ADVANCES IN LATTICE GAUGE THEORY

ADVANCES IN LATTICE GAUGE THEORY

Editors
D. W. DUKE
J. F. OWENS

DEPARTMENT OF PHYSICS
THE FLORIDA STATE UNIVERSITY
TALLAHASSEE, FLORIDA

World Scientific

Published by

World Scientific Publishing Co Pte Ltd.
P. O. Box 128, Farrer Road, Singapore 9128
242, Cherry Street, Philadelphia PA 19106-1906, USA

Library of Congress Cataloging-in-Publication Data is available.

ADVANCES IN LATTICE GAUGE THEORY

ISBN 9971-50-030-2

Printed in Singapore by Kim Hup Lee Printing Co Pte Ltd.

PREFACE

During the period April 10-13, 1985 over eighty scientists from around the world participated in the conference *Advances in Lattice Gauge Theory* held at The Florida State University in Tallahassee, Florida. The papers collected in this volume represent the state of the art in lattice gauge theory as of mid-1985. It seems clear that the subject promises great excitement and continued progress for the forseeable future.

This conference was sponsored by the Supercomputer Computations Research Institute (SCRI) of The Florida State University. The SCRI is a joint venture of The Florida State University, the State of Florida, Control Data Corporation and the United States Department of Energy. Inaugurated in October, 1984, the SCRI is the nation's first university-industry-government cooperative effort into research in computational science, combining modern supercomputers and a large staff of dedicated research scientists.

We wish to thank Professor Joseph E. Lannutti, Director of the SCRI, and the SCRI administrative staff, headed by Susan Fell and Pat Meredith, for all of their help and cooperation in arranging the conference. The effort and assistance of Bernd Berg, Gyan Bhanot, Anna Hasenfratz and Tony Kennedy in arranging the speakers and the program was invaluable and is gratefully acknowledged. Finally, we wish to especially thank our Conference Coordinator, Susan Lampman of The Florida State University Center for Professional Development, for her splendid organization of the conference.

The Florida State University
Tallahassee, Florida
September, 1985

Dennis Duke
Jeff Owens

CONTENTS

HADRON OPERATORS FOR STAGGERED FERMIONS

Jan Smit
Institute for Theoretical Physics, Valckenierstraat 65
1018 XE Amsterdam, The Netherlands

ABSTRACT

We report on recent work in which simple restframe hadron operators for staggered fermion lattice QCD are constructed and classified according to the lattice symmetry group. The connection with continuum hadronic states is also given.

INTRODUCTION

The flavor and spin interpretation of staggered fermions applies to the continuum limit at weak coupling. At coupling values used in numerical simulations, the symmetries of the continuum limit may be valid approximately, but only the lattice symmetry group is valid exactly. In classifying the particle spectrum and the construction of hadron fields on the lattice, it seems therefore safe to rely at first only on group theory, as was done for glueballs [1]. This requires the following steps to be taken:

1. Determine the lattice symmetry group. For hadron masses the restframe group is relevant.
2. Determine the irreducible representations (irreps). This gives a classification of particle states in Hilbert space.
3. Construct hadron fields transforming in such irreps. These serve to create the particle excitations out of the ground state.
4. For the determination of particle masses from correlation functions, the lattice analogue of the Kählén-Lehman representation is needed.
5. Decompose the irreps of the continuum symmetry with respect to the irreps of the lattice symmetry group. This tells us which lattice states may be considered as candidates for which continuum states. Provided we know enough about the continuum limit we can go beyond group theory and construct lattice fields which go over in continuum fields in the weak coupling limit.

6. Assemble lattice irrep fields into continuum limit fields. This will help in making a more definite one to one correspondence between hadrons and lattice states.

Progress on the steps 2-6 has recently been achieved [2,3]. For related work in the same spirit see [4,5].

SYMMETRIES

In the following we restrict ourselves to the four flavor theory with degenerate mass term,

$$S_{\bar{\chi}\chi} = -\sum_{x,\mu} \tfrac{1}{2}\eta_\mu(x) \left[\bar{\chi}(x)U_\mu(x)\chi(x+a_\mu) - \bar{\chi}(x+a_\mu)U_\mu^+(x)\chi(x)\right]$$

$$-\sum_x m\,\bar{\chi}(x)\chi(x) \ . \tag{1}$$

The symmetries of this theory have been discussed in [6] (see [7] for the two flavor theory). We recall the transformations on the fermion fields

- shifts S_μ :

$$\chi(x) \rightarrow \zeta_\mu(x)\chi(x+a_\mu) \tag{2}$$

- 90° rotations $R^{(\kappa\lambda)}$:

$$\chi(x) \rightarrow S_R(R^{-1}x)\ \chi\ (R^{-1}x)\ ,\ R^{(12)}x = (x_2, -x_1, x_3, x_4) \tag{3}$$

- inversion I_s :

$$\chi(x) \rightarrow \eta_4(x)\zeta_4(x)\chi(I_s x)\ ,\ I_s x = (-x_1, -x_2, -x_3, x_4) \tag{4}$$

- charge conjugation C_o :

$$\chi(x) \rightarrow \varepsilon(x)\bar{\chi}(x),\ \bar{\chi}(x) \rightarrow -\varepsilon(x)\chi(x) \tag{5}$$

- U(1) transformations:

$$\chi(x) \rightarrow e^{i\alpha}\ \chi(x)\ ,\ \bar{\chi}(x) \rightarrow e^{-i\alpha}\ \bar{\chi}(x)\ . \tag{6}$$

In (1)-(5):

$\eta_1(x) = 1$ $\qquad\qquad$ $\zeta_1(x) = (-1)^{x_2+x_3+x_4}$

$\eta_2(x) = (-1)^{x_1}$ $\qquad\qquad$ $\zeta_2(x) = (-1)^{x_3+x_4}$

$\eta_3(x) = (-1)^{x_1+x_2}$ $\qquad\qquad$ $\zeta_3(x) = (-1)^{x_4}$

$\eta_4(x) = (-1)^{x_1+x_2+x_3}$ $\qquad\qquad$ $\zeta_4(x) = 1$

$$\varepsilon(x) = (-1)^{x_1+x_2+x_3+x_4}$$

$$S_R(x) = \tfrac{1}{2}[1 \pm \eta_\kappa(x)\eta_\lambda(x) \mp \zeta_\kappa(x)\zeta_\lambda(x) + \eta_\kappa(x)\eta_\lambda(x)\zeta_\kappa(x)\zeta_\lambda(x)], \quad \kappa \lessgtr \lambda \tag{7}$$

The transformations (2)-(6) generate the defining representation of the lattice symmetry group. The interpretation of these transformations is best seen in momentum space [6]:

$$\chi(x) \to \psi(p) \ , \ -\frac{\pi}{2} < p_\mu \leq \frac{\pi}{2} \tag{8}$$

$$S_\mu \quad : \psi(p) \to e^{ip_\mu} \xi_\mu \, \psi(p) \tag{9}$$

$$R^{(\kappa\lambda)} \quad : \psi(p) \to \exp(\frac{\pi}{4} \gamma_\kappa\gamma_\lambda)\exp(\frac{\pi}{4} \xi_\kappa\xi_\lambda) \, \psi(R^{-1}p) \tag{10}$$

$$I_s \quad : \psi(p) \to \gamma_4\xi_4 \, \psi(I_s p) \tag{11}$$

$$C_o \quad : \psi(p) \to (\bar{\psi}(p)C_o)^T \ , \ \bar{\psi}(p) \to -(C_o^+ \psi(p))^T$$

$$C_o = C^\gamma C^\xi \ , \ C^\gamma \gamma_\mu C^{\gamma+} = -\gamma_\mu^{\ T} \ , \ C^\xi \xi_\mu C^{\xi+} = -\xi_\mu^{\ T} \tag{12}$$

The field ψ is constructed out of the fourier transform of χ by dividing momentum space in 16 blocks in the usual way; γ_μ and ξ_μ are hermitian Dirac matrices acting respectively on the spin and flavor components of ψ. The S_μ are a mixture of translations (e^{ip_μ}) and flavor transformations (ξ_μ). Note that the S_μ anticommute. The $R^{(\kappa\lambda)}$ are simultaneous rotations in spin and flavor space. Besides parity ($\to \gamma_4$), I_s also involves a flavor (ξ_4) transformation. The transformation differs from the usual charge conjugation $C = C^\gamma$ by a conjugation C^ξ in flavor space.

These symmetry operations are represented by operators in Hilbertspace which we denote by a ^ .In particular, the

$$\hat{T}_\mu \equiv \hat{S}_\mu^{\ 2} \tag{13}$$

are interpreted as translation operators; $\hat{T} = \hat{T}_4$ is the transfermatrix [7,8]. Their action on eigenstates is given by

$$\hat{T}\,|E\,\vec{p}\rangle = e^{-2E}\,|E\,\vec{p}\rangle ,$$

$$\hat{T}_k\,|E\,\vec{p}\rangle = e^{i2p_k}\,|E\,\vec{p}\rangle . \qquad (14)$$

We define unitary operators $\hat{\Xi}_\mu$ by (cf. (9))

$$\hat{S}_4\,|E\,\vec{p}\rangle = e^{-E}\,\hat{\Xi}_4\,|E\,\vec{p}\rangle ,$$

$$\hat{S}_k\,|E\,\vec{p}\rangle = e^{ip_k}\,\hat{\Xi}_k\,|E\,\vec{p}\rangle . \qquad (15)$$

The $\hat{\Xi}_\mu$ are interpreted as lattice flavor transformations. They generate a group which is isomorphic to the 32 element group generated by the Dirac γ-matrices. It is a subgroup of the continuum flavor group $SU(4)_F$.

THE RESTFRAME $\vec{p} = 0$

The restframe group RF is generate by the transformations which commute with the transfermatrix*): $R^{(\kappa\ell)}$, Ξ_μ, I_s, C_o, $U(1)$, in the subspace $T_k = 1$. We denote this group by

$$RF = G(R^{(k\ell)}, \Xi_m, \Xi_4, I_s, C_o, U(1)). \qquad (16)$$

Note that $\pm\,\Xi_m$ transform as a vector under $R^{(k\ell)}$, while Ξ_4 and I_s are scalars, with (cf. (11))

$$[\Xi_4, I_s] = 0 , \quad \Xi_k\,I_s = -I_s\,\Xi_k . \qquad (17)$$

We define parity by (cf. (11))

$$P = \Xi_4\,I_s . \qquad (18)$$

This P commutes with flavor transformations Ξ_μ and satisfies

*) For convenience we omit the ^ when indicating group elements.

$C_o P = -P C_o$.

Since C_o and U(1) are easy to understand we concentrate on $R^{(k\ell)}$, Ξ_μ, I_s and since P commutes with these it is useful to factor it out. We define the geometrical restframe group GRF, the proper restframe group PRF and the geometrical time slice group GTS by

$$GRF = G(R^{(k\ell)}, \Xi_m, \Xi_4, I_s) , \tag{19}$$

$$PRF = G(R^{(k\ell)}, \Xi_m, \Xi_4) , \tag{20}$$

$$GTS = G(R^{(k\ell)}, \Xi_m, I_s) . \tag{21}$$

The geometrical restframe group GRF is just RF without the C_o and U(1) transformations. The PRF and GTS are obtained from GRF by extracting P in two different ways,

$$GRF = PRF \times \{1, P\} = GTS \times \{1, P\} . \tag{22}$$

(Note that PRF is isomorphic to GTS). The group GRF is used to classify the states $|E, \vec{p} = o\rangle$ and GTS to classify the hadron fields $\phi(t)$. We want the fields to be local in time (involving lattice sites separated by only a finite number of shifts in the time direction). Therefore we cannot use GRF or PRF which involve Ξ_4 whose definition is non-local in time. The fields $\phi(t)$ cannot carry a definite parity P but we can choose them to possess the next best thing, namely a definite I_s - parity. The irreps of GRF are denoted by r^σ, where r denotes the irreps of PRF $\simeq$ GTS and σ the eigenvalue of P. Choosing Ξ_4 diagonal, we write

$$\hat{S}_4 \; |E \; r \; \sigma; \; \sigma_t\rangle = e^{-E} \; \sigma_t \; |E \; r \; \sigma; \; \sigma_t\rangle$$

$$\hat{P} \; |E \; r \; \sigma; \; \sigma_t\rangle = \sigma \; |E \; r \; \sigma; \; \sigma_t\rangle$$

$$\hat{I}_s \; |E \; r \; \sigma; \; \sigma_t\rangle = \sigma \; \sigma_t \; |E \; r \; \sigma; \; \sigma_t\rangle , \tag{23}$$

where $\sigma, \sigma_t = \pm 1$.

Suppose now we have constructed fields $\phi(t)$ and $\bar{\phi}(t)$ which are translation invariant (commute with $\hat{T}_k$), transform in the irrep r under C^{GTS}, have a definite I_s parity σ_s and are related by C_o ($\phi \overset{o}{\leftrightarrow} \bar{\phi}$, for mesons we assume $\bar{\phi} = \pm\phi$) . Then it is possible to derive the following spectral decomposition for the time-time correlation functions [2]:

$$C(t) \equiv \langle\phi(t)\ \bar{\phi}\ (0)\rangle - \langle\phi\rangle\ \langle\bar{\phi}\rangle$$

$$= \sum_{E\sigma} R_\sigma\ (E)\ [(\sigma\sigma_s)^t\ e^{-Et} + \sigma_c K(\sigma_c\sigma\sigma_s)^t\ e^{-E(2L-t)}]\ . \tag{24}$$

Where

$$R_\sigma(E) = \sum \langle 0\,|\hat{\phi}(0)\,|E r\sigma;\sigma_t\rangle\ \langle E r\sigma;\sigma_t|\ \hat{\bar{\phi}}(0)\ |0\rangle\ ,\ \sigma_s = \sigma\sigma_t \tag{25}$$

and

$\sigma_c = +\ 1(-1)$ for mesons (baryons)
$K = +1((-1)^q)$ for antiperiodic (periodic) boundary conditions in time; q = quark number. (26)

In (24) the time extension of the lattice was taken as [-L+1, L] and t is assumed ≥ 0; negative t is incorporated by allowing t = 0, 1, ..., 2L-1. In (25) $\sum$ denotes a summation over contributing states. Only states with GRF irrep $r\sigma$ and with σ_t such that $\sigma\,\sigma_t = \sigma_s$ contribute to $R_\sigma(E)$. Of course, the baryon number and (for mesons) the C_o parity of the states are also determined by the fields. For example, in the narrow width approximation the contribution of particles with masses m_+ and m_- in parity sectors + and -, respectively, is given, in case $\sigma_s = +1$, by

$$R_+[e^{-m_+t} + e^{-m_+(2L-t)}] + R_-(-1)^t\ [e^{-m_-t} + e^{-m_-(2L-t)}] \tag{27}$$

for mesons, and

$$R_+ [e^{-m_+ t} - K(-1)^t e^{-m_+(2L-t)}] + R_-(-1)^t [e^{-m_- t} - K(-1)^t e^{-m_-(2L-t)}] \quad (28)$$

for baryons. Hence, by comparing even and odd times we can distinguish the parities of the particles.

THE GEOMETRIC TIMESLICE GROUP GTS

The cubic rotation group $O = G(R^{(k\ell)})$ has 24 elements, the γ-matrix group $G(\Xi_\mu)$ has 32 elements and it can be shown that $GTS \simeq PRF = G(R^{(k\ell)}, \Xi_\mu)$ has $24 \times 32 = 768$ elements. It has 43 irreps: 3 fermionic and 40 bosonic*). The defining representation is carried by the quark fields, at zero momentum, i.e. the translation invariant combination

$$\hat{\chi}_A = \sum_{\vec{m}} \hat{T}_1^{-m_1} \hat{T}_2^{-m_2} \hat{T}_3^{-m_3} \hat{\chi}(\vec{a}_A) \hat{T}_3^{m_3} \hat{T}_2^{m_2} \hat{T}_1^{m_1}$$

$$= \sum_{\vec{m}} \hat{\chi}(\vec{a}_A + 2\vec{m}a) \equiv \sum_{\vec{x}, x_k \text{ even}} \hat{\chi}(\vec{a}_A + \vec{x}) \ . \quad (29)$$

Here $\vec{a}_A$ is one of the 8 vectors,

$$\vec{0},\ \vec{a}_1,\ \vec{a}_2,\ \vec{a}_3,\ \vec{a}_2 + \vec{a}_3,\ \vec{a}_1 + \vec{a}_3,\ \vec{a}_1 + \vec{a}_2,\ \vec{a}_1 + \vec{a}_2 + \vec{a}_3 \ . \quad (30)$$

The eight components (for a given color) χ_A form the fundamental irrep $\underset{\sim}{8}$. The reduction with respect to the group $O_h = O \times \{1, I_s\}$ is given by

$$\underset{\sim}{8} \rightarrow A_1^+ + A_1^- + F_1^+ + F_1^- \ , \quad (31)$$

where the $\pm$ denote the I_s parity σ_s . We recall the irreps of O and the decomposition of the first few SO(3) irreps with respect to

*) By fermionic (bosonic) we mean that the group element −1 is represented by multiplication with the number −1(+1).

O:

O irrep	dimension	SO(3) irrep		O irrep
A_1	1	ℓ		
A_2	1	0	$\rightarrow$	A_1
E	2	1	$\rightarrow$	F_1
F_1	3	2	$\rightarrow$	$E+F_2$
F_2	3	3	$\rightarrow$	$A_2+F_1+F_2$

The decomposition (31) is illustrated in fig. 1. Note the particular definition of $R^{(k\ell)}$ which we employ (cf. (3)). One may also consider rotations about the center of a cube: these differ from our $R^{(k\ell)}$ by shifts and generate a different subgroup of GTS. So there is an ambiguity in what one chooses as the definition of O. However, the complete group GTS is unambiguous.

The other fermionic irreps of GTS are obtained from the reduction of $\underset{\sim}{8} \times \underset{\sim}{8} \times \underset{\sim}{8}$. They are 8 and 16 dimensional, denoted by $\underset{\sim}{8}'$ and $\underset{\sim}{16}$, with O_h content

$$\underset{\sim}{8}' \rightarrow A_2^+ + A_2^- + F_2^+ + F_2^- \quad , \tag{32}$$

$$\underset{\sim}{16} \rightarrow E^+ + E^- + F_1^+ + F_1^- + F_2^+ + F_2^- \; . \tag{33}$$

For the bosonic irreps the group element $-1 = \Xi_k \Xi_\ell \Xi_k \Xi_\ell$, $(k \neq \ell)$ is represented by unity, so Ξ_μ and $-\Xi_\mu$ have the same representation, say X_μ. The X_μ commute amongst themselves and with I_s, it follows that $X_1 X_2 X_3$ and I_s commute with all group elements in a bosonic irrep of GTS and we can write $r = r^{-\sigma_s \sigma_{123}}$, where σ_{123} denotes the eigenvalue of $X_1 X_2 X_3$. For the states we use the PRF group $(I_s \rightarrow X_4)$ with irreps $r = r^{-\sigma_t \sigma_{123}}$. The bosonic irreps with their O content are listed below (the number indicates the dimension):

$\bar{r}$		O irrep	$\bar{r}$		O irrep
$\underset{\sim}{1}$	→	A_1	$\underset{\sim}{3}''$	→	F_1
$\underset{\sim}{1}'$	→	A_2	$\underset{\sim}{3}'''$	→	F_2
$\underset{\sim}{2}$	→	E	$\underset{\sim}{3}'^{v}$	→	A_1+E
$\underset{\sim}{3}$	→	F_1	$\underset{\sim}{3}^{v}$	→	A_2+E
$\underset{\sim}{3}'$	→	F_2	$\underset{\sim}{6}$	→	F_1+F_2

BARYON FIELDS

To find suitable baryon operators it is convenient to first introduce symmetric shift operators D_A , defined by

$$D_0\ \phi(\vec{x}) = \phi(\vec{x}) ,$$
$$D_k\ \phi(\vec{x}) = \tfrac{1}{2}[\phi(\vec{x}+\vec{a}_k) + \phi(\vec{x}-\vec{a}_k)] , \quad k=1,2,3,$$
$$D_{k\ell}\ \phi(\vec{x}) = D_k\ D_\ell\ \phi(\vec{x}) \quad , \quad k\neq\ell ,$$

etc. (34)

Note that (cf. (29))

$$\chi_A = \sum_{\vec{x},x_k \text{even}} D_A\chi(\vec{x}) \quad . \tag{35}$$

Baryon fields are introduced by

$$\tilde{B}_{ABC} = \sum_{\vec{x},x_k \text{even}} \frac{1}{6}\ \varepsilon_{abc}\ D_A\chi^a(\vec{x})D_B\chi^b(\vec{x})D_C\chi^c(\vec{x}) \quad , \tag{36}$$

where a, b and c are color indices. The $\tilde{B}_{ABC}$ form a reducible representation $(\underset{\sim}{8} \times \underset{\sim}{8} \times \underset{\sim}{8})_{symm}$ of GTS, which decomposes as

$$(\underset{\sim}{8} \times \underset{\sim}{8} \times \underset{\sim}{8})_{symm} = 5.\underset{\sim}{8} + 2.\underset{\sim}{8}' + 4.\underset{\approx}{16} \quad . \tag{37}$$

The $\tilde{B}_{ABC}$ may be simplified by leaving out of (36) all terms with

χ fields separated by two lattice spacings in the same direction. The so obtained B_{ABC} have the same transformation properties. From (31)-(33) it follows that we can obtain all components of the $\underset{\sim}{8}$, $\underset{\sim}{8}'$ and $\underset{\sim}{16}$ by first identifying components of A_1, A_2 and E, respectively, and then applying rotations and shifts. Fig. 2 shows a list of such components. The fields are made locally gauge invariant by introducing the gauge fields $U_\mu(x)$ on the links between the χ's in the obvious way. It is necessary to average over various equivalent ways of introducing the U_μ's in order to maintain covariance under GTS. For more details, see [2].

MESON FIELDS

Meson fields may be introduced by

$$\tilde{M}_{AB} = \sum_{\vec{x}, x_k \text{even}} D_A \bar{\chi}_a(\vec{x}) D_B \chi^a(\vec{x}) . \tag{38}$$

These transform as

$$\underset{\sim}{8} \times \underset{\sim}{8} = \sum_{\substack{\sigma_s=\pm 1 \\ \sigma_{123}=\pm 1}} (\underset{\sim}{1}^{\sigma_s \sigma_{123}} + \underset{\sim}{3}^{\sigma_s \sigma_{123}} + \underset{\sim}{3}''^{\sigma_s \sigma_{123}} + 3'^{v\, \sigma_s \sigma_{123}} + \underset{\sim}{6}^{\sigma_s \sigma_{123}}) . \tag{39}$$

As in the case for baryon fields, we may drop terms in (38) which do not fit into a lattice cube. The reduction (39) leads to table 1. Only one value of the C_o parity τ_o occurs for each irrep. We can obtain the opposite value by separating the $\bar{\chi}$ and χ fields by one lattice unit in the time direction. This does not affect the GTS transformation properties. For the detailed construction see the revised version of [3].

LATTICE HADRONS VERSUS CONTINUUM HADRONS

The lattice restframe group RF is a subgroup of the continuum restframe group generated by spin rotations

$\exp(\frac{1}{4}\phi_{k\ell}\gamma_k\phi_\ell) \in SU(2)_S$, flavor transformations

$\exp(i\alpha + i\alpha_\mu\xi_\mu + \frac{1}{4}\alpha_{\mu\nu}\xi_\mu\xi_\nu + \alpha_\mu^5\xi_\mu\xi_5 + i\alpha_5\xi_5) \in U(4)_F$, parity

$P \to \gamma_4$ and charge conjugation C^{γ}, where we used the notation (9)-(12).

We want to reduce the continuum irreps, which correspond to hadrons, with respect to the lattice irreps. Consider first the baryons, which can be classified into irreps (J^{σ}, R) of $SU(2)_S \times U(4)_F \times \{1, P\}$. Here R is one of the $U(4)_F$ irreps $20_S = \square\square\square$, $20_M = \begin{smallmatrix}\square\square\\ \square\end{smallmatrix}$ or $4_A = \begin{smallmatrix}\square\\ \square\\ \square\end{smallmatrix}$ and J^{σ} denotes the spin-parity. We found in (31)-(33) that the baryonic RF irreps $\underset{\sim}{8}$, $\underset{\sim}{8}'$ and $\underset{\sim}{16}$ are identified by their 0 content. Hence all we have to do is to reduce along the chain

$$SU(2)_S \times U(4)_F \supset SU(2)_S \times SU(2)_F \supset SU(2)_{diag} \supset 0 \ ,$$

where

$$SU(2)_F \leftarrow \exp(\tfrac{1}{4}\alpha_{k\ell}\xi_k\xi_\ell); \ SU(2)_{diag} \leftarrow \exp\,[\tfrac{1}{4}\alpha_{k\ell}(\gamma_k\gamma_\ell + \xi_k\xi_\ell)];$$

$$0 \leftarrow \exp[\tfrac{1}{4}\pi(\gamma_k\gamma_\ell + \xi_k\xi_\ell)] \ .$$

The reduction of the first few irreps is given in the following table.

$SU(2)_S \times U(4)_F \times \{1,P\}$ irrep		GRF representations
$(\frac{1}{2}^{\pm}, 4_A)$	$\to$	$\underset{\sim}{8}^{\pm}$
$(\frac{1}{2}^{\pm}, 20_M)$	$\to$	$3\cdot\underset{\sim}{8}^{\pm} + \underset{\sim}{16}^{\pm}$
$(\frac{1}{2}^{\pm}, 20_S)$	$\to$	$\underset{\sim}{8}^{\pm} + 2\cdot\underset{\sim}{16}^{\pm}$
$(\frac{3}{2}^{\pm}, 4_A)$	$\to$	$\underset{\sim}{16}^{\pm}$
$(\frac{3}{2}^{\pm}, 20_M)$	$\to$	$\underset{\sim}{8}^{\pm} + \underset{\sim}{8}'^{\pm} + 4\cdot\underset{\sim}{16}^{\pm}$
$(\frac{3}{2}^{\pm}, 20_S)$	$\to$	$2\cdot\underset{\sim}{8}^{\pm} + 2\cdot\underset{\sim}{8}'^{\pm} + 3\cdot\underset{\sim}{16}^{\pm}$

From the table we can make a picture of how the experimentally observed baryon resonances should fit in the spectra of the irreps $\underset{\sim}{8}^{\pm}$, $\underset{\sim}{8}'^{\pm}$ and $\underset{\sim}{16}^{\pm}$ in the continuum limit. See fig 3. The $\Lambda(1405)$

and $\Lambda(1520)$ are assumed to be mainly SU(3) singlets, so we assigned them to $(\frac{1}{2}^-, 4_A)$ and $(\frac{3}{2}^-, 4_A)$, respectively. In the figure their masses are corrected downward with 170 MeV (= $\Lambda(1690)$-N(1520)) to compensate for SU(3) flavor breaking.

Consider next the mesons, which can be classified into irreps $(J^{\sigma\tau}, R)$ of the continuum restframe group. Here R = S or A, the $SU(4)_F$ singlet or adjoint irreps, and τ is the charge conjugation parity. Reductions to RF are given in [3]. The situation is more favourable than in the baryon case, because there are many more bosonic irreps.

Actually, the meson fields given in the table can be understood from their classical continuum limit [6]. In this limit the link variables which render the fields gauge invariant effectively approach 1 and then the correspondence [6]

$$\eta_\mu(x)\ D_\mu\chi(x) \to \gamma_\mu\psi(p)\ \cos p_\mu\ ,$$

$$\zeta_\mu(x)D_\mu\chi(x) \to \xi_\mu\psi(p)\ \cos p_\mu\ ,$$

$$\eta_\mu(x)\zeta_\mu(x)\chi(x) \to \gamma_\mu\xi_\mu\psi(p)\ , \tag{40}$$

allows for easy interpretation (also $\cos p_\mu \to 1$ in the continuum limit). For example, for

$$\phi(x) = \bar{\chi}(x)\varepsilon(x)\zeta_\mu(x)D_\mu\chi(x) \tag{41}$$

the continuum limit is

$$\phi(x) \to \bar{\psi}(x)\gamma_5\xi_k\xi_5\psi(x)\ , \tag{42}$$

since $\varepsilon(x) \to \gamma_5\xi_5$. Similarly

$$(-1)^{x_4}\phi(x) \to \bar{\psi}(x)\gamma_4\xi_k\xi_4\psi(x)\ , \tag{43}$$

since $(-1)^{x_4} = \varepsilon(x)\eta_4(x)\zeta_4(x) \to \gamma_4\gamma_5\xi_4\xi_5$. For low momenta and energies $\phi(x)$ excites only $\sigma_t=+1$ states and $(-1)^{x_4}\phi(x)$ only $\sigma_t = -1$ states. Indeed, from (24) follows $(\sigma_s = -1)$

$$\sum_{t=0}^{2L-1} e^{-ip_4 t} C(t) = \sum_E 2 \sinh E\left[\frac{R_-(E)}{\cosh E - \cos p_4} + \frac{R_+(E)}{\cosh E + \cos p_4}\right] , \qquad (44)$$

as $L \to \infty$. Hence, for $p_4 \sim 0$ ($\leftarrow$ (42)) $R_+(E)$ drops out of the low energy contributions and for $p_4 \sim \pi$ ($\leftarrow$ (43)) $R_-(E)$ drops out. We can establish an alternative labelling for those states $|E r \underline{\sigma} \sigma_t\rangle$ which are in 1-1 correspondence with the quantum numbers of $\bar{\psi}\gamma\xi\psi$, as given by the classical continuum limit. See table 2; more extensive tables are in [3].

CONCLUSION

The group theoretical properties of staggered fermions are now much better understood. Simple meson fields which reduce to a form $\bar{\psi}\gamma\xi\psi$ in the classical continuum limit have been put in a group theoretical frame work. The continuum limit of the baryon field proposed here is yet to be explored (step 6 in the introduction). For example, the amplitudes for creation of the higher spin states in fig. 3 with the fields in fig. 2 may well vanish in the continuum limit.

ACKNOWLEDGEMENT

I would like to thank M.F.L. Golterman for a pleasant collaboration.

References

[1] R.C. Johnson, Phys.Lett. 114B (1982) 147
B. Berg and A. Billoire, Phys.Lett. 114B (1982) 324; Nucl.Phys. B221 (1983) 109.
[2] M.F.L. Golterman and J. Smit, Nucl.Phys. B255 (1985) 328.
[3] M.F.L. Golterman, Staggered Mesons, Amsterdam preprint ITFA-85-5, submitted to Nucl. Phys. B.
[4] G. Parisi and Y.C. Zhang, Nucl.Phys. B230 [FS10] (1984) 97.
[5] D. Verstegen, Nucl.Phys. B249 (1985) 685.
[6] M.F.L. Golterman and J. Smit, Nucl.Phys. B245 (1984) 61.
[7] C.P. van den Doel and J. Smit, Nucl.Phys. B228 (1983) 122.
[8] H.S. Sharatchandra, H.J. Thun and P. Weisz, Nucl.Phys. B192 (1981) 205.

class	irrep	operator	C_0 parity (τ_0)
0	1^{++}	$\sum_{\vec{x}} \bar{\chi}(\vec{x})\chi(\vec{x})$	+
	1^{+-}	$\sum_{\vec{x}} \eta_4(x)\zeta_4(x)\bar{\chi}(\vec{x})\chi(\vec{x})$	+
	$3^{iv\,+-}$	$\sum_{\vec{x}} \eta_k(x)\varepsilon(x)\zeta_k(x)\bar{\chi}(\vec{x})\chi(\vec{x})$	+
	$3^{iv\,++}$	$\sum_{\vec{x}} \eta_4(x)\zeta_4(x)\eta_k(x)\varepsilon(x)\zeta_k(x)\bar{\chi}(\vec{x})\chi(\vec{x})$	+
1	3^{-+}	$\sum_{\vec{x}} \bar{\chi}(\vec{x})\eta_k(x)D_k\chi(\vec{x})$	-
	3^{--}	$\sum_{\vec{x}} \eta_4(x)\zeta_4(x)\bar{\chi}(\vec{x})\eta_k(x)D_k\chi(\vec{x})$	+
	$3''^{--}$	$\sum_{\vec{x}} \bar{\chi}(\vec{x})\varepsilon(x)\zeta_k(x)D_k\chi(\vec{x})$	+
	$3''^{-+}$	$\sum_{\vec{x}} \eta_4(x)\zeta_4(x)\bar{\chi}(\vec{x})\varepsilon(x)\zeta_k(x)D_k\chi(\vec{x})$	-
	6^{--}	$\sum_{\vec{x}} \eta_k(x)\varepsilon(x)\zeta_k(x)\bar{\chi}(\vec{x})\eta_l(x)D_l\chi(\vec{x})\vert_{k\neq l}$	-
	6^{-+}	$\sum_{\vec{x}} \eta_4(x)\zeta_4(x)\eta_k(x)\varepsilon(x)\zeta_k(x)\bar{\chi}(\vec{x})\eta_l(x)D_l\chi(\vec{x})\vert_{k\neq l}$	+
2	3^{++}	$\sum_{\vec{x}} \bar{\chi}(\vec{x})\eta_k(x)D_k\{\eta_l(x)D_l\chi(\vec{x})\}\varepsilon_{klm}$	-
	3^{+-}	$\sum_{\vec{x}} \eta_4(x)\zeta_4(x)\bar{\chi}(\vec{x})\eta_k(x)D_k\{\eta_l(x)D_l\chi(\vec{x})\}\varepsilon_{klm}$	-
	$3''^{++}$	$\sum_{\vec{x}} \bar{\chi}(\vec{x})\zeta_k(x)D_k\{\zeta_l(x)D_l\chi(\vec{x})\}\varepsilon_{klm}$	-
	$3''^{+-}$	$\sum_{\vec{x}} \eta_4(x)\zeta_4(x)\bar{\chi}(\vec{x})\zeta_k(x)D_k\{\zeta_l(x)D_l\chi(\vec{x})\}\varepsilon_{klm}$	-
	6^{++}	$\sum_{\vec{x}} \eta_m(x)\zeta_m(x)\bar{\chi}(\vec{x})\eta_k(x)D_k\{\zeta_l(x)D_l\chi(\vec{x})\}_{k\neq l\neq m\neq k}$	+
	6^{+-}	$\sum_{\vec{x}} \eta_4(x)\zeta_4(x)\eta_m(x)\zeta_m(x)\bar{\chi}(\vec{x})\eta_k(x)D_k\{\zeta_l(x)D_l\chi(\vec{x})\}_{k\neq l\neq m\neq k}$	+
3	1^{-+}	$\sum_{\vec{x}} \bar{\chi}(\vec{x})\eta_1(x)D_1\{\eta_2(x)D_2\{\eta_3(x)D_3\chi(\vec{x})\}\}$	+
	1^{--}	$\sum_{\vec{x}} \eta_4(x)\zeta_4(x)\bar{\chi}(\vec{x})\eta_1(x)D_1\{\eta_2(x)D_2\{\eta_3(x)D_3\chi(\vec{x})\}\}$	-
	$3^{iv\,--}$	$\sum_{\vec{x}} \eta_k(x)\varepsilon(x)\zeta_k(x)\bar{\chi}(\vec{x})\eta_1(x)D_1\{\eta_2(x)D_2\{\eta_3(x)D_3\chi(\vec{x})\}\}$	-
	$3^{iv\,-+}$	$\sum_{\vec{x}} \eta_4(x)\zeta_4(x)\eta_k(x)\varepsilon(x)\zeta_k(x)\bar{\chi}(\vec{x})\eta_1(x)D_1\{\eta_2(x)D_2\{\eta_3(x)D_3\chi(\vec{x})\}\}$	+

table 1 Irreducible meson operators contained in $\tilde{M}_{AB}$.

operator	lattice states $\sigma_t = +1$	lattice states $\sigma_t = -1$	corresponding continuum states	J_R^{PC}
$\bar{\chi}\chi$	1	$\gamma_4\gamma_5\xi_4\xi_5$	0_S^{++}	0_A^{-+}
$\eta_4\zeta_4\bar{\chi}\chi$	$\gamma_4\xi_4$	$\gamma_5\xi_5$	0_A^{+-}	0_A^{-+}
$\eta_k\varepsilon\zeta_k\bar{\chi}\chi$	$\gamma_k\gamma_5\xi_k\xi_5$	$\gamma_k\gamma_4\xi_k\xi_4$	1_A^{++}	1_A^{--}
$\eta_4\zeta_4\eta_k\varepsilon\zeta_k\bar{\chi}\chi$	$\gamma_l\gamma_m\xi_l\xi_m$	$\gamma_k\xi_k$	1_A^{+-}	1_A^{--}
$\bar{\chi}\eta_k D_k\chi$	γ_k	$\gamma_l\gamma_m\xi_4\xi_5$	1_S^{--}	1_A^{+-}
$\eta_4\zeta_4\bar{\chi}\eta_k D_k\chi$	$\gamma_k\gamma_4\xi_4$	$\gamma_k\gamma_5\xi_5$	1_A^{--}	1_A^{++}
$\bar{\chi}\varepsilon\zeta_k D_k\chi$	$\gamma_5\xi_k\xi_5$	$\gamma_4\xi_k\xi_4$	0_A^{-+}	0_A^{+-}
$\eta_4\zeta_4\bar{\chi}\varepsilon\zeta_k D_k\chi$	$\gamma_4\gamma_5\xi_l\xi_m$	ξ_k	0_A^{-+}	0_A^{++}
$\eta_k\varepsilon\zeta_k\bar{\chi}\eta_l D_l\chi\vert_{k\neq l}$	$\gamma_m\gamma_4\xi_k\xi_5$	$\gamma_m\gamma_5\xi_k\xi_4$	1_A^{--}	1_A^{++}
$\eta_4\zeta_4\eta_k\varepsilon\zeta_k\bar{\chi}\eta_l D_l\chi\vert_{k\neq l}$	$\gamma_m\xi_l\xi_m$	$\gamma_k\gamma_l\xi_k$	1_A^{--}	1_A^{+-}
$\bar{\chi}\eta_k D_k\{\eta_l D_l\chi\}\varepsilon_{klm}$	$\gamma_k\gamma_l$	$\gamma_m\xi_4\xi_5$	1_S^{+-}	1_A^{--}
$\eta_4\zeta_4\bar{\chi}\eta_k D_k\{\eta_l D_l\chi\}\varepsilon_{klm}$	$\gamma_m\gamma_5\xi_4$	$\gamma_m\gamma_4\xi_5$	1_A^{++}	1_A^{--}
$\bar{\chi}\zeta_k D_k\{\zeta_l D_l\chi\}\varepsilon_{klm}$	$\xi_k\xi_l$	$\gamma_4\gamma_5\xi_m$	0_A^{++}	0_A^{-+}
$\eta_4\zeta_4\bar{\chi}\zeta_k D_k\{\zeta_l D_l\chi\}\varepsilon_{klm}$	$\gamma_4\xi_m\xi_5$	$\gamma_5\xi_m\xi_4$	0_A^{+-}	0_A^{-+}
$\eta_m\zeta_m\bar{\chi}\eta_k D_k\{\zeta_l D_l\chi\}_{k\neq l\neq m\neq k}$	$\gamma_k\gamma_m\xi_l\xi_m$	$\gamma_l\xi_k$	1_A^{+-}	1_A^{--}
$\eta_4\zeta_4\eta_m\zeta_m\bar{\chi}\eta_k D_k\{\zeta_l D_l\chi\}_{k\neq l\neq m\neq k}$	$\gamma_l\gamma_5\xi_k\xi_5$	$\gamma_l\gamma_4\xi_k\xi_4$	1_A^{++}	1_A^{--}
$\bar{\chi}\eta_1 D_1\{\eta_2 D_2\{\eta_3 D_3\chi\}\}$	$\gamma_4\gamma_5$	$\xi_4\xi_5$	0_S^{-+}	0_A^{++}
$\eta_4\zeta_4\bar{\chi}\eta_1 D_1\{\eta_2 D_2\{\eta_3 D_3\chi\}\}$	$\gamma_5\xi_4$	$\gamma_4\xi_5$	0_A^{-+}	0_A^{+-}
$\eta_k\varepsilon\zeta_k\bar{\chi}\eta_1 D_1\{\eta_2 D_2\{\eta_3 D_3\chi\}\}$	$\gamma_k\gamma_4\xi_k\xi_5$	$\gamma_k\gamma_5\xi_k\xi_4$	1_A^{--}	1_A^{++}
$\eta_4\zeta_4\eta_k\varepsilon\zeta_k\bar{\chi}\eta_1 D_1\{\eta_2 D_2\{\eta_3 D_3\chi\}\}$	$\gamma_k\xi_l\xi_m$	$\gamma_l\gamma_m\xi_k$	1_A^{--}	1_A^{+-}

table 2 Operators of table 1 with the lattice states excited by them and the corresponding continuum states with lowest spin.

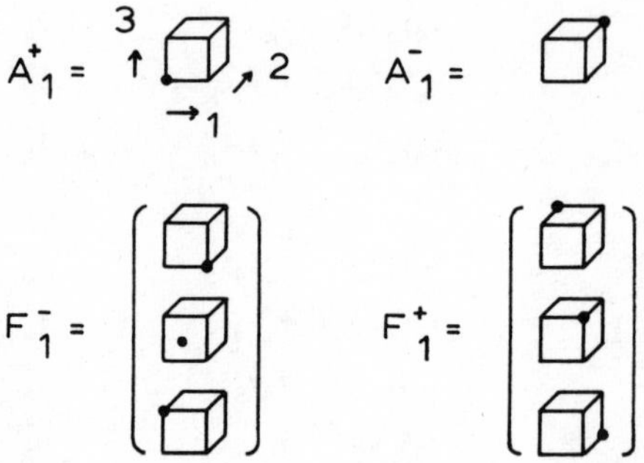

fig. 1 Pictorial notation for the decomposition (31) of χ_A .

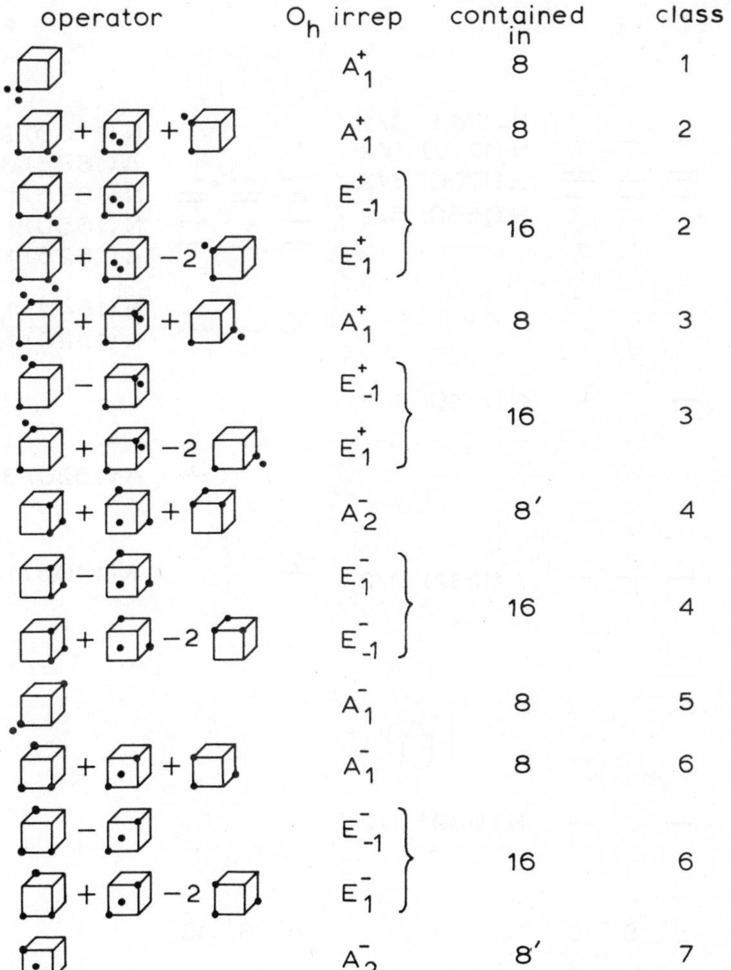
operator
O_h irrep
contained in
class
A_1^+ 8 1
A_1^+ 8 2
E_{-1}^+
16 2
E_1^+
A_1^+ 8 3
E_{-1}^+
16 3
E_1^+
A_2^- 8′ 4
E_1^-
16 4
E_{-1}^-
A_1^- 8 5
A_1^- 8 6
E_{-1}^-
16 6
E_1^-
A_2^- 8′ 7

fig. 2 Irreducible baryon operators contained in $\tilde{B}_{ABC}$. A dot at site $\vec{a}_A$ indicates $D_A\chi(\vec{x})$. Components are shown by the irreps $A_1^{\pm}$, $A_2^{\pm}$ and $E^{\pm}$ of the O_h subgroup of GTS. The subscript of E indicates the eigenvalue of $R^{(12)}$. All other components can be obtained by the application of shifts.

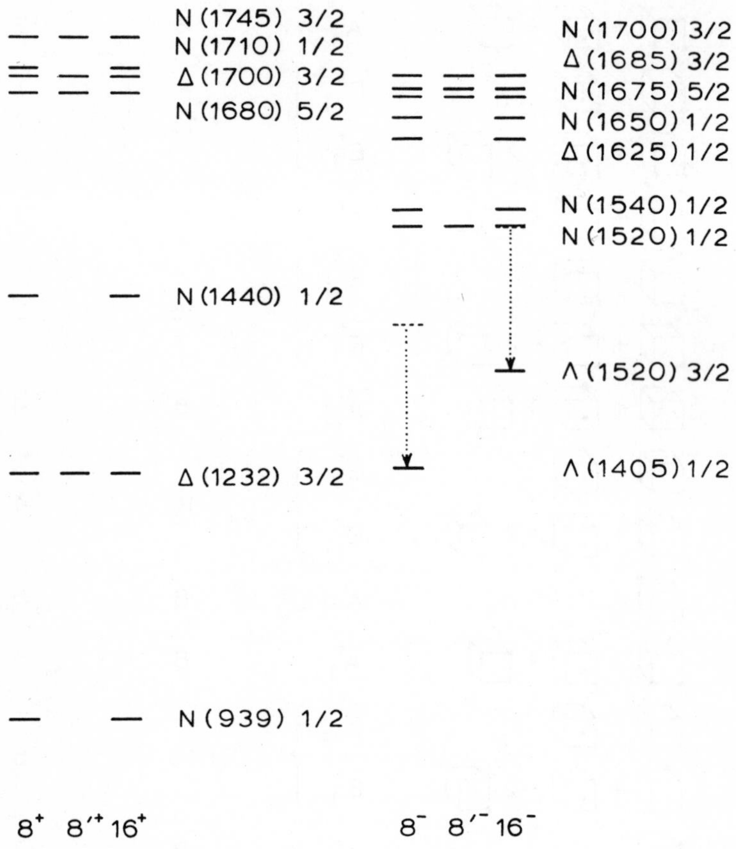

fig. 3 Tentative assignment of baryon resonances to the spectra of $\underset{\sim}{8}^{\pm}$, $\underset{\sim}{8}'^{\pm}$ and $\underset{\approx}{16}^{\pm}$. The symbols N, Δ and Λ refer to the flavor irreps 20_M, 20_S and 4_A, respectively.

SIMULATING LATTICE GAUGE THEORY WITH FERMIONS

John B. KOGUT
Department of Physics, University of Illinois at Urbana-Champaign, 1110 West Green Street, Urbana, IL 61801

ABSTRACT

We present a progress report in lattice gauge theory computer simulations which include the effects of light, dynamical fermions. Microcanonical and hybrid microcanonical-Langevin alogrithms are presented and discussed. Physics applications such as the thermodynamics of Quantum Chromodynamics, hierarchal energy scales in unified gauge theories, and the phase diagram of theories with many fermion species are discussed. Prospects for future research are assessed.

I. LATTICE GAUGE THEORY WITH FERMIONS

The four dimensional Euclidean Action density S for lattice gauge theory with fermions reads generically,

$$S = \sum_{ij} \bar{\psi}_i [\not{D}(u)+m]_{ij} \psi_j + S_o(u) \tag{1.1}$$

where ψ_i is a Grassman field at site i, $A_{ij} = [\not{D}(u)+m]_{ij}$ is the gauge covariant Dirac operator and $S_o(u)$ is the pure gauge field Action on the lattice.[1] The precise form of the gauge covariant discrete difference operator $\not{D}(u)$ depends on the lattice fermion method employed. We will be considering staggered fermions[2] in this article so ψ_i will be one component objects and the fermion contribution to Eq. (1.1) reads,

$$\sum_n \bar{\psi}(n)\{\frac{1}{2}\sum_{\mu=1}^{4} \eta_\mu(n)[U_\mu(n)\psi(n+\mu)-U_\mu^+(n-\mu)\psi(n-\mu)]+ m\psi(n)\} \quad (1.2)$$

where $\eta_\mu(n)$ are phase factors that carry the spin-1/2 character of the continuum Dirac field and $U_\mu(n)$ is the SU(3) rotation matrice residing on the link between sites n and n + μ. For the purposes of this discussion all these details are not essential. Suffice it to say that Eq. (1.2) has the good feature of describing four species of Dirac fermions which become massless when $m \to 0$ in a natural fashion. $\langle\bar{\psi}\psi\rangle$ is a good order parameeter for chiral symmetry, one of the two basic quantities (confinement is the other) of interest here.

Since the subject of this talk is the status of computer simulations of lattice gauge theory with fermions, our interest focuses on the partition function,

$$Z = \int \prod_i d\psi_i \prod_j d\bar{\psi}_j \prod_{n,\mu} dU_\mu(n)\exp(-S) \quad (1.3)$$

Since the ψ_i are anti-commuting numbers a direct simulation of Eq. (1.3) is not practical. Instead the fermions can be integrated out of Eq. (1.3) since Eq. (1.1) is a quadratic form in ψ,

$$Z = \int \prod dU_\mu(n) \det[\not{D}(u)+m] \exp(-S_o(u))$$

$$= \int \prod dU_\mu(n) \exp(-S_o(u) + \mathrm{tr}\, \ell n[\not{D}(u)+m]) \quad (1.4)$$

It is not so clear, however, that this step represents real progress since tr $\ell n[\not{D}(u)+m]$ is an effective, non-local interaction among the U-variables. Such Actions are not well studied and classified in the context of traditional statistical mechanics approaches to critical phenomena. At least the determinant in Eq. (1.4) is positive semi-definite for staggered fermions.

We all recognize the physical origin for the determinant here. It represents closed fermion loops, virtual quark-antiquark pairs, and the plus sign, +tr $\ell n[\not{D}(u)+m]$, in Eq. (1.4) is responsible for the perturbation theory rule: -1 for each closed fermion loop.

Various numerical approaches to evaluating Eq. (1.4) and physically relevant matrix elements have been proposed. Monte Carlo methods, the so-called pseudo-fermion algorithms[3], are being studied as well as microcanonical[4,5] and Langevin equations.[5] I will concentrate on the latter two methods in this review. At this time all such algorithms are controversial -- we have not studied enough cases with enough computer power to delineate the clear successes and limitations of each method. However, such studies are being vigorously pursued at this time and solid answers concerning the reliability, scope and error estimates in each method should be forthcoming.

II. THE MICROCANONICAL ENSEMBLE AND MOLECULAR DYNAMICS

We begin by reviewing the molecular dynamics approach[4] to problems in equilibrium statistical mechanics. Consider a boson field ϕ which might be defined on a lattice. The theory has an action $S(\phi)$ which determines its Path Integral and equilibrium statistical mechanics properties. This system has no natural dynamics which would govern its approach to equilibrium. However, it can be given dynamics in several ways -- the molecular dynamics and the Langevin equations are two alternatives. In the molecular dynamics approach we associate $S(\phi)$ with a potential $V(\phi) \equiv \beta^{-1}S(\phi)$ and construct a fictitious Hamiltonian,

$$H = T + V = \sum_i \frac{1}{2} p_i^2 + V(\phi) \tag{2.1}$$

where i labels lattice sites and p_i will soon be interpreted as the momentum conjugate to ϕ_i. Using Eq. (2.1) we could consider the classical statistical mechanics based on the invariant phase space $\prod_i dp_i d\phi_i$ and the Boltzmann factor $\exp(-\beta H)$. Since the p_i-integrals are trivial, ,this formulation reduces to the original Path Integral formulation of the boson field theory.

To give this approach some meat, we identify p_i with the momentum conjugate to ϕ_i by introducing a 5th dimension τ into the problem,

$$p_i = d\phi_i/dt \tag{2.2}$$

Then the ensemble given by the phase space measure $\prod_i dp_i d\phi_i$ and the Boltzmann factor $\exp(-\beta H)$ defines the usual canonical ensemble of classical statistical mechanics. There is still no advantage in all this until one passes to the microcanonical ensemble. Now the energy is fixed $H = E$ and the measure in phase space is $\prod_i dp_i d\phi_i \delta(H-E)$. Observables in the system $\theta(p,\phi)$ have expectation values,

$$\langle\theta\rangle = \frac{1}{z} \int \prod_i dp_i d\phi_i \delta(H-E)\ \theta(p,\phi) \tag{2.3}$$

If θ is just a function of ϕ, then standard arguments apply to show that $\langle\theta\rangle$ calculated in the microcanonical ensemble is the same as $\langle\theta\rangle$ calculated in the canonical ensemble in the large volume $V \to \infty$ limit.[6]

But $\langle\theta\rangle$ can also be calculated from the time evolution of the classical system. This is the molecular dynamics approach to the problem. Let $(\phi(\tau),p(\tau))$ describe the phase space point of the physical system. Then a time-average of θ can be calculated,

$$\langle\theta\rangle = \lim_{T\to\infty} \frac{1}{T} \int_0^T \theta(p(\tau),\phi(\tau))d\tau \tag{2.4}$$

This time average reproduces the expectation value Eq. (2.3) if the Ergodic Hypothesis works for this physical system. Roughly speaking, one must assume that the Hamiltonian dynamics of the system carries the phase space point $(\phi(\tau),p(\tau))$ uniformly over the energy shell $H = E$.

The final ingredient in this molecular dynamics approach is the computation of the coupling β given the system's fixed energy. The necessary correspondence follows from the equi-partition theorem,

$$\langle T \rangle = \frac{1}{2} \beta^{-1} N \qquad (2.5)$$

where N is the number of independent, excited degrees of freedom in the system.

Eq. (2.4) and (2.5) coupled with the Hamilton equations of motion following from Eq. (2.1) and (2.2) represent a clear alternative to Monte Carlo simulation procedures of pure bose systems. This formulation has several interesting points: (1) It is fully deterministic, (2) It involves ordinary coupled differential equations and (3) It generalizes to a practical method for fermions. Let's review the fermion method before discussing its strengths and weaknesses further.

Now we wish to invent a classical system in 4+1 dimensions involving only complex numbers whose molecular dynamics generates the Path Integral Eq. (1.4) with the infamous fermion determinant. Consider the Lagrangian[5],

$$L = -S_o(u) + \frac{1}{2} \sum_{n,\mu} \dot{U}^{\dagger}_{\mu}(n)\hat{P}\dot{U}_{\mu}(n) + \sum_{ij} \dot{\phi}^{\dagger}_{i}[A^{\dagger}A]_{ij}\dot{\phi}_{j} - \omega^2 \sum_{i} \phi^{\dagger}_{i}\phi_{i} \qquad (2.6)$$

where A is the lattice Dirac operator defined earlier and $\hat{P}$ is a projection operator. Thus L consists of kinetic energy terms for the gauge fields and the pseudo-fermions, and potential terms for both. $A^{\dagger}A$ appears in L rather than A itself to insure positivity. This unusual form for the pseudo-fermion kinetic energy will generate the fermion determinant with the correct sign. Note that this L is local because A couples only nearest neighbors.

It is straight-forward to identify the canonical momenta of this physical system,

$$p_{\mu}(n) = \dot{U}_{\mu}(n) \qquad P_i = [\dot{\phi}^{\dagger}A^{\dagger}A]_i \qquad (2.7)$$

and construct the Hamiltonian

$$H = \frac{1}{2} \sum p^2 + \sum P^{\dagger}(A^{\dagger}A)^{-1}P + S_o(u) + \omega^2 \sum \phi^{\dagger}\phi \tag{2.8}$$

and consider the Hamiltonian equations of motion,

$$\dot{P}^{\dagger} = \frac{d}{d\tau}\left(A^{\dagger}(u)A(u)\dot{\phi}\right) = -\omega^2 \phi$$

$$\dot{p} = \ddot{U} = -\frac{\partial}{\partial u^{\dagger}} S_o(u) + \dot{\phi}^{\dagger} \frac{\partial}{\partial u^{\dagger}}\left(A^{\dagger}(u)A(u)\right)\dot{\phi} \tag{2.9}$$

These equations are generic in character. The real equations which are simulated choose a convenient parametrization for the $U_\mu(n)$ matrices and incorporate constraints appropriately.[7] But the point to be stressed here is simply that Eq. (2.9) is a tractable set of coupled ordinary differential equations. The fermions introduce the complication of requiring the solution of a sparse set of linear equations for $\dot{\phi}$ of the form $A^{\dagger}A\dot{\phi} = \ldots$ for <u>each</u> time step. This is done very efficiently with good control of errors by standard methods such as the conjugate-gradient algorithm. As the bare quark mass approaches zero, these iterative sparse matrix algorithms require more computer time, but they prove to be quite practical.[8]

Our last task is to check that L really gives the original Path Integral. The canonical ensemble based on Eq. (2.8) reads,

$$Z = \int DuDpD\phi D\phi^{\dagger}DPDP^{\dagger} \exp(-H/T) \tag{2.10a}$$

All the variables except U enter H quadratically, so the integrals can be done,

$$Z = \text{const.} \int DU \det{}^2 A(u) \exp(-S_o(u)/T) \tag{2.10b}$$

which is the required answer except for the second power of the determinant. However, since $A^{\dagger}A$ in the staggered fermion method does not couple nearest neighbor pseudo-fermion fields, ϕ can be set to zero on every other lattice site.[8] In this final scheme $\det^2 A$ is replaced by $\det A$.

Now we see clearly the character of the tricks in Eq. (2.6) and (2.8). The pseudo-fermion kinetic energy in L is "$\frac{1}{2} mv^2$" with $m \sim A^\dagger A$. When the H is constructed we have "$p^2/2m$" and the $(A^\dagger A)^{-1}$ here was responsible for the positive power of $\det A^\dagger A$ in Eq. (2.10b). The nice feature of this scheme is that the full non-local character of the determinant is avoided by the algorithm. In each time step $A^\dagger A\dot{\phi} = \ldots$ is solved for $\dot{\phi}$ -- this is a local operation since $A^\dagger A$ only couples nearby degrees of freedom.

The last ingredient in the algorithm is the calculation of the coupling constant β. If we identify the number of active, independent degrees of freedom N^* of the system, this can be done using the equipartition theorem,

$$\frac{1}{2} \beta^{-1} N^* = \langle T \rangle = \langle \dot{\phi} A^\dagger A \dot{\phi} + \frac{1}{2} \sum \dot{U}^2 \rangle \tag{2.11}$$

The calculation of N^* for particular parametrizations of the U matrices is discussed in ref. 7.

III. MOLECULAR DYNAMICS APPROACH TO THE CANONICAL ENSEMBLE

The "naive" microcanonical fermion + gauge field algorithm of Sec. 2 can be generalized and improved in many ways. Let's discuss a variation on the original method which has three interesting features:

1. It is completely deterministic.
2. It simulates the canonical ensemble.
3. It treats β as an input rather than an output variable.

The idea here is to add one degree of freedom s which will act as a heat bath for the original microcanonical system. If its kinetic and potential energies can be chosen appropriately, properties 1-3 follow. Since the new variable changes the system from one at fixed energy to one at fixed temperature, we will call it a "demon" following a similar, but different, idea used for the Ising model.[9]

Let's illustrate the idea for a set of N point particles[10],

$$L = \sum_i \frac{1}{2} m_i \dot{\underset{\sim}{r}}_i^2 - \phi(\{\underset{\sim}{r}\}) \tag{3.1}$$

which could be simulated by the usual molecular dynamics equation. Instead, introduce a demon s and a Lagrangian describing the system of N + 1 particles,

$$L = \sum_i \frac{1}{2} m_i s^2 \dot{r}_i^2 - \phi(\{\underset{\sim}{r}\}) + \frac{1}{2} Q\dot{s}^2 - (N+1)\, T \,\ell n\, s \tag{3.2}$$

and simulate the equations of motion here. To see that the new system describes the original N point particles at temperature T, form the Hamiltonian from Eq. (3.2),

$$H = \sum_i \underset{\sim}{p}_i^2/2m_i s^2 + \phi(\{r\}) + p_s^2/2Q + (N+1)T\, \ell n\, s \tag{3.3}$$

and consider the microcanonical ensemble,

$$Z = \int dp_s ds \prod_i d\underset{\sim}{p}_i d\underset{\sim}{r}_i\, \delta(H-E) \tag{3.4}$$

Rescale $\underset{\sim}{p} \to \underset{\sim}{p}/s$, do the s integral using the delta function and do the p_s Gaussian integral trivially, to find

$$Z = \int \prod_i d\underset{\sim}{p}_i d\underset{\sim}{r}_i \exp(-[\sum_i \underset{\sim}{p}_i/2m_i + \phi(\{\underset{\sim}{r}\})]/T) \tag{3.5}$$

which the desired answer.

In retrospect we see that the logarithmic potential for the demon was essential to generate the Boltzmann factor in Eq. (3.5).

The nice features of the molecular dynamics simulation of Eq. (3.2) are (1.) T can be chosen as an input variable, (2.) 1/V effects which distinguish the microcanonical and canonical ensembles are suppressed and (3.) Equi-partition can be monitored clearly through $\langle p_s^2/2Q\rangle = T/2$.

This approach to field theory simulations has been tested on a number of systems. The two dimensional planar spin model was simulated by Monte Carlo, naive microcanonical and the demon algorithms. The average Action, topological charge (the theory has a vortex driven phase transition first described by Kosterlitz and Thouless), spin-spin correlation function and demon kinetic energies were monitored. The Monte Carlo and demon simulations were in excellent agreement on 15^2 and 30^2 lattices and their expectation values differed only at 1/V (V = volume) effects from the naive microcanonical results.

The demon trick is easily generalized to gauge theories with fermions. The Lagrangian becomes,

$$L = \frac{1}{2} \sum s^2 \operatorname{tr} \dot{U}^\dagger \hat{P} \dot{U} - \beta S_o(u) + \dot{\phi}^\dagger A^\dagger(u) A(u) \dot{\phi} - \omega^2 \phi^\dagger \phi / s^2 + \frac{1}{2} Q \dot{s}^2 - (N+1)\, T \ln s \tag{3.6}$$

Long runs (15,000 sweeps) have been carried out at $\beta = 6/g^2 = 5.512$ on a 4×8^3 lattice at fermion mass of $m = 0.10$ where there is extensive microcanonical data. The two algorithms were in very good agreement but the demon results showed less severe long time correlations as one would hope.

IV. FACING ERGODICITY BREAKING SQUARELY

Chemical physicists have considerable experience with molecular dynamic simulations of systems containing 5-50 degrees of freedom. These systems can be mapped onto a polymer of 5-50 monomers which interact through strong nearest neighbor harmonic forces perturbed by weaker anharmonic effects. In these cases the simplest molecular dynamic algorithms fail badly because the normal modes do not exchange energy on short enough time scales for practical simulations.[11] In fact, if the anharmonic forces in the system are weak enough the KAM theorem applies which implies that the system will not sample the energy surface uniformly. In these cases of relatively few degrees of freedom ergodicity breaking is easily monitored in the simulation.

The breaking is clear and obvious. In our field theory applications it is harder to monitor potentially disasterous effects such as these. However, in asymptotically free theories where the ultraviolet fixed point lies at vanishing coupling, we must expect trouble with ergodicity as the continuum limit of the lattice theory is made. In addition, at strong coupling where correlation lengths are small ergodicity breaking is also expected. In simulations at intermediate couplings obvious failures of ergodicity have not been formed in SU(2) and SU(3) gauge theories on "large" lattices (8^4, $8^3 \times 16$, 6×12^3, for example), but some observables have shown dangerous long time correlations. Certainly the microcanonical algorithm should be improved. In fact, the physical chemists have adopted the molecular dynamics technique to physical systems which are not ergodic. The "quick fix" they use is simply to "refresh" the velocities in the system from time to time, i.e. the velocities $v_i(\tau)$ are put into a Boltzmann distribution at τ_o, the system is evolved from τ_o by molecular dynamics to the time $\tau_o + \tau^*$ where the velocities are "refreshed" again, etc. This method has become the standard for many chemistry problems and has been discussed by B. Berne, H. Andersen and others extensively.[10,11]

Luckily there is more to such "quick fixes" than just guesswork. They are closely related to Langevin dynamics and can be placed on a solid theoretical footing.[12] We will refer to such schemes as "hybrids" -- they combine the strong points of the naive microcanonical and the Langevin algorithms into a new, improved method.

Consider a simple example: an Action S(q) and a bose variable q.[12] We want to calculate an expectation value,

$$\langle F(q)\rangle = \frac{1}{Z}\int dq\ F(q)\ \exp(-S(q)) \tag{4.1}$$

In the microcanonical approach, the system is given dynamics,

$$\ddot{q}(\tau) = -\partial S(q)/\partial q \tag{4.2}$$

and expectation values are replaced by time averages,

$$\langle F(q)\rangle = \lim_{T\to\infty} \frac{1}{T} \int_0^T dt\ F(q(\tau)) \tag{4.3}$$

In the Langevin approach, the system is given dynamics with explicit white noise,

$$\dot{q}(\tau) = -\partial S/\partial q + \eta(\tau)$$

$$\langle \eta(\tau)\eta(\tau')\rangle = 2\delta(\tau-\tau') \tag{4.4}$$

and Eq. (4.3) is applied again. The stochastic differential equation Eq. (4.4) causes $q(\tau)$ to execute a forced random walk such that it covers phase space with the weight $\exp(-S(q))dq$.

There is an intimate relation between Eq. (4.2) and (4.4) which is best seen by replacing the differential equations by discrete difference equations. The Langevin system becomes,

$$q_{n+1} = q_n + \Delta\xi_\mu - \frac{1}{2}\Delta^2 S'(q_\mu)$$

$$\xi_n \xi_{n'} = \delta_{nn'}$$

$$\tau_{n+1} - \tau_n = \frac{1}{2}\Delta^2 \tag{4.5}$$

which the microcanonical reads,

$$q_{n+1} = 2q_n - q_{n-1} - \Delta^2 S'(q_n)$$

$$\tau_{n+1} - \tau_n = \Delta \tag{4.6}$$

which can be written more suggestively as,

$$q_{n+1} = q_n + \frac{1}{2}(q_{n+1} - q_{n-1}) - \frac{1}{2}\Delta^2 S'(q_n) \tag{4.7}$$

So, we have the correspondences,

Langevin	Microcanonical
noise	velocity
$\tau_{n+1} - \tau_n = \frac{1}{2}\Delta^2$	$\tau_{n+1} - \tau_n = \Delta$

Each scheme has the following features. (1.) Langevin has explicit noise, so it is ergodic by construction. However, it samples the phase space very slowly in many cases becuase $q(\tau)$ executes a forced random walk which, if the noise term dominates, fills space at a rate $\sqrt{N}$, N = number of time steps of Eq. (4.5). (2.) Microcanonical dynamics follows the classical equations of motion so it is as efficient as possible in probing the important regions of phase space locally. Also, its time step is large, so it moves along its trajectories rapidly. However, it may not be ergodic, i.e. for long times it may get trapped into regions of phase space which are only accessible to the Langevin simulation because of its explicit noise.

All this suggests a hybrid method which combines the best features of both algorithms.[12] In the hybrid scheme a time step will be either Langevin or microcanonical with a certain probability $p\Delta$:

$$q_{n+1} = q_n + \Delta v_n - \frac{1}{2}\Delta^2 S'(q_n)$$

with

$$v_n = \begin{cases} (q_{n+1} - q_{n-1})/2\Delta & , \text{ probability } p\Delta \\ \xi_n & , \text{ otherwise} \end{cases}$$

$$\tau_{n+1} = \tau_n + \Delta$$

One can attempt to optimize this algorithm by choosing p appropriately. The generic best choice is to set p to the frequency of the "slowest mode" in the system.[12] Inspecting the observables of a typical microcanonical run one can find slowly relaxing modes and estimate p. The idea here is the following: The microcanonical algorithm permits the system to sample a finite region of phase space efficiently for a fixed time interval. That time should be as long as the period of the slowest mode in the system. Then it is best to "refresh" the system, so it can move to a completely new region of phase space. This is done with the (relatively unlikely) Langevin step.

These ideas can be implemented for lattice gauge theory with fermions. In fact, there is considerable freedom here and various hybrid simulation methods are being tested. The results are encouraging and suggest that improved algorithms will be forthcoming for this difficult system.[13]

V. QUANTUM CHROMODYNAMICS SIMULATIONS

Now let's discuss the status of large scale simulations of SU(2) and SU(3) gauge theories with four light, dynamical Dirac fermions -- simulations close to the real theory QCD. Various projects are in progress.

First is the thermodynamics of the continuum field theory. Here one wants to understand QCD at finite temperature and study the transition from hadronic matter to a quark-gluon plasma. One wants to know if there are true non-analyticities in the thermodynamic quantities of interest such as the entropy and internal energy densities. The SU(2) and SU(3) theories without fermion feedback showed such non-analyticities and their behavior is well-understood in the context of traditional statistical mechanics. The situation is relatively unclear when fermion feedback is accounted for and the subject is quite controversial. A chiral restoring transition is certainly present, but fermion screening may be qualitatively similar for all temperatures rendering the thermodynamics of the "transition" smooth. This is a particularly interesting question for the groups developing fermion algorithms, because they are completely dependent on numerical methods for the answers to these physics questions.

The SU(2) theory has been studied on a 6×12^3 lattice at 6 β values and three fermion mass values (0.10, 0.075 and 0.050) with $5 \cdot 10^3$ to 10^4 sweeps of the microcanonical algorithm for each point. The SU(2) spectrum has been studied similarly on a $8^3 \times 16$ lattice and an analogous SU(3) project is underway.

In Fig. 1 I show the scaling regions of the pure SU(2) theory and the theory with $N_f = 4$ species of fermions. The agreement with asymptotic freedom for $\langle\bar{\psi}\psi\rangle$ is quite nice. Note that fermion feedback shifts the $\langle\bar{\psi}\psi\rangle$ curve toward stronger coupling as N_f is increased <u>and</u> the slope of $\ell n\langle\bar{\psi}\psi\rangle$ vs. β changes appropriately. It appears that the scaling region for the $N_f = 4$ theory begins at $\beta = 1.85$.

In Fig. 2 I show $\langle\bar{\psi}\psi\rangle$ extrapolated to zero mass and the Wilson line for a 6×12^3 lattice in the SU(2), $N_f = 4$ theory. It appears that the transition from hadron to quark-gluon matter is abrupt. It is crucial to confirm or refute this result with other algorithms.

VI. HIERARCHY PROBLEMS IN UNIFIED GAUGE THEORIES

I want to illustrate that lattice methods can be applied to theories "beyond QCD" which might have interesting mass scales at arbitrarily high energies. Unfortunately, the most interesting schemes involve chiral fermions and these cannot be attacked by lattice methods because we cannot place a

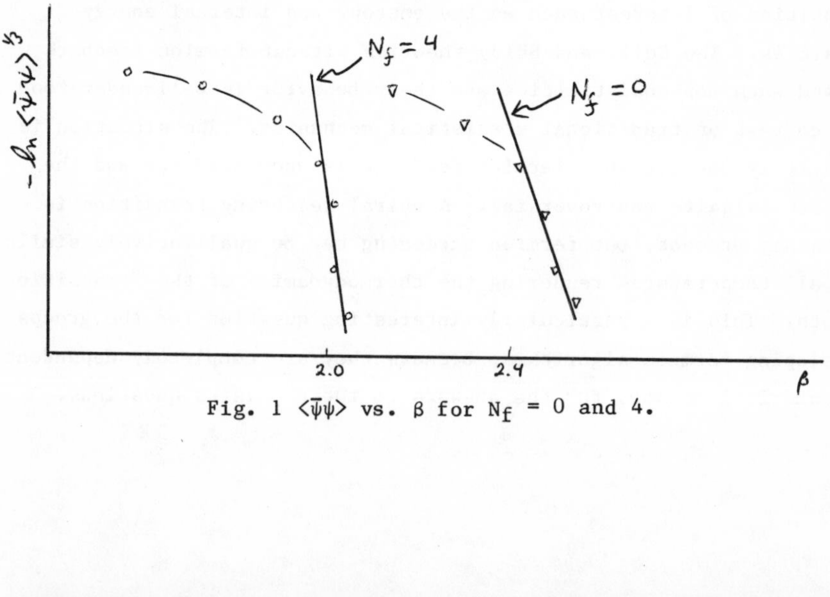

Fig. 1 $\langle\bar{\psi}\psi\rangle$ vs. β for $N_f = 0$ and 4.

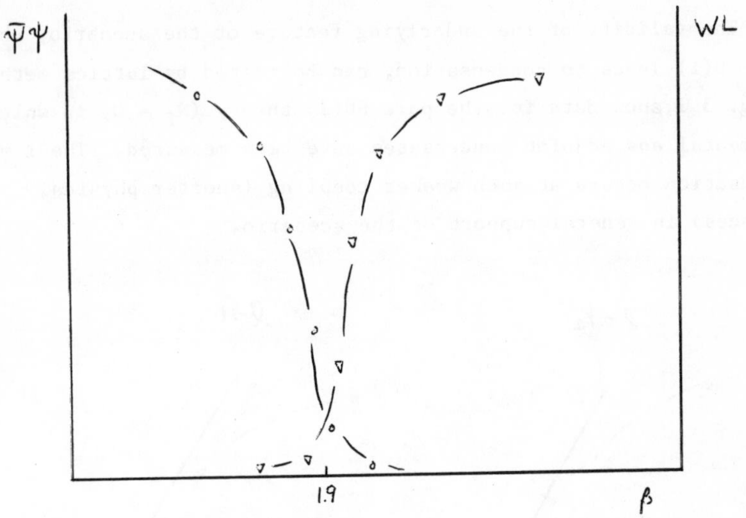

Fig. 2 $\langle\bar{\psi}\psi\rangle$ and the Wilson line for SU(2), $N_f = 4$, 6×12^3 lattice.

single neutrino on the lattice with a conventional Action. Anyway, in the realm of vector theories we can ask whether a theory can support disparate mass scales without the need to fine tune a fundamental parameter. Chiral symmetry breaking and asymptotic freedom can conspire to do this, as suggested in the present context by Raby, Dimopolous and Susskind.[14] By considering single gluon exchange they suggest that when $C_f g^2 \sim 0(1)$ massless fermions of color charge C_f will condense into a chiral condensate. By asymptotic freedom, this criterion leads to an exponential sensitively of the characteristic energy scale of the condensate to the fermion's color charge. Changes of scale of 10^{5-10} are possible in such "technicolor" schemes although realistic models do not exist.

The validity of the underlying feature of the scenerio, that $C_f g^2 \sim O(1)$ leads to condensation, can be tested by lattice methods. In Fig. 3 I show data for the pure SU(2) theory ($N_f = 0$) in which fundamental and adjoint condensates have been measured. The $\ell = 1$ condensation occurs at much weaker coupling (shorter physical distances) in general support of the scenerio.

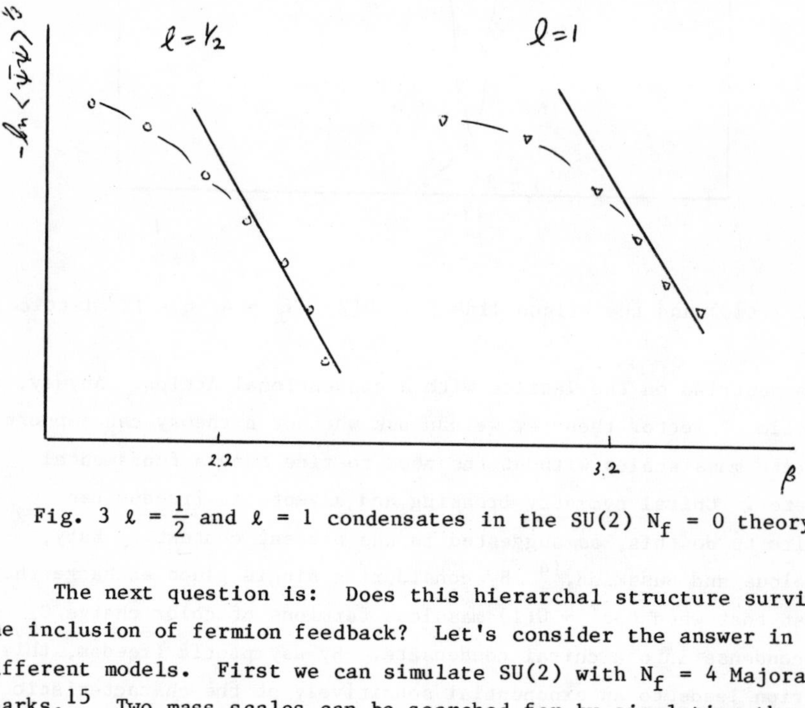

Fig. 3 $\ell = \frac{1}{2}$ and $\ell = 1$ condensates in the SU(2) $N_f = 0$ theory.

The next question is: Does this hierarchal structure survive the inclusion of fermion feedback? Let's consider the answer in two different models. First we can simulate SU(2) with $N_f = 4$ Majorana quarks.[15] Two mass scales can be searched for by simulating the theory of finite temperature and measuring $\langle\bar{\psi}\psi\rangle$ for the $\ell = 1$ quarks and the string tension for $\ell = 1/2$ static quarks. In Fig. 4 I show data from a 4×8^3 simulation depicting $\langle\bar{\psi}\psi\rangle$ and the Wilson line. Clearly the deconfinement and the chiral symmetry restoration temperatures are distinct. This is an encouraging result.

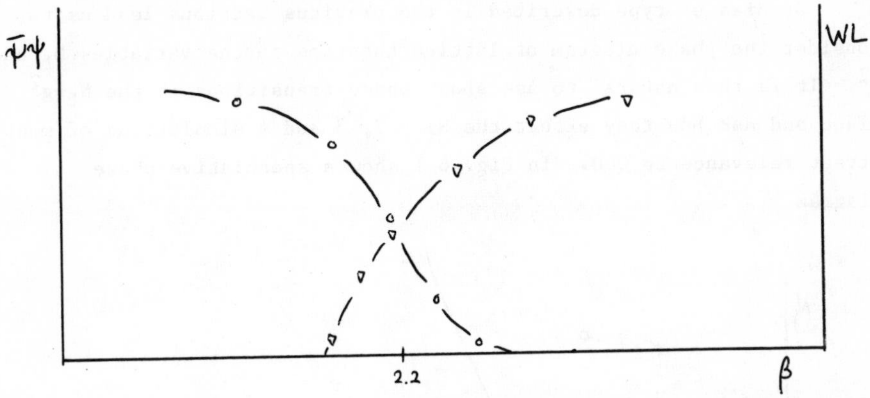

Fig. 4 $\langle\bar{\psi}\psi\rangle$ and the Wilson line for SU(2) theory with adjoint quarks.

And finally we can simulate a model with so much fermion feedback that asymptotic freedom is lost but the problem of multiple energy scales can be posed in the cutoff theory. Consider SU(2) with 4 flavors of fundamental Dirac fermions and 4 flavors of adjoint Majorana fermions. The results of a finite temperature simulation (4×8^3 lattice) are shown in Fig. 5 and support the hierarchy picture.

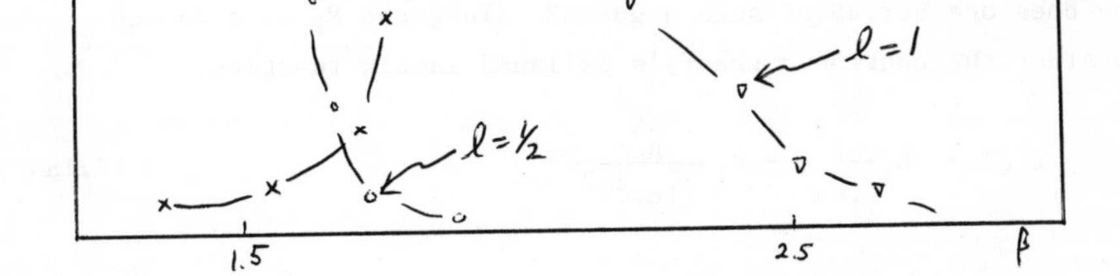

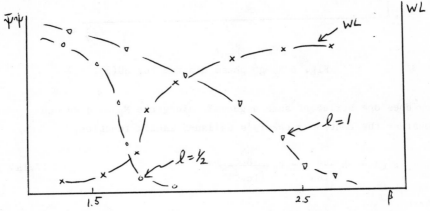

Fig. 5 $\ell = \frac{1}{2}$ and 1 condensates and the Wilson line vs. β.

VII. LATTICE GAUGE THEORY WITH MANY FLAVORS

Studies of type described in the previous Sections lead us to consider the phase diagram of lattice theories in the variables N_f and g^2. It is then natural to ask about phase transitions in the N_f-g^2 plane and ask how they effect the N_f = 2, 3 and 4 simulations of most direct relevance to QCD. In Fig. 6 I show a speculative phase diagram.[16]

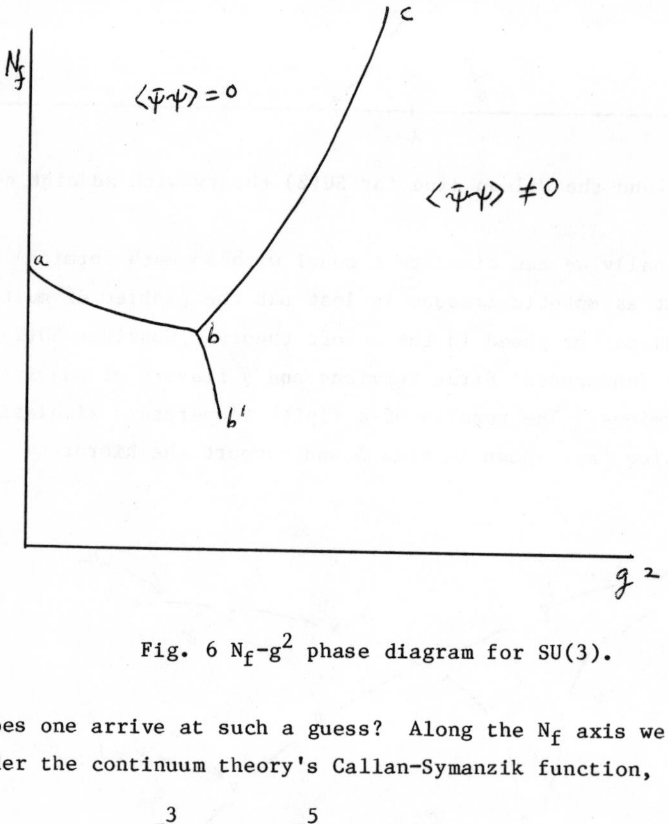

Fig. 6 N_f-g^2 phase diagram for SU(3).

How does one arrive at such a guess? Along the N_f axis we can consider the continuum theory's Callan-Symanzik function,

$$\beta(g) = -\beta_o \cdot \frac{g^3}{16\pi^2} - \beta_1 \cdot \frac{g^5}{(16\pi^2)^2} \tag{7.1a}$$

where

$$\beta_o = 11-2N_f/3 \qquad \beta_1 = 102-38N_f/3 \tag{7.1b}$$

Note that β_o changes sign at $N_f = 16.50$ and β_1 changes sign at $N_f = 8.05$. For $N_f > 16.50$ and $g^2 \approx 0$ the theory is not asymptotically free, so a small g^2 at short distances gives rise to a yet smaller g^2 at larger distances. This strongly suggests that $\langle\bar{\psi}\psi\rangle = 0$ in this region of the phase diagram. However, for large N_f and large g^2 strong coupling expansions imply that $\langle\bar{\psi}\psi\rangle \neq 0$. These last two observations suggest that the N_f-g^2 phase diagram separates into two parts labelled with the order parameter $\langle\bar{\psi}\psi\rangle$.

Note that if the $\langle\bar{\psi}\psi\rangle = 0$ region dips as shown in Fig. 6 the crossover from strong to weak coupling of the $N_f = 3$ theory will be effected by the nearby structure. It is natural to speculate that very abrupt crossover phenomena is present in the $N_f = 3$ theory as a consequence of the rich N_f-g^2 phase diagram. Such effects may obscure the approach to continuum behavior of asymptotically free theories with fermions.

Some evidence for the line of transitions on Fig. 6 has already appeared in computer simulations.[17] Fig. 7 shows a microcanonical simulation of SU(3) with $N_f = 12$ on a small lattice. A clear signal for a first order transition as a function of β is seen.

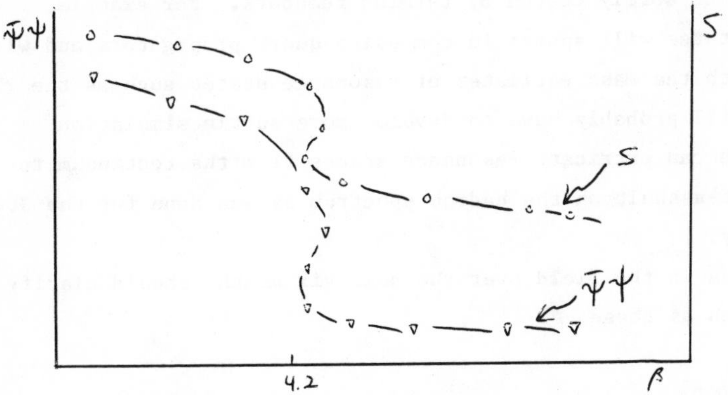

Fig. 7 $\langle\bar{\psi}\psi\rangle$ and the Action for SU(3), $N_f = 12$.

In fact this figure reminds us of a useful feature of microcanonical simulations: the absence of energy fluctuations can stabilize metastable states on finite systems leading to particularly clear evidence for first order transitions.[7]

Further studies of this type should elucidate Fig. 6. Simulations at variable N_f should reach the small g^2, $N_f \sim 8$–16 region of the diagram where little theoretical insight is available. For large N_f and large g^2 the transition line is expected to be first order[16] as found in Fig. 7.

VIII. FUTURE DIRECTIONS

The field of computer simulations of fermion systems is in its infancy. Algorithm development, testing and error analysis, are crucial projects here. This is a controversial field at the moment and direct comparisons of different fermion algorithms are needed before proceeding to additional applications.

With a reliable algorithm in hand it will be particularly interesting to study the mass spectrum of QCD with fermion feedback. The validity of the naive quark model, and the role of the axial anomaly and topology in the spectrum and thermodynamics will become topics of study. The brute force measurement methods of present day studies will be sorely tested by fermion feedback. For example, multi-pion states will appear in composite quark propagators and will interfere with the mass estimates of resonance states such as the rho meson. We will probably have to develop more subtle simulation methods which can extricate resonance states from the continuum to make a direct assault on the hadron spectrum as was done for the SU(3) $N_f = 0$ theory.

Progress in the field over the next six months should clarify questions such as these.

ACKNOWLEDGEMENT

This work is partially supported by the National Science Foundation under grant number NSF-PHY82-01948.

REFERENCES

1) Wilson, K. G., Phys. Rev. D14, 2455 (1974).

2) Kogut, J.B. and Susskind, L., Phys. Rev. D9, 3501 (1974); Phys. Rev. D11, 395 (1975).
Susskind, L., Phys. Rev. D16, 3031 (1977).

3) Weingarten, D. and Petcher, D., Phys. Lett. B99, 33 (1981).
F. Fucito, E. Marinari, G. Parisi and G. Rebbi, Nucl. Phys. B180[FS2], 369 (1981).

4) Callaway, D. and Rahman, A., Phys. Rev. Lett. 49, 613 (1982).

5) Polonyi, J. and Wyld, H.W., Phys. Rev. Lett. 51, 2257 (1983).
Polonyi, J., Wyld, H.W., Kogut, J.B., Shigemitsu, J. and Sinclair, D.K., Phys. Rev. Lett. 53, 644 (1984).

6) Guha, A. and Lee, S.-C., Phys. Rev. D27, 2412 (1983).
Ukawa, A. and Fukugita, M., University of Tsukuba preprint, March, 1985.
Batrouni, R., seminar at the Workshop on Lattice Gauge Theory, Florida State University, April 1985.

7) Kogut, J., Polonyi, J., Shigemitsu, J, Sinclair, D.K. and Wyld, H.W., Nucl. Phys. B251[FS13], 311 (1985).

8) See the second reference in item 5 above, for example.

9) Creutz, M., Phys. Rev. Lett. 50, 411 (1983).

10) Nosé, S., Molecular Physics 52, 255 (1984).

11) Hall, R.W. and Berne, B.J., J. Chem. Phys. 81, 3641 (1984).

12) Duane, S., University of Illinois preprint, April, 1985.

13) Duane, S. and Kogut, J.B., in preparation.

14) Raby, S., Dimopoulos, S. and Susskind, L., Nucl. Phys. B169, 373 (1980).

15) Kogut, J., Polonyi, J., Sinclair, D.K. and Wyld, H.W., Phys. Rev. Lett. 54, 1980 (1985).

16) Banks, J. and Zaks, A., Nucl. Phys. B196, 189 (1982).

17) Fucito, F., Solomon, S. and Hamber, H., Phys. Lett. 150B, 285 (1985).
Kogut, J., Polonyi, J., Sinclair, D.K. and Wyld, H.W., Phys. Rev. Lett. 54, 1475 (1985).

SIMULATION OF LATTICE QCD WITH THE LANGEVIN EQUATION

Ghassan George Batrouni
Newman Laboratory of Nuclear Studies
Cornell University, Ithaca, NY 14853
U.S.A.

ABSTRACT

I present an algorithm for simulating lattice QCD using the Langevin equation. Systematic errors are analyzed and shown to be mainly renormalization effects and under control. Then I demonstrate how critical slowing down can be greatly reduced by updating in momentum space. Fermions are introduced into the Langevin equation in a new way, by including a bilinear noise term. This allows the simulation of any number of flavors, even or odd. Analytic arguments are backed by several numerical experiments.

1. INTRODUCTION

In this talk I will report on work done at Cornell with G.R. Katz, A.S. Kronfeld, G.P. Lepage, B. Svetitsky and K.G. Wilson[1)]. We are working on an algorithm to simulate lattice QCD using the Langevin equation.

Before describing the algorithm, I will present some motivation for taking this approach. One very attractive feature of the Langevin equation is that the entire lattice field configuration is updated at the same time. This enables us to include the fermion loops very efficiently (as will be discussed below) without the need for a Monte Carlo within a Monte Carlo. We can also fight critical slowing down by updating the field configurations in momentum space. This would not be practical had we needed to perform a Fast Fourier Transform (FFT) after updating each individual site or link variable. Here, since the entire lattice is updated simultaneously, we need only two FFT's per sweep. In the case of momentum space updating of the gauge fields, large gauge differences between neighboring links will be mistaken for high momentum fluctuations. This is remedied by fixing to a smooth gauge, thus restoring the relationship between short distance and high momentum. This gauge fixing is very easy and inexpensive to implement in the Langevin equation, and does not introduce long correlations in the data. Finally, systematic errors introduced by discretizing the differential Langevin equation are easy to compute, understand, and control.

2. DISCRETE LANGEVIN EQUATION

Before discussing the discrete time Langevin equation, I will briefly review stochastic quantization with continuous time[2]. Consider a generic field theory with an action $S[\phi]$. The Langevin equation that describes the evolution of the fields in the fictitious Langevin time, t, is given by

$$\dot{\phi}(x,t) \equiv \frac{d}{dt}\phi(x,t) = -\frac{\delta S}{\delta\phi(x,t)} + \eta(x,t) \tag{1a}$$

where $\eta(x,t)$ is a Gaussian random number with a two-point function given by

$$<\eta(x,t)\,\eta(x',t')> = 2\delta(x-x')\,\delta(t-t') \tag{1b}$$

and where $< >$ stands for stochastic averaging, i.e., averaging over an ensemble of η configurations. It can be shown that the probability distribution for the fields satisfying Eqs. (1a and 1b) satisfies the Fokker-Planck equation[3]

$$\frac{d}{dt}P[\{\phi\},t] = \sum_x \frac{\delta}{\delta\phi(x,t)}\left[\frac{\delta S}{\delta\phi(x,t)}P[\{\phi\},t] + \frac{\delta}{\delta\phi(x,t)}P[\{\phi\},t]\right] \tag{2}$$

Clearly, in the stationary limit $(\dot{P}=0)$ the probability distribution is given by exp(-S). This means that if $\phi_\eta(x,t)$ is a solution of the Langevin equation (1a,b) and $F[\phi]$ is some operator, then

$$<F[\phi_\eta]> \xrightarrow[t\to\infty]{} \overline{F[\phi]} \tag{3}$$

where $\overline{F[\phi]}$ is the usual field theoretic expectation value. For a gauge theory, the operator F must be gauge invariant, or the longitudinal component will undergo a random walk and thus give a divergent term in t.

A more general Langevin equation, which I will mention for future reference, is[4]

$$\begin{aligned}\dot{\phi}(x,t) &= -\sum_y Q(x-y)\frac{\delta S}{\delta\phi(y,t)} + \eta(x,t)\\ <\eta(x,t)\,\eta(y,t')> &= 2\delta(t-t')\,Q(x-y)\end{aligned} \tag{4}$$

where Q(x-y) is an arbitrary function. Clearly, if Q(x-y) is replaced by $\delta(x-y)$ we regain Eqs. (1a and b). As before, one can show, using the

Fokker-Plank equation, that Eq. (4) leads to the same expectation values in the infinite t limit as Eq. (1a and b).

This is the idea which we want to implement numerically. To do this we need to discretize the Langevin equation. The simplest such equation for a scalar theory is

$$\phi(x,\tau_{n+1}) = \phi(x,\tau_n) - f_x[\phi,\eta]$$

$$f_x[\phi,\eta] = \sum_y \epsilon_{x,y} \frac{\delta S}{\delta\phi(y,\tau_n)} + \sqrt{\epsilon_{x,y}}\,\eta\,(y,\tau_n) \tag{5}$$

where I am using the general form, Eq. (4), after rescaling $\eta \rightarrow \sqrt{\epsilon_{x,y}}$. $\epsilon_{x,y}$ is Q(x-y) multiplied by a small parameter $\overline{\epsilon}$. To find the equilibrium action for this discrete Langevin process, we use the discrete Fokker-Planck equation[1] and find to $O(\epsilon)$

$$P[\{\phi\},t\rightarrow\infty] = e^{-\overline{S}} \tag{6a}$$

$$\overline{S} = S[\phi] + \frac{1}{4}\sum_{x,y}\epsilon_{x,y}\left\{2\frac{\delta^2 S[\phi]}{\delta\phi(x)\delta\phi(y)} - \frac{\delta S[\phi]}{\delta\phi(x)}\,\frac{\delta S[\phi]}{\delta\phi(y)}\right\} \tag{6b}$$

Clearly it is $\overline{S}$ rather than S that determines the physics of the simulations. Consequently, we must understand the effect of the $O(\epsilon)$ terms in $\overline{S}$ to see how it differs from S. To this end note that, at least to $O(\epsilon)$, Eq. (6b) can be simplified by the change of variable

$$\phi(x)\rightarrow\phi(x) + \frac{1}{4}\sum_y \epsilon_{x,y}\frac{\delta S}{\delta\phi(y)}$$

$$\prod_x d\phi(x)\rightarrow\prod_x d\phi(x)\,\exp\left[\frac{1}{4}\sum_{x,y}\epsilon_{x,y}\,\frac{\delta^2 S}{\delta\phi(x)\delta\phi(y)}\right] \tag{7}$$

leading to the new equilibrium action (to $O(\epsilon)$)

$$\overline{S} = S[\phi] + \frac{1}{4}\sum_{x,y}\epsilon_{x,y}\,\frac{\delta^2 S}{\delta\phi(x)\delta\phi(y)} \tag{8}$$

For $\epsilon_{x,y} = \overline{\epsilon}\delta_{x,y}$ this becomes

$$\overline{S} = S[\phi] + \frac{\overline{\epsilon}}{4}\sum_x \frac{\delta^2 S}{\delta\phi^2(x)} \tag{9}$$

For most theories the only effect of the $O(\epsilon)$ term is to renormalize bare coupling constants in S. For example, for a $\lambda\phi^4$ theory, $m^2\rightarrow m^2 + \overline{\epsilon}\lambda/4$. If $\epsilon_{x,y}$ is not diagonal, the $O(\epsilon)$ correction from the $\lambda\phi^4$ term before changing variables Eq. (6b), includes the term

$$\lambda^2 \sum_{x,y} \phi(x)^3 \epsilon_{x,y} \phi(y)^3 \tag{10}$$

This is the type of term one gets if there were another scalar "particle" $\varsigma(x)$ coupled to the ϕ field by terms like $\lambda\phi^3(x)\varsigma(x)$. From this point of view taking $\epsilon_{x,y} = \overline{\epsilon}\delta_{x,y}$ corresponds to the infinite mass limit of the ς. I will demonstrate later that choosing $\epsilon_{x,y}$ to be proportional to the free propagator is optimal for Fourier acceleration.

We can easily write a Langevin equation for SU(N) gauge theories[5)]. The above arguments and error analysis will apply here as well when this equation is discretized. First, we define our convention for differentiation within the group manifold

$$f(e^{i\delta\cdot T}U) = f(U) + \delta^i \partial_i f + O(\delta^2) \tag{11}$$

where the generators T_i are normalized by $[T^i, T^j] = iC_{ijk}T^k$ and $tr(T^iT^j) = \delta_{ij}/2$. Then the simplest discretized Langevin equation is

$$U_\mu(x, \tau_{n+1}) = e^{-if\cdot T} U_\mu(x, \tau_n) \tag{13a}$$

$$f_j(U_\mu, \eta_j) = \overline{\epsilon}\partial_j S[U] + \sqrt{\overline{\epsilon}}\eta_j. \tag{13b}$$

For the standard plaquette action

$$S[U] = -\frac{\beta}{2N} \sum_P tr(P + P^\dagger), \tag{14}$$

where P is the product of links around a plaquette, we have

$$if\cdot T = \frac{\overline{\epsilon}\beta}{4N} \sum_{P\supset U_\mu} (P-P^\dagger) - \frac{\overline{\epsilon}\beta}{4N^2} \sum_{P\supset U_\mu} tr(P-P^\dagger) + i\sqrt{\overline{\epsilon}}H \tag{15a}$$

where H is a traceless Hermitian random noise matrix obeying

$$< H_{ik}(x,\tau) H_{lm}(x',\tau') > = (\delta_{il}\,\delta_{km} - \frac{1}{N}\delta_{ik}\delta_{lm})\delta_{x,x'}\ \delta_{\tau,\tau'} \tag{15b}$$

Use of the Fokker-Planck equation leads to the equilibrium action

$$\overline{S}[U] = (1 + \frac{\overline{\epsilon}}{12}N)\, S[U] + \frac{\overline{\epsilon}}{4} \sum_j \left\{ 2\partial_j{}^2 S[U] - (\partial_j S[U])^2 \right\} + O(\epsilon^2) \tag{16}$$

where care was taken in manipulating the nonabelian derivatives, since

$$[\partial_i, \partial_j] = -C_{ijk}\partial_k. \tag{17}$$

Shifting variables

$$U \rightarrow e^{-i\frac{\overline{\epsilon}}{4}\partial_j S T^j}\, U \tag{18}$$

and using the plaquette action Eq. (14) yields[1)]

$$\overline{S}[U] = \frac{\overline{\beta}}{2N}\sum_{P} tr(P + P^{\dagger}) \tag{19a}$$

$$\overline{\beta} = \beta(1-\overline{\epsilon}\frac{5N^2-6}{12N}) \tag{19b}$$

The above change of variables also means

$$<f[U]> \rightarrow \left(1+\frac{\overline{\epsilon}(N^2-1)}{8N}\right)^{\#\text{ links}} <f[U]> + O(\overline{\epsilon}^{\,2}) \tag{20}$$

where f[U] is any linear functional of the U's.

Once again we see that the only effect of the $O(\epsilon)$ terms was to renormalize coupling constants. To test these results we iterated Eqs. (13) and (15) for SU(3) on a 2×5^3 lattice at $\beta = 4.9$. Fig. 1 shows a plot of the average plaquette vs ϵ compared with the Metropolis result. Notice that even at $\epsilon = 0.1$ the error is no worse than about 11%. Next we iterated the same equations but instead of $\beta = 4.9$, we rescaled β to give $\overline{\beta} = 4.9$, to test the renormalization arguments given above. Those results are plotted in Fig. 2 and are in much better agreement with Metropolis, even up to the relatively large value of $\epsilon = 0.05$. The main effect of $O(\epsilon)$ corrections is to renormalize β.

Eq. (13b) is the simple Euler discretization of the differential Langevin equation. We can use more efficient integration schemes to decrease the error. In particular, there is a whole class of Runge-Kutta algorithms[6)] that lead to errors of order higher than ϵ. The first order Runge-Kutta algorithm, which leads to errors of $O(\epsilon^2)$, is

$$f_j = \frac{\overline{\epsilon}}{2}(1 + \frac{\overline{\epsilon}N}{6})\{\partial_j S[U(\tau_n)] + \partial_j S[\widetilde{U}\,(\tau_{n+1})]\} + \sqrt{\overline{\epsilon}}\eta_j(\tau_n) \tag{21}$$

where $\{\widetilde{U}\,(\tau_{n+1})\}$ is obtained from the $\{U(\tau_n)\}$ configuration by using the simple Euler algorithm Eq. (13b). This algorithm was tested for SU(3) at $\beta = 4.9$ on a 2×5^3 lattice, and the results are plotted in Fig. 3. Agreement with Monte Carlo improved dramatically and is quite good, even at $\epsilon = 0.1$. Also, for the same number of sweeps, this algorithm requires only about half again as much time as the simple Euler algorithm, but results in much higher precision. Its main drawback is that the noise matrices must be stored because they are used twice; once in Eq. (21) and once to compute $\widetilde{U}\,(\tau_{n+1})$. In effect this means that another copy of the lattice must be stored, thus putting more stringent limits on the size of the lattice that can be simulated.

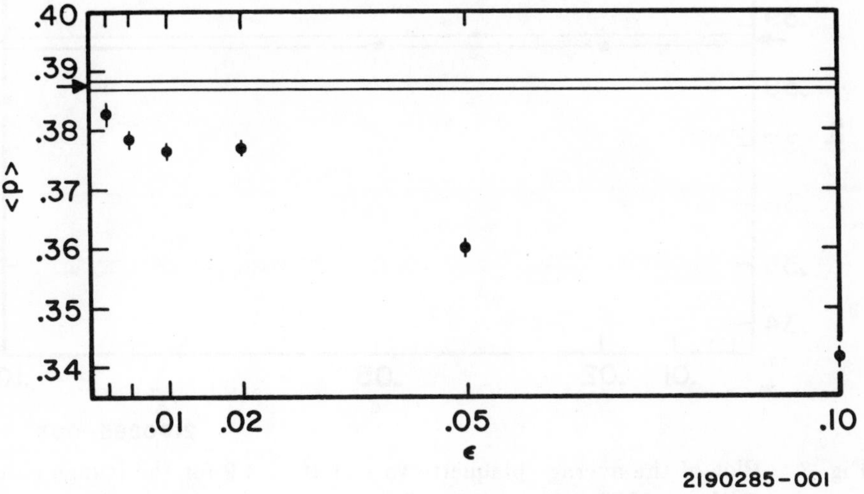

Fig. 1 Plot of the average plaquette vs ϵ at $\beta = 4.9$ for the simple Euler algorithm. The horizontal lines near $<p> = 0.39$ indicates the Metropolis values. 1200 sweeps at $\epsilon = 0.1$.

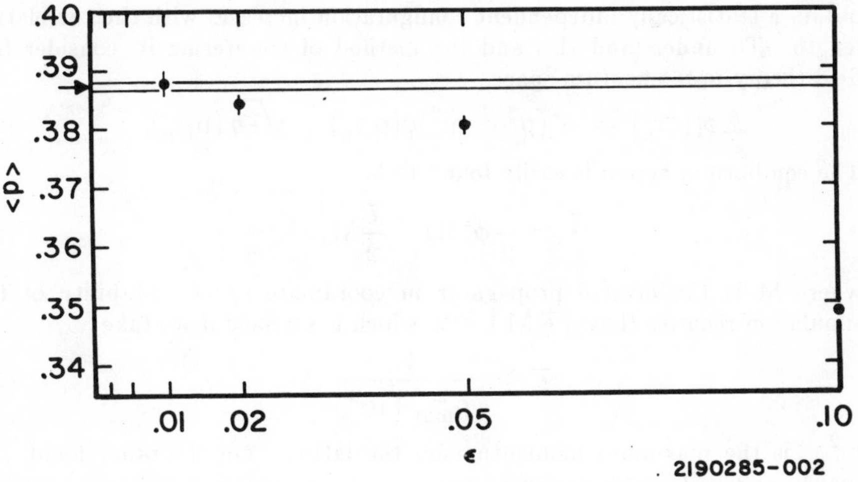

Fig. 2 Plot of the average plaquette vs ϵ at $\beta = 4.9$ for the simple Euler algorithm with shifted coupling. 3200 sweeps at $\epsilon = 0.1$.

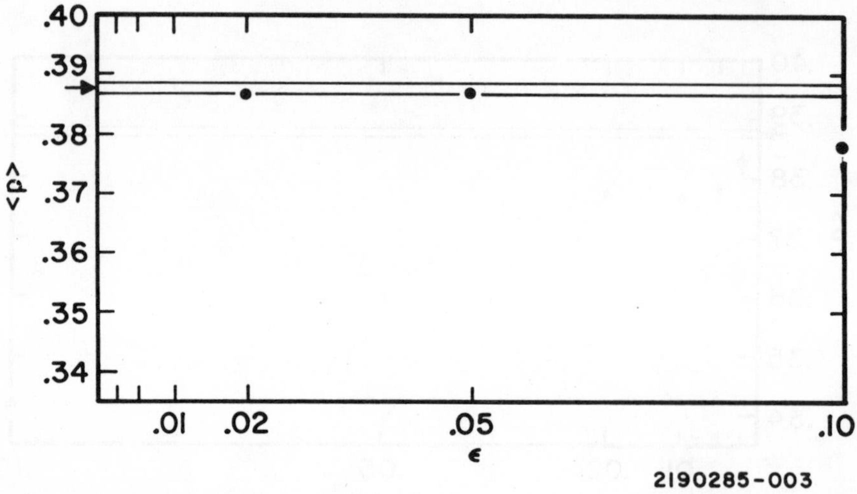

Fig. 3 Plot of the average plaquette vs ϵ at $\beta = 4.9$ for the Runge-Kutta method. 1000 sweeps at $\epsilon = 0.1$.

3. CRITICAL SLOWING DOWN AND FOURIER ACCELERATION

When simulating a system in the weak coupling limit, the system experiences critical slowing down, i.e., the number of sweeps needed to obtain a statistically independent configuration increases with the correlation length. To understand this and the method of countering it, consider free field theory in momentum space,

$$\Delta\phi(p,\tau_n) = -\overline{\epsilon}\,(p^2 + m^2)\phi(p,\tau_n) + \sqrt{\overline{\epsilon}}\eta\,(p,\tau_n). \tag{22}$$

The equilibrium action is easily found to be

$$\overline{S} = \frac{1}{2}\phi M(1 - \frac{\overline{\epsilon}}{2}M)\phi \tag{23}$$

where M is the inverse propagator in coordinate space. Stability of the simulation requires that $|\overline{\epsilon}M| < 2$, which is satisfied if we take

$$\overline{\epsilon} \lesssim \frac{1}{p^2_{max}+m^2} \tag{24}$$

p^2_{max} is the maximum momentum on the lattice. On the other hand, Eq. (22) is easily solved for $\phi(p,\tau_n)$, giving

$$\phi(p,\tau_n) = \sqrt{\overline{\epsilon}}\sum_{j=1}^{n}(1 - \overline{\epsilon}\,[p^2 + m^2])^{n-j}\eta(p,\tau_{j-1}) \tag{25}$$

From Eq. (24) we see that $(1 - \overline{\epsilon}\,[p^2 + m^2]) < 1$, which means that as (n–j) increases, $\phi(p,\tau_n)$ will depend less and less on $\eta(p,\tau_j)$. One can then define a correlation time, $N_c(p)$, by demanding

$$\ln\left\{(1-\overline{\epsilon}\,[p^2+m^2])^{N_c(p)}\right\}\sim -1 \tag{25a}$$

leading to

$$N_c(p)\sim\frac{1}{\overline{\epsilon}\,(p^2+m^2)}. \tag{25b}$$

From Eq. (24), we see that $N_c(p=p_{max})\sim 1$, whereas

$$N_c(p=0)\sim\frac{1}{m^2}=\xi^2 \tag{26}$$

Thus, the correlation time grows quadratically with the longest correlation length (in lattice units). The physical origin of this is Eq. (24): ultraviolet stability requirements lead to the condition on $\overline{\epsilon}$ given by Eq. (24), thus holding back the evolution of the infrared structure of the theory.

The way around this critical slowing down is clear from Eq. (25b)[7]; we use

$$\epsilon(p)\sim\frac{1}{(p^2+m^2)} \tag{27}$$

This means that we are resolving different length scales and treating them separately. In coordinate space this corresponds to using a nonlocal $\epsilon_{x,y}$, which is the reason it was introduced to begin with. So, from Eqs. (25b) and (27), we find $N_c(p)\sim 1$ for all momenta, not just p_{max}. This means that the number of passes needed for a simulation is reduced by a factor of ξ^{-2}, and no critical slowing down is experienced.

Implementing this momentum space updating for an interacting theory is straightforward and inexpensive in the Langevin algorithm. Eq. (5) is replaced by

$$\Delta\phi(x,\tau_n)=-\hat{F}\left\{\epsilon(p)\hat{F}^{-1}\frac{\delta S}{\delta\phi(x,\tau_n)}+\sqrt{\epsilon(p)}\,\eta(p,\tau_n)\right\} \tag{28}$$

where $\hat{F}$ is a Fourier transformation operator. This algorithm was tested for the two dimensional XY model[8],

$$S=-\beta\sum_{n,\mu}\cos\,(\theta_n-\theta_{n+\mu})-h\sum_n\cos(\theta_n). \tag{29}$$

We shifted β, h, and the field θ_n so that errors of $O(\epsilon)$ due to discrete Langevin updating cancelled (Eqs. 7-8). We ran on a 16×16 lattice, at β

= 1.5 and h = 0.23, both for p-dependent and p-independent ϵ,

$$\epsilon(\mathrm{p}) = \begin{cases} 0.02 \\ 0.02\dfrac{[\beta\sum(1-\cos(2\pi \mathrm{p}_\mu/16))+\mathrm{h}]_{\max}}{\beta\sum(1-\cos(2\pi \mathrm{p}_\mu/16))+\mathrm{h}} \end{cases} \tag{30}$$

i.e. the p-dependent ϵ is proportional to the lattice propagator.

Our results for the momentum space Green function $<|\theta(\mathrm{p})|^2>$ are given in Table 1. We see that, as argued from the free field theory case, the performance of the algorithm is greatly improved by Fourier acceleration. For small momenta the statistical errors with FFT are smaller for the sample of 400 sweeps than for a sample of 6000 sweeps with no FFT. As the momentum increases, i.e., as we start probing the short distance structure, both algorithms become the same, as expected from Eq. (30). Clearly, if one is only interested in measuring short distance properties of the system, Fourier acceleration is not needed. It is in the evolution of the infrared structure that one gains in employing such acceleration. To show these gains more dramatically, we plot in Fig. 4 the square of the ratio of the standard deviations with and without FFT versus momentum. At high momentum the ratio is almost unity, as expected. However, at low momentum, the simulation was much faster with FFT, in particular at p $\sim$ 0 it was 20 to 25 times faster than without FFT. We also measured the correlation time $\mathrm{N_c(p)}$, which is plotted in Fig. 5. As expected $\mathrm{N_c(p)}$ is small and independent of p when Fourier acceleration is used. Without Fourier acceleration $\mathrm{N_c(p)}$ increases rapidly at low momenta.

Table 1 Values of the momentum space propagator for the two dimensional XY model on a 16$\times$16 lattice. Runs were done at β = 1.5 and h = 0.23.

TABLE 1. $\langle|\theta(p)|^2\rangle$ **for the** xy **action of Eq. (29).**

	Perturbative		FFT		No FFT	
$p^2/p^2_{\max}$	0 order	1 loop	400 steps	6000 steps	400 steps	6000 steps
1/32	2.093	2.512	2.372(172)	2.568(36)	2.373(710)	2.654(180)
1/16	0.798	0.958	0.971 (49)	0.973(11)	1.023(127)	0.998 (35)
1/8	0.338	0.405	0.443 (17)	0.421 (6)	0.356 (37)	0.402 (10)
1/4	0.190	0.228	0.257 (18)	0.239 (3)	0.263 (19)	0.232 (4)
1/2	0.105	0.126	0.131 (7)	0.129 (2)	0.125 (8)	0.131 (2)
1	0.082	0.098	0.095 (18)	0.097 (4)	0.067 (19)	0.098 (4)

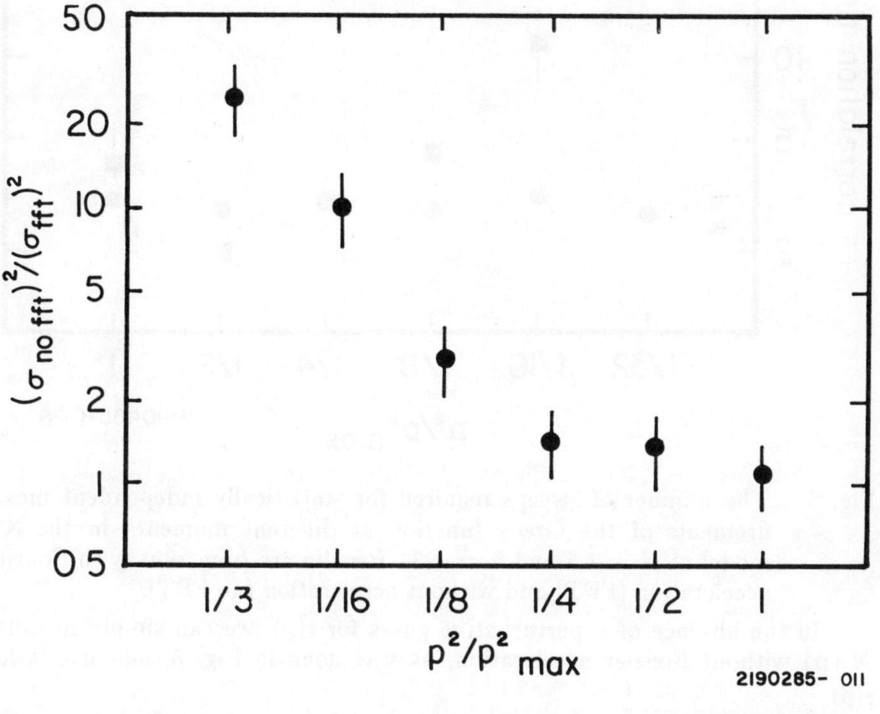

Fig. 4 Improvement due to Fourier acceleration in measurements of the momentum space Green functions for the XY model at $\beta = 1.5$ and h = .23. The squares of the standard deviations with and without acceleration are compared for different momenta.

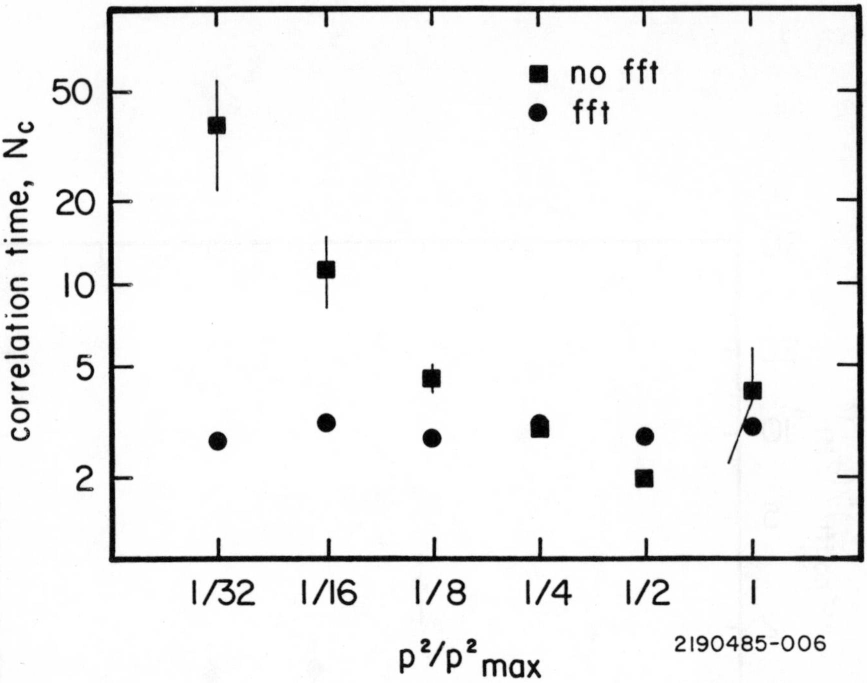

Fig. 5 The number of sweeps required for statistically independent measurements of the Green function at different momenta in the XY model at $\beta = 1.5$ and h = .23. Results are from runs with Fourier acceleration (FFT) and without acceleration (no FFT).

In the absence of a perturbative guess for $\epsilon(p)$, we can simply measure $N_c(p)$ without Fourier acceleration, as was done in Fig. 5, and use it for $\epsilon(p)$.

We also studied the effect of topological excitations on Fourier acceleration by running in the nonperturbative phase of the XY model (at $\beta = 1.0$ and 0.7), where there is a condensate of vortex-antivortex pairs. We found that, again, Fourier acceleration speeds up the simulation by factors of ξ^2 although in this phase ξ is not very large and gets smaller as β decreases.

In using Fourier acceleration for the XY model, we Fourier transformed the field configuration $\{\theta_n\}$. The analog of this for nonabelian theories is to Fourier transform the configuration of driving functions $\{f_j(U_\mu, \eta_j)\}$ (Eqs. 13a and b), and not the link variables U_μ. Remember that there is a term $f_j(U_\mu, \eta_j)$ for every link on the lattice. Fourier transforming U_μ and updating it in momentum space would destroy its unitarity. However, there is a problem in updating gauge theories in momentum space. The $O(\epsilon)$ corrections to the equilibrium action break gauge invariance and thus give a mass term to the gluons. To avoid this we can fix to an axial gauge[1] or use gauge invariant variables[9].

4. FERMIONS IN THE LANGEVIN EQUATION

When quarks are included, the full action for the theory becomes (for one flavor)

$$S = S_g[U] - \mathrm{Tr}\ln M[U] \tag{31}$$

where $S_g[U]$ is the gauge action and $M[U] = \gamma_5(D\cdot\gamma + m)$. This action can be simulated simply and efficiently by using the Langevin equation.

$$U_\mu(x,\tau_{n+1}) = e^{-if\cdot T}U_\mu(x,\tau_n) \tag{32a}$$

$$f_j = \bar{\epsilon}\left\{\partial_j S_g - \frac{1}{2}\mathrm{Re}(\eta_q^\dagger \frac{1}{M}\partial_j M\eta_q)\right\} + \sqrt{\bar{\epsilon}}\eta_j \tag{32b}$$

where η_j and η_q are gaussian random noise terms normalized to $<\eta_q^\dagger\eta_q> = <\eta_j\eta_j>=2, <\eta_q\eta_j>=0$. The new bilinear noise term in Eq. (32b) generates the fermion contribution. This can be seen by using the Fokker-Planck equation, which requires, to $O(\epsilon)$, only the stochastic average of f_j with respect to η_q,

$$<f_j>_{\eta_q} = \bar{\epsilon}\left\{\partial_j S_g - \mathrm{Tr}(\frac{1}{M}\partial_j M)\right\} + \sqrt{\bar{\epsilon}}\eta_j \tag{33a}$$

$$= \bar{\epsilon}\partial_j(S_g - \mathrm{Tr}\ln M) + \sqrt{\bar{\epsilon}}\eta_j. \tag{33b}$$

This is just the simplest driving term one would write down for the action S, if one wanted only one noise term in the Langevin equation. The cost in Eq. (32) is that we need to evaluate $\psi=M^{-1}\eta_q$ once per sweep. This is done by solving

$$M[U]\psi=\eta_q \tag{34}$$

using the conjugate gradient method[10].

The most striking feature of this algorithm is that any number of flavors, even or odd, can be simulated simply by including a bilinear noise term for each flavor, or if the masses are equal, by multiplying the bilinear noise term by the number of flavors. Furthermore, no Monte Carlo within a Monte Carlo is needed here as is needed in most pseudofermion algorithms. The analysis of systematic errors proceeds exactly as before[1], and $O(\epsilon)$ corrections are under control and pose no problems.

We tested this algorithm on a 2^4 lattice with Wilson fermions, for which

$$\gamma_5 M\psi = \psi(n) - \kappa \sum_{\mu} [(1-\gamma_\mu)U_\mu(n)\psi(n+\hat{\mu}) + (1+\gamma_\mu)U_\mu^\dagger(n-\hat{\mu})\psi(n-\hat{\mu})]. \quad (35)$$

Our results for β=4 and κ = 0.15 are presented in Table 2 and compared with recent results of Weingarten.[11] Even though our algorithm seems to require the same number of sweeps as Weingarten's Metropolis algorithm, it is much faster because we need to solve equations like Eq. (34) only once per lattice sweep, whereas Weingarten has to do that once per link update. Note that our boundary conditions are skew while Weingarten's are periodic. Consequently, results of the two algorithms should differ slightly.

Table 2 Data from three different algorithms for simulating QCD: Langevin (Eq. 32), Metropolis and pseudofermion (Ref. 11). Runs were done on a 2^4 lattice, β = 5, κ = 0.15.

	Number of Sweeps	Tr[line]	Tr[plaquette]
Eqs. 4.2			
ϵ = 0.05	6000	-0.15(2)	0.29(1)
ϵ = 0.01	12000	-0.15(2)	0.31(1)
ϵ = 0.005	24000	-0.15(2)	0.30(1)
Ref. 11			
Metropolis	2400	-0.16(2)	0.36(2)
Pseudofermion	85000	-0.15(1)	0.31(3)

Finally, I mention that the conjugate gradient method for solving Eq. (34) suffers from critical slowing down for very small quark masses. The number of iterations needed to obtain a fixed accuracy grows like ξ, the longest correlation length in the theory. This factor multiplies the obvious growth rate which is linear in the lattice volume. The remedy for this, as in the updating case, is to invert in momentum space, thus eliminating critical slowing down. For a more detailed argument and illustration with the Jacobi method, see Ref. 1.

We tested Fourier acceleration of the conjugate gradient algorithm for QCD on a 4^4 lattice. Using Wilson fermions with n_f = 2 and β = 6 we solved Eq. (34) for ψ (actually, we multiply both sides of the equation by M, since the conjugate gradient requires a positive matrix). Near the critical $\kappa \sim 0.158$, Fourier acceleration reduced the number of sweeps needed by a factor of a third, roughly as expected for this size lattice. We also examined the residual vector, $M^2\psi - M\eta_q$, as a function of momentum. As expected, the error was the same for all momenta when Fourier acceleration was used, and peaked for small momenta when no such acceleration was used. This is the analog of what happened with the correlation time, $N_c(p)$, for the XY model. Finally, we found that Fourier acceleration had no effect if the gauge is not fixed, confirming the earlier remark that without gauge fixing there will be no relation between short distance and high momentum.

5. CONCLUSIONS

In this talk, I have explained the Langevin approach to simulating lattice field theories. I examined systematic errors and argued that they are understood and under control. This control of systematic errors is a very attractive feature of the Langevin equation. Furthermore, I showed how Runge-Kutta algorithms increase the accuracy of the simulation and allow larger step size. I also demonstrated critical slowing down and how it can be removed by using Fourier acceleration both for updating and conjugate gradient. This should result in an algorithm whose computational needs grow linearly with lattice volume instead of $V^{7/4}$ (assuming $\xi^4 = V$). Finally, fermions were introduced very simply by introducing a new bilinear noise term. This method of introducing fermion loops avoids a Monte Carlo within a Monte Carlo, and allows the simulation of any number of flavors, even or odd.

6. ACKNOWLEDGEMENTS

I thank all my collaborators and also Pietro Rossi for many illuminating discussions and comments on this paper. This work was supported in part by a grant from the NSF.

REFERENCES

1. Batrouni, G.G., Katz, G.R., Kronfeld, A.S., Lepage, G.P., Svetitsky, B., and Wilson, K.G., Cornell Preprint CLNS 85/651 (May 1985).
2. Parisi, G. and Wu, Y., Sci. Sin. 24, 483 (1981).
3. See for example van Kampen, N.G., "Stochastic Processes in Physics and Chemistry", (North-Holland, Amsterdam, 1981).
4. See for example Damgaard, P.H. and Tsokos, K., Nucl. Phys. B235, 75 (1984).
5. Halpern, M.B., Nucl. Phys. B228, 173 (1983); Drummond, I.T., Duane, S. and Horgan, R.R., Nucl. Phys. B220, [F58] 119 (1983); Guha, A. and Lee, S.C., Phys. Rev. D27, 2412 (1983).

6. Helfand, E., The Bell System Technical Journal, 58, 2289 (1979); Greenside, H.S. and Helfand, E., ibid., 60, 1927 (1981).

7. Similar ideas are discussed by Parisi, G. in "Progress in Gauge Field Theory", edited by G. 't Hooft, et al. (Plenum, New York, 1984).

8. See for example Tobochnik, J. and Chester, G.V., Phys. Rev. B20, 3761 (1977).

9. Bars, I., Nucl. Phys. B148, 445 (1979); ibid., B149, 39 (1979); Batrouni, G.G. and Halpern, M.B., Phys. Rev. D30, 1782 (1984).

10. Hestenes, M.R. and Stiefel, E., J. Res. Natl. Bur. Stand. 49, 409 (1952); Hageman, L.A. and Young, D.M., "Applied Iterative Methods", (Academic, New York, 1981).

11. Weingarten, D., IBM report (January 1985), unpublished, and his contribution to this conference.

APPLICATION OF THE LANCZOS ALGORITHM IN LATTICE QCD

I. M. Barbour

Department of Natural Philosophy
University of Glasgow
Glasgow G12 8QQ

Two applications of the Lanczos Algorithm are discussed. First a fast and efficient method is proposed for Metropolis updating of the lattice with dynamical fermions via Lanczos block inversion on a hypercube and rank annihilation. Secondly we present the results of a high statistics study of the chiral condensate in quenched lattice QCD. We see clear evidence for deviation from asymptotic scaling in the range of β considered.

A. UPDATING WITH DYNAMICAL FERMIONS

The inclusion of the effects of quark loops in the Monte Carlo calculations of lattice QCD has been a longstanding problem. In an earlier paper[1)] we have described how the Lanczos algorithm can be used to either diagonalize large matrices or invert large matrices row by row. Indeed the convergence of the latter method is superior to that of the conjugate gradient and will converge even at zero quark mass (for Kogut-Susskind fermions). We describe here[2)] a generalization of the single row inversion to simultaneous multi-row inversion and its application to fermionic updating.

For example, a hypercube on the lattice contains 16 sites and 32 links and for SU(3) a 48 x 48 block of the inverse of the fermion matrix is sufficient to update any link in the hypercube with the inclusion of the ratio of the fermion determinants. Block Lanczos[3)] (described below) will invert 24 rows (or more) simultaneously and more efficiently than single row by row inversion. Rank annihilation[4] then allows us to update the 48 x 48 block to give the inverse for the new configuration without any further inversion. This means that all the links in the hypercube can be updated one at a time as many times as desired without much more than the computation of 48 rows of the inverse.

This procedure then increases the efficiency of the updating procedure by a factor of 4 relative to a link by link sweep of the lattice. In addition the number of sweeps for thermalisation will be reduced since hypercubes (or larger objects) are brought to local equilibrium in each sweep. Combining these ideas together with the use of block Lanczos gives an overall time saving of at least 100 (possibly more if larger objects are considered) over link by link updating.

We use the Susskind[5] fermionic scheme in which the fermionic part of the action has the for m

$$S_F = \bar{\psi}(\not{D} + m)\psi \tag{1}$$

where m is the quark mass in lattice units and $\not{D}$ is the antihermitian fermion matrix given by

$$\bar{\psi}\not{D}\psi = \tfrac{1}{2} \sum_{x,\mu} \bar{\psi}(x)\, U_\mu(x)\, f(x)\psi(x+\hat{\mu}) - \text{h.c.} \tag{2}$$

with f(x) containing the fermionic signs and antiperiodic boundary conditions in the spin diagonalized form.

In order to perform the Metropolis updating we must calculate the ratio of the determinants of $\not{D}$ + m when a change is made to one link

$$R = \frac{\det(H + \Delta)}{\det\,(H)} \qquad\qquad H = i(\not{D} + m) \tag{3}$$

$$= \det\,(1 + H^{-1}\Delta)$$

where Δ is the change in the fermiom matrix. Δ is non-zero in the 6 x 6 block corresponding to the two end points of the link. The only elements of H^{-1} which contribute are in the same block. Knowledge of the 6 x 6 block of the inverse is obviously sufficient to enable updating of the same link as many times as desired.

Now consider the change in one like of the hypercube just prior to the update of another link. It makes a change ΔH in the fermion matrix with 18 non-zero elements which we separate into 18 consecutive changes each of the form

$$\Delta H_i = a_i U_i V_i^+ \qquad i = 1 \; .... \; 18 \tag{4}$$

where a_i is the change to the element and U_i and V_i are unit column vectors, each zero in all elements except one. Then rank annihilation gives

$$(H + a_i U_i V_i^+)^{-1} = H^{-1} - \frac{a H^{-1} U_i V_i^+ H^{-1}}{1 + a V_i^+ H^{-1} U_i} \tag{5}$$

It is easily seen that this gives the new 48 x 48 block of the inverse without knowledge of the rest of its elements.

We now consider how we may use the Lanczos algorithm to obtain a block of the inverse of the fermion matrix prior to updating the links of a hypercube say. The method is a straightforward generalization of the single row inversion described in ref. 1). As an illustration we discuss a hypercube at m = 0 which is algebraically simpler.

Ordering sites as say even then odd the fermion matrix has the block structure

$$H = \begin{pmatrix} im & M \\ M^+ & im \end{pmatrix} \tag{6}$$

For block Lanczos the alphas and betas become LxL (24 x 24) matrices (for a hypercube it is equally efficient to invert on the even and odd sites separately) and the Lanczos vectors NxL arrays, $N = 3n_s^3 n_t$, each half empty. Remembering $m = 0 \Rightarrow \alpha_i = 0$ ($\alpha_i = m$) the iterative scheme is

$$\begin{aligned} Hx_1 &= x_2 B_1 & B_1^+ B_1 &= x_1^+ H^+ H x_1 \\ Hx_2 &= x_1 B_1^+ + x_3 B_2 & B_2^+ B_2 &= (Hx_2 - x_1 B_1^+)^+ (Hx_2 - x_1 B^+) \end{aligned} \tag{7}$$

(the B_i can be chosen to be triangular) transforming H to block tridiagonal form. The initial vector x_1 is a NxL array of L column vectors each projecting out in $\mathbf{H}^{-1} x_1$ a particular column of the inverse relevant to the hypercube. We use the Lanczos scheme to calculate $H^{-1} x_1$ as a series

$$H^{-1} x_1 = x_1 C_1 + x_2 C_2 + \; .... \tag{8}$$

For the case m = 0 we need only every alternate Lanczos equation starting with the second

$$H^{-1}x_1 = x_2(B_1^+)^{-1} - H^{-1}x_3B_2(B_1^+)^{-1} \qquad (9)$$

The other Lanczos equations are used in sequence to eliminate the remainder term

$$H^{-1}x_1 = x_2(B_1^+)^{-1} - x_4(B_3^+)^{-1}B_2(B_1^+)^{-1} + x_6(B_5^+)^{-1}B_4(B_3^+)^{-1}B_2(B_1^+)^{-1} + \ldots. \qquad (10)$$

As in Lanczos applied to single row inversion, this series proceeds without any sign of convergences until the smallest eigenvalues of H converge at which point the series converges rapidly to about machine precision. Convergence criteria for the block inversion are similar to the single row case. At small quark masses the block inversion is about $^L/2$ times faster than the equivalent single row inversion for Susskind fermions.

To summarise, the Metropolis algorithm is carried out as follows: To cover all links we use $^1/_8$ of the possible hypercubes, i.e. only those with no links in common. The hypercubes can be taken in sequence or at random. Block Lanczos gives the 48 x 48 block of the inverse required to update its links. For each link extract the 6 x 6 block. Updating the link requires the calculation of 6 x 6 determinants. Before proceeding to the next link in the hypercube, update the 48 x 48 block by rank annihilation for the overall change in the link. It is worthwhile going round the hypercube ~10 times until it is close to equilibrium within itself.

Studies are underway on varying sizes of lattice to investigate the overall efficiency of the method.

B. HIGH STATISTICS STUDY OF THE CHIRAL CONDENSATE IN QUENCHED LATTICE QCD.

We[6)] have measured the chiral condensate on a 8^4 lattice at β = 5.4, 5.5, 5.6, 5.7 and 5.8 for 8, 8, 16, 32 and 48 configurations respectively. The configurations were generated by a modified heat-bath algorithm[7)] and separated by 600 sweeps with about 2,400 sweeps for equilibrium at each β-value. We also measured the condensate at β = 6 to check our expectation that the symmetry be restored on an 8^4

lattice.

For Susskind[5] fermions the chiral condensate at zero mass is given by[8] $\lim_{m\to 0}\lim_{V\to\infty}\langle\bar{\psi}\psi(m)\rangle = 3\pi\rho(0)$ where $\rho(0)$ is the fermion matrix spectral density at $\lambda = 0$ which was determined by using the Lanczos algorithm to find the small eigenvalues and then fitting $\rho(\lambda)$ as described in refs. (8,9). The spectral density is given by

$$\rho(\lambda) = \frac{1}{N}\frac{\Delta n}{\Delta\lambda} \tag{11}$$

where Δn is the number of eigenvalues in the range $\Delta\lambda$. Figure 1 shows the results for the lowest eigenvalues for 2 configurations at various β values

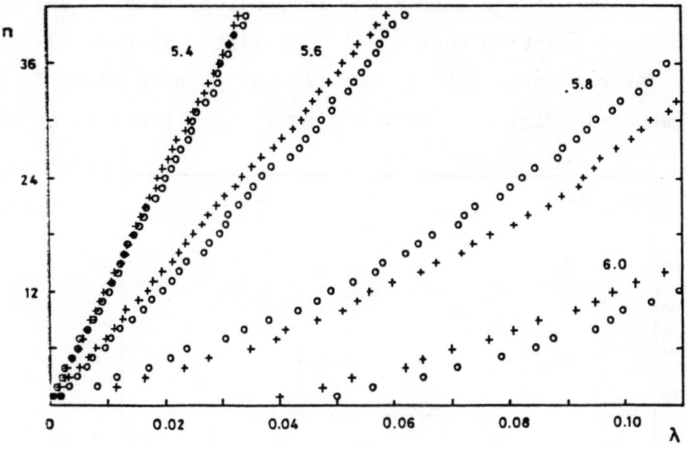

Figure 1: Lowest Eigenvalues of the fermion matrix at various β values.

Fits to the theoretically expected form

$$n(\lambda) = \tfrac{1}{2} + a\lambda + b\lambda^3 \tag{12}$$

give the values for the spectral density at $\lambda = 0$ as given in the Table.

Table 1: Values of the normalised spectral density $\rho(0)$, the chiral condensate, $\langle\psi\psi(0)\rangle$ and the asymptotically invariant quantity $(\langle\psi\psi\rangle_{inv})^{1/3}/\Lambda_L$.

β	$\rho(0)$	$\langle\psi\psi(0)\rangle$	$\{\langle\psi\psi\rangle_{inv}\}^{1/3}/\Lambda_L$
5.4	.0942 ± .0008	0.888 ± 0.008	109.0 ± 0.5
5.5	.0754 ± .0004	0.710 ± 0.004	112.5 ± 0.3
5.6	.0521 ± .0008	0.490 ± 0.008	110.6 ± 0.6
5.7	.0269 ± .0012	0.254 ± 0.012	98.9 ± 2.0
5.8	.0145 ± .0008	0.138 ± 0.008	89.8 ± 3.3

The values of $\langle\bar{\psi}\psi(0)\rangle$ are also given in lattice units. The asymptotically invariant quantity $\langle\psi\psi\rangle_{inv}^{1/3}$ as calculated in ref. 9) for Λ_{mom} = 200 MeV and normalised to one flavour is also given in the Table and shown in fig. 2. We see clearly that there is no evidence

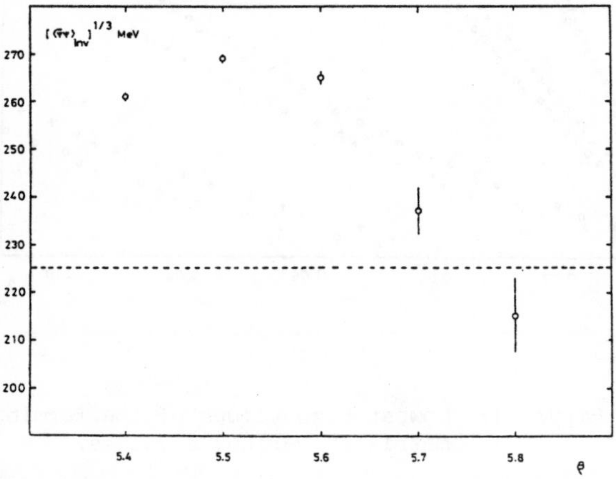

Figure 2: Comparison of measured values of the condensate with the expected continuum behaviour

for asymptotic scaling in the range of β studied. The earlier low statistics analysis of ref. 8) was not able to discriminate between asymptotic scaling behaviour and our high statistics data points.

The change $\Delta\beta(\beta)$,[10)] i.e. the change in β corresponding to a change in length scale by a factor of 2, can be estimated from the table by scaling $\langle\bar{\psi}\psi(0)\rangle$ at $\beta = 5.4$ with the cube of the lattice spacing and comparing with a simple linear extrapolation of the data at $\beta = 5.7$ and 5.8. We find

$$\Delta\beta(\beta = 5.82 \pm 0.03) = 0.42 \pm 0.03 \qquad (13)$$

consistent with the MCRG measurements of Ref. 10). Also chiral symmetry restoration observed at $\beta = 6$ suggests that the inverse restoration coupling is very close to 6 on an 8^4 lattice. Equivalent measurements on a $12^3 \times 4$ lattice[11)] give $\beta_r \simeq 5.675$ and hence

$$\Delta\beta(\beta = 6.0) \simeq 0.33 \qquad (14)$$

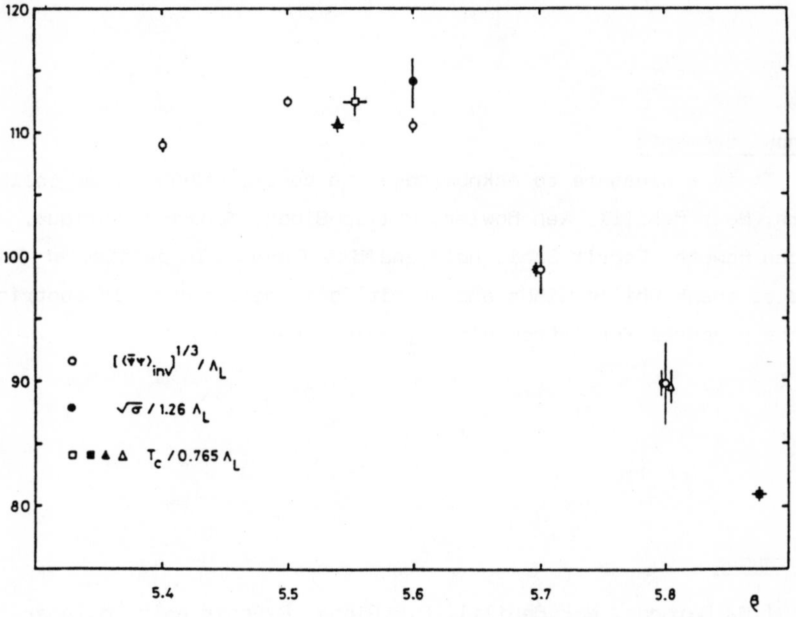

Figure 3: Comparison of the scaling behaviour of $\{\langle\psi\psi\rangle_{inv}\}^{1/3}\,\Lambda_L$ (o), this work, with that of $\sqrt{\sigma}/1.26\Lambda_L$ (●). Barkai et al. and of $T_c/0.765\Lambda_L$ (■), Kennedy et al. (▲), Svetitsky and Fucito, (□), Celik et al., Karsch and Petronzio (△).

In Fig. 3 we compare the scaling behaviour of the condensate with that of other physical quantities. The string tension data of Barkai Moriarty and Rebbi has been scaled at 5.8 to agree with our measured value of $\langle\psi\psi\rangle^{1/3}_{inv}/\Lambda_L$. The deconfinement temperature data of Ref. 12) has been scaled at $\beta \simeq 5.7$.

Given the small number of data points at which comparison is possible the data is consistent with a common scaling behaviour for $<\psi\psi>^{1/3}$, Λ_{inv}, T_c and $\sqrt{\sigma}$. On the other hand the recent data of de Forcrand et al.,[13)] on the 0^{++} glueball mass suggests a markedly different scaling behaviour in this range of β. A universal pre-asymptotic scaling regime may well not exist.

Acknowledgements

It is a pleasure to acknowledge the contributions of my collaborators, Nasr Behilil, Ken Bowler, Philip Gibbs, Muhammad Rafique, Duncan Roweth, Gerrit Schierholz and Mike Teper. In particular I would like to thank Philip Gibbs and Gerrit Schierholz for their contribution to the proposal for introducing dynamical fermions.

References

1) I.M. Barbour, N-E.Behilil, P.E.Gibbs, G.Schierholz, M.Teper, DESY Preprint 84-087, LAPP-TH 118 (1984) (to be published in Lecture Notes on Physics "The Recursion Method and its Applications" Springer Verlag).

2) Glasgow Preprint, in preparation.

3) D.S.Scott, in "Sparse Matrices and Their Uses", 1981, Ed. I.S. Duff (Academic Press).

4) H.S.Wilf, Mathematical Methods for Digital Computers

5) L.Susskind, Phys. Ref. D16 (1976) 3031

6) I.M.Barbour, K.C. Bowler, P.E.Gibbs and D. Roweth, Phys. Lett. to

be published.

7) N.Cabibbo and E. Marinari, Phys. Lett. 119B (1982) 387
K.C. Bowler and B.J.Pendleton, Nucl. Phys. B230 (FS10) (1984) 109

8) I.M.Barbour, P.E.Gibbs, J.Gilchrist, H.Schneider, G.Schierholz, M.Teper, Phys. Lett. 136B (1984) 80

9) I.M.Barbour, in "Gauge Theory on a Lattice: 1984", Proc. Argonne National Laboratory Workshop (ANL 1984) 120

10) A.Hasenfratz, P.Hasenfratz, U.Heller and F. Karsch, Phys. Lett. 76 (1984) 1408
K.C.Bowler, A. Hasenfratz, P.Hasenfratz, U.Heller, F.Karsch, R.D. Kenway, H.Meyer-Ortmanns, I.Montvay, G.S.Pawley and D.J.Wallace, Nucl. Phys. B(FS) (1985) to appear

11) I.M.Barbour, P.Gibbs, G.Schierholz and M.Teper, CERN Preprint in preparation, 1985

12) T.Celik, J.Engels and H.Satz, Phys. Lett. 129B (1983) 323
F.Karsch and R.Petronzio, Phys. Lett. 139B (1984) 403
A.D.Kennedy, J.Kuti, S. Meyer and B.J.Pendleton, Phys. Rev. Lett. 54 (1985) 87
B.Svetitsky and F.Fucito, Phys. Lett. 131B (1983) 165

13) Ph. de Forcrand, G.Schierholz, H.Schneider and M.Teper, Phys. Lett. 152B (1985) 107.

THE PSEUDOFERMION METHOD AND ITS APPLICATIONS IN LATTICE QCD*

F. Fucito** and S. Solomon†

California Institute of Technology, Pasadena, CA 91125

ABSTRACT

We show how the pseudofermion method can be efficiently implemented on a parallel machine. Results obtained for finite temperature and zero temperature QCD are discussed.

1. Introduction

Lattice QCD has been quite successful, up to now, in describing some of the properties of strong interactions. Until now all of this information was obtained in the so-called quenched approximation, which amounts to neglecting the effects of dynamical quarks. In the last year some computations have appeared in which proper care of these effects was taken [1]. These results seem to be in good agreement among themselves. In this talk I will describe one of these techniques, called the pseudofermion method [2]. I will show how this method can be efficiently implemented on a parallel computing machine and I will discuss some physical results obtained for $T = 0$ and $T \neq 0$ QCD, where T is the physical temperature.

* Work supported in part by the U.S. Department of Energy under contract DEAC 03-81-ER40050.

** Weingart Fellow in Theoretical Physics.

† Bantrell Fellow in Theoretical Physics

2. The Parallel Implementation of the Pseudo Fermion Algorithm

The actual implementation of our algorithm was performed on the 32 and 64 processors cosmic cube at Caltech. In the rest of this section we will make an effort to show that this implementation does not depend on the details of the machine on which we worked, but holds in general for all parallel machines built up of processors which communicate only with their nearest neighbors and have periodic boundary conditions in all the directions of the space in which they are embedded.

The partition function of lattice QCD is

$$Z = \int \prod_{x,\mu} \mathrm{D}U_{x,\mu} \prod_{x} \mathrm{D}\psi_x \mathrm{D}\overline{\psi}_x e^{-S(\psi,\overline{\psi},U)} \tag{2.1}$$

where:

$$S(U,\psi,\overline{\psi}) = S_G(U) + S_F(\overline{\psi},\psi,U) \,. \tag{2.2}$$

$S_G(U)$ is the usual Wilson action and

$$S_F(\overline{\psi},\psi,U) = \sum_{x,x'} \overline{\psi}_x^A Q_{xx'}^{AB} \psi_{x'}^B \,. \tag{2.3}$$

$Q_{xx'}^{AB}$ is defined as

$$\begin{aligned} Q_{xx'}^{AB} &\equiv \frac{1}{2} \sum_{\mu=1}^{4} \Gamma_\mu(x) [\, U_{x,\mu}^{AB} \delta_{x',x+\mu} + (U_{x',\mu}^{+})^{AB} \delta_{x',x-\mu}] + m\, \delta_{xx'} \delta^{AB} \\ &\equiv D_{xx'}^{AB} + m\, \delta_{xx'} \delta^{AB} \,. \end{aligned} \tag{2.4}$$

m is the fermion bare mass and $\Gamma_x(x) \equiv (-)^{x_3}$, $\Gamma_y(x) \equiv (-)^{x_1}$, $\Gamma_z(x) \equiv (-)^{x_2}$, $\Gamma_t(x) \equiv (-)^{x_1+x_2+x_3}$, with $x \equiv (x_1, x_2, x_3, x_4) \equiv (\vec{x}_1, x_4)$. Integrating out the fermionic degrees of freedom we can define an effective action S_{eff}:

$$\begin{aligned} e^{-S_{eff}(U)} &= \int \prod_{y,c} \mathrm{D}\overline{\psi}_y^c \mathrm{D}\psi_y^c e^{-\overline{\psi}_x^A Q_{xx'}^{AB} \psi_x^B} \\ &= \det Q(U) \,. \end{aligned} \tag{2.5}$$

The partition function (2.1) can now be rewritten as:

$$Z = \int \prod_{x,\mu,A} \mathrm{D}\, U_{x,\mu}^A e^{-S_G(U) - S_{eff}(U)} \,. \tag{2.6}$$

The variation of the action $S(U,\overline{\psi},\psi)$ is:

$$\delta S = \delta S_G + \delta S_{eff} \tag{2.7}$$

where

$$\delta S_{eff}(U) = \delta Tr \ln Q(U) = Tr\, Q^{-1}(U)\delta Q(U) + O(\delta U^2)$$

$$= Tr(Q^{-1}\frac{\delta Q}{\delta U})\delta U + O(\delta U^2)\,. \tag{2.8}$$

$\delta S_{eff}(U)$ can be also rewritten as

$$\delta S_{eff}(U) = tr(G\delta U) \tag{2.9}$$

with G defined as

$$G = (Q^+Q)^{-1}Q^+\frac{\delta Q(U)}{\delta U}\,. \tag{2.10}$$

$(Q^+Q)^{-1}$ can be obtained using the formula:

$$(Q^+Q)^{-1}{}^{BA}_{xx} = \langle \varphi^{+A}_x \varphi^B_x \rangle = \frac{\int \prod_y \mathrm{D}\varphi_y \mathrm{D}\varphi^+_y \varphi^{+A}_x \varphi^B_x e^{-\varphi^+ M \varphi}}{\int \prod_y \mathrm{D}\varphi_y \mathrm{D}\varphi^+_y e^{-\varphi^+ M\varphi}}\,. \tag{2.11}$$

The ratio in the right-hand-side can now be computed statistically and the auxiliary complex bosonic variables φ's are called pseudofermions. The combination Q^+Q is used in (2.11) to insure the positivity of the action.

The interaction in (2.11) is clearly second neighbor because there the matrix Q appears squared. We can now transform the action in (2.11) in one with nearest neighbors interaction, by defining a suitable field h^A_x:

$$h^A_x = m\,\varphi^A_x + \frac{1}{2}\sum_{\mu=1}^{4}\Gamma_\mu(x)[\,U^{AB}_{x,\mu}\delta_{x+\mu,y}\varphi^B_y - (U^+)^{AB}_{x-\mu,\mu}\delta_{x-\mu,y}\varphi^B_y] \equiv Q^{AB}_{xy}\varphi^B_y\,. \tag{2.12}$$

The φ's can now be computed (e.g., by a heat bath algorithm) employing the field h:

$$(\varphi^{new})^A_x - \varphi^A_x = \frac{Gauss}{\sqrt{2(m^2+2)}} + \frac{1}{m^2+2}[mh^A_x - \frac{1}{2}\sum_{\mu=1}^{4}\Gamma_\mu(x)(U^{AB}_{x,\mu}h^B_{x+\mu} - U^+)^{AB}_{x-\mu,\mu}h^B_{x-\mu}] \tag{2.13}$$

where Gauss is a random number with Gaussian distribution of width equal to one.

In turn, the G's can be computed as

$$G^{AB}_{x,\mu} = \langle (h^B_x)^* \varphi^A_{x+\mu} - h^A_{x+\mu} (\varphi^B_x)^* \rangle \tag{2.14}$$

where $\langle\rangle$ means average over an ensemble of configurations generated according to (2.12). Now that we have δS_{eff}, it is easy to compute (2.7) adding δS_{eff} to the variation of the Wilson action δS_G, obtained by well known and standard methods.

If we had to compute δS_{eff} every time we upgrade a link of our lattice, the pseudofermions algorithm would still be very slow. Our approximation will consist of computing δS_{eff} only at the end of one full sweep through the entire lattice. This approximation seems to be well under control for $\delta U \ll 1$.

Now let's see how the parallel algorithm actually works: if our 2^d-processors machine has a geometry of a d-hypercube we can transform via a Gray code [3] this geometry to a hypercube living in a four-dimensional space, having D_1, D_2, D_3, D_4 processors in each one of the four directions of our space. Now our space-time lattice $N_1 \times N_2 \times N_3 \times N_4$ will be divided among each processor. The amount of lattice points contained in each processor will be

$$v = \frac{N_1}{D_1} \times \frac{N_2}{D_2} \times \frac{N_3}{D_3} \times \frac{N_4}{D_4} = n_1 \times n_2 \times n_3 \times n_4 \, . \tag{2.15}$$

Each processor will store:

i) the φ and h fields residing on the sites belonging to the processor;

ii) the U's and G's of the links originating in those points;

iii) the $U^{AB}_{x-\mu,\mu}$'s for $x = 0$.

Storing the matrices of iii) represents an increase in the amount of storage required that is largely compensated by the saving in communication time. In fact, if we store the matrices iii), we won't need to communicate any matrix U during the entire evaluation of the G's.

It is now straightforward to describe the flow of the parallel algorithm. Given a set of U's and φ's (2.12) is computed. Every time we need a φ field which does not belong to the processor we have a communication among processors. As all of them perform the same tasks each processor acts both like a sender and a receiver of information at the same time.

Then we compute $(\Delta\varphi)^A_x$ from (2.13). From this one gets:

$$(\varphi^{new})^A_x = \varphi^A_x + (\Delta\varphi)^A_x$$

$$(h^{new})^A_{x+\mu} = h^A_{x+\mu} - \frac{1}{2}\Gamma_\mu(x)(U^+)^{AB}_{x,\mu}(\Delta\varphi)^B_x$$

$$(h^{new})^A_x = h^A_x + m(\Delta\varphi)^A_x$$

$$(h^{new})^A_{x-\mu} = h^A_{x-\mu} + \frac{1}{2}\Gamma_\mu(x)\, U^{AB}_{x-\mu,\mu}(\Delta\varphi)^B_x\,. \tag{2.16}$$

The efficiency of this algorithm is related to

$$\varepsilon^{comm} \approx 1 - \frac{T_{communication}}{T_{computation}} \quad (T \text{ stands for time})$$

In ref. [4] it was computed $\varepsilon^{comm} \approx .97$, which is a very good result.

3. The Phase Diagram of Finite Temperature QCD.

In this section I will summarize the results obtained, with the pseudo-fermions method, for finite temperature QCD. Finite temperature QCD is very interesting for both practical and theoretical reasons:

i) it gives us information about the early universe;

ii) it can help us in understanding the mechanisms of confinement and chiral symmetry breaking;

iii) as a first application of methods including dynamical quarks, we can rely on very precise data obtained in the pure gauge approximation. From these data it seems that volume effects, even for relatively small lattices are negligible.

With respect to the computations of finite temperature QCD in the pure gauge approximation, the presence of dynamical quarks creates some problems. Dynamical quarks break explicitly the Z_3 symmetry of the lattice action and as a direct consequence of this the Wilson thermal line, defined as

$$Wl = Tr\sum_{\vec{x}} Wl(\vec{x}) = \frac{1}{N_\sigma^3} Tr\Big(\prod_{t=1}^{N_\beta} U_{\vec{x},t;t}\Big) \tag{3.1}$$

is not an order parameter anymore (N_β, N_σ is the number of lattice sites in the time, space directions). Luckily enough, as we shall see later, we can still use it to study the properties of our system. A relevant parameter, even in the presence of dynamical quarks, is the thermodynamical potential:

$$\varepsilon_g \simeq \frac{36}{a^4}\Big(\sum_{\{space\}}{}' Tr\,U - \sum_{\{time\}} Tr\,U \Big) . \tag{3.2}$$

which represents the gluonic internal energy. In (3.2) a stands for the lattice spacing and U for the product of the matrices residing on the links of the elementary plaquette.

In all the computations I'm going to talk about, we used Kogut-Susskind fermions. Discretizing the fermion field according to this method, leaves a continuous chiral symmetry in the action; this symmetry, in turn, will assure the presence of Goldstone bosons in case of spontaneous symmetry breaking. Kogut-Susskind fermions have also practical advantages: their chiral limit is at $m = 0$ and, furthermore, Dirac indices do not appear in the fermionic part of the action.

Let me now present these results. We tried to build hysteresis cycles starting from ordered and disordered runs. We ran for various values of the bare quark mass $m = .1, .2, .5, 1., 2$. In fig. 1 we present $\mathcal{W}l$ vs. β ($\beta = \frac{6}{g^2}$ where g is the coupling constant of QCD) for various masses. In fig. 2 we have ε_g vs. β. In fig. 3 we show some metastable states obtained at the transition temperatures for different m's. This metastability might indicate that the transition is first order as in the pure gauge case. At last, in fig. 4, we draw a tentative diagram for finite temperature QCD. The entire body of these results was obtained on $6^3 \times 4$ and $8^3 \times 4$ lattices for three flavors of dynamical quarks.

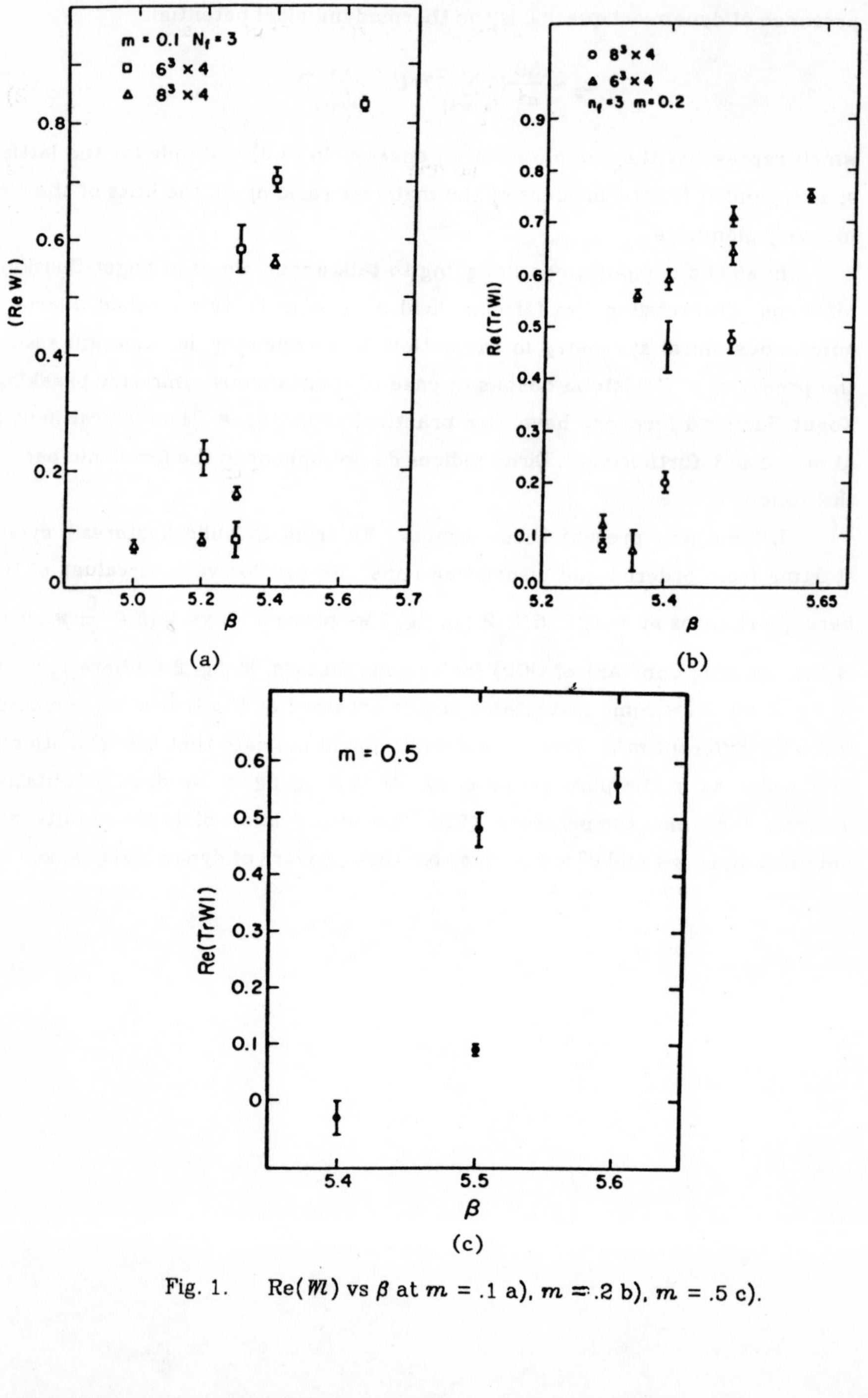

Fig. 1. Re($\mathcal{W}$) vs β at m = .1 a), m = .2 b), m = .5 c).

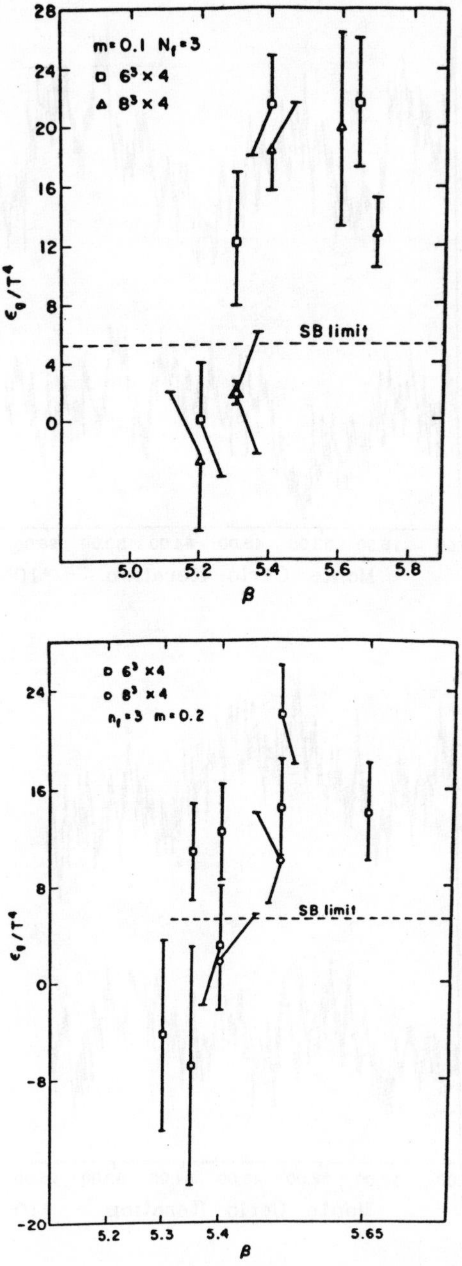

Fig. 2. E_g vs. β at m = .1 a), m = .2 b).

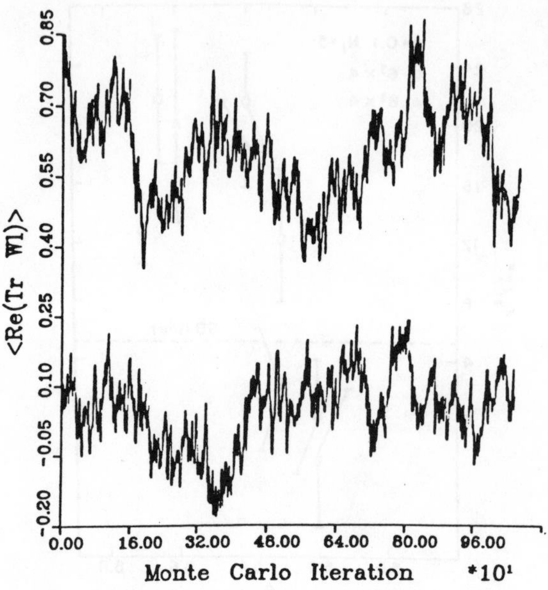

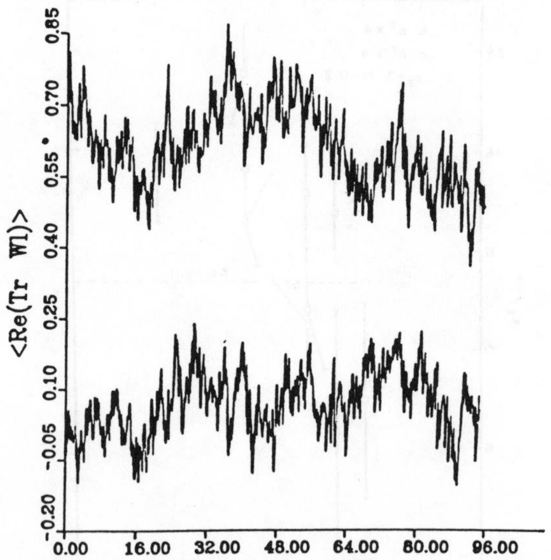

Fig. 3. Metastable states at $\beta = 5.3$ $m = .1$ a), $\beta = 5.35$ $m = .2$ b).

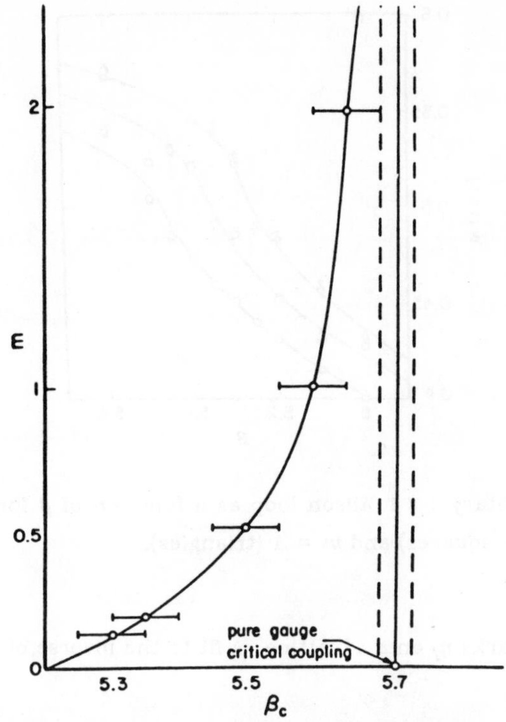

Fig. 4. The tentative phase diagram of QCD at finite temperature.

4. Crossover in the Presence of Dynamical Fermions [5]

I will now discuss QCD at zero physical temperature, still in presence of dynamical fermions discretized according to the Kogut-Susskind scheme.

It is a well-known fact that around $\beta = 5.55$, QCD, in the pure gauge approximation, shows and abrupt "transition" between the so-called strong coupling and weak coupling regions. The location of the "transition" is important to the study of the continuum limit of the theory. In the presence of dynamical quarks the "transition" region should be shifted to smaller values of β. In fig. 5 we see this phenomenon. The data were obtained for three flavors

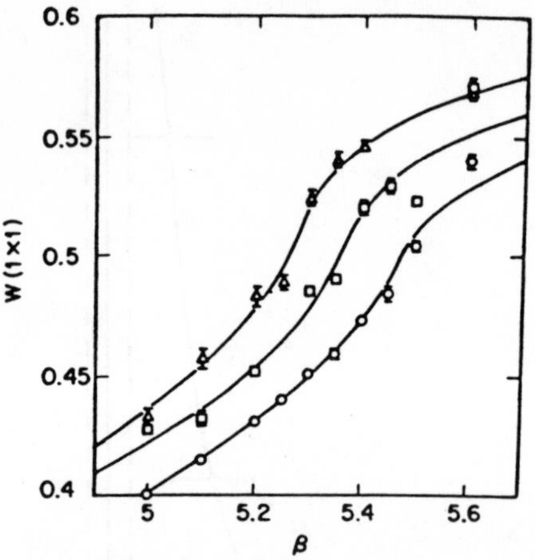

Fig. 5. Elementary 1 × 1 Wilson loop as a function of β for $m = \infty$ (circles), $m = .2$ (squares) and $m = .1$ (triangles).

of dynamical quarks n_f on a 4^4 lattice. A fit to the inverse of the specific heat of the theory

$$C_V^{-1} = A + B\beta + C\beta^2 + \cdots \tag{4.1}$$

gives $\beta \approx 5.0 \div 5.2$ for $m = 0$.

Let's try to interpret this fact. At the position of the peak of C_V I should have

$$\xi(\beta = 5.1, n_f = 3, m = 0) \approx \xi(\beta = 5.55, n_f = 0, m = \infty) . \tag{4.2}$$

If ξ scales like $\Lambda_{\overline{MS}}$ with a proportionality constant weakly dependent on n_f, then

$$\frac{\Lambda_{\overline{MS}}(\beta = 5.1, n_f = 3, m = 0)}{\Lambda_{\overline{MS}}(\beta = 5.55, n_f = 0, m = \infty)} = .83 \tag{4.3}$$

which is not far from the value one. From fig. 5 and its fit can also be deduced that the peak in C_V gets sharper in the presence of dynamical fermions. This, in turn, would imply that the transition line in the mixed SU(3) - SU(3)/Z_3 model gets closer to the pure SU(3) Wilson axis.

5. Conclusions

In this talk I presented the pseudofermion algorithm and some of its applications to lattice QCD. After the first studies on the effectiveness of this method on toy models we are finally able to challenge the full QCD theory. The results of the studies I presented you here seem to be in agreement, at the moment, with those performed by other groups with totally different methods. This gives us the hope that in a short period we will be able to reliably perform computation for the full theory of lattice QCD and hopefully (in the case of finite temperature QCD) to confront them with experiment.

References

[1] F. Fucito, S. Solomon, Phys. Lett. 140B (1984) 387.
R. V. Gavai, M. Lev, B. Petersson, Phys. Lett. 140B (1984) 397.
F. Fucito, S. Solomon, C. Rebbi, Nucl. Phys. B248 (1984) 615.
F. Fucito, R. Kinney, S. Solomon, Nucl. Phys. B253 (1985) 727.
J. Polonyi, H. W. Wyld, J. Kogut, J. Shigenitsu, D. K. Sinclair, Phys. Rev. Lett. 53 (1984) 644.

[2] F. Fucito, E. Marinari, G. Parisi, C. Rebbi, Nucl. Phys. B180 (1981) 369.

[3] J. Salmon, Caltech preprint (CCP Hm-53

[4] F. Fucito, S. Solomon, to appear in J. Comp. Phys.

[5] F. Fucito, H. Hamber, S. Solomon, Phys. Lett. 150B (1985) 285.

HADRON MASS CALCULATIONS ON LARGE LATTICES

K.H. Mütter

Physics Department
University of Wuppertal
Gauss-Str. 2o, D-56oo Wuppertal 1

ABSTRACT

I want to report on two computations of the hadron mass spectrum, done in collaboration with A. König, K. Schilling and J. Smit. The calculations were performed with Wilson fermions on two lattices $16^3 \times 28$ and $16^3 \times 56$ at $\beta = 6$. We used the quenched approximation in the updating of the background fields and the approximate block diagonalization scheme for the computation of quark propagators.

1. INTRODUCTION

The prediction of the hadron mass spectrum is one of the crucial tests for establishing QCD as the correct theory of strong interactions. Nowadays, the lattice formulation appears to be the most promising method to compute the hadron mass spectrum within the framework of QCD. Needless to say that Monte-Carlo simulations on large lattices are needed in order to approximate in a realistic way a field theory with its infinite number of degrees of freedom. The main obstacle to a realistic evaluation of lattice QCD arises from the inclusion of dynamical fermions. Therefore, the so-called quenched approximation - where one neglects the effect of dynamical fermions - has been used in all hadron mass calculations, done so far on large lattices. If you accept the quenched approximation the main limitation on the lattice size comes from the computation of quark propagators, i.e. the inversion of the large fermion matrix. Two years ago Klaus Schilling and I have proposed an approximation scheme for the computation of quark propagators, which we call "approximate block diagonalization"[1]. The basic idea of this approach is easily explained: For the computation of the lowest lying hadron masses, we need the quark propagators for large time separations, that means its infrared behaviour. The latter is governed by the small eigenvalue modes of the fermion matrix. Therefore, we pursued the strategy to find an effective fermion matrix with effective gluon fields defined on a coarser "blocked" lattice, which keeps the small eigenvalue modes almost unchanged.

In the first part of my talk I will review the approximate block diagonalization method. In the second and third part I will present the results of our Monte-Carlo simulations on lattices $16^3 \times 28$ and $16^3 \times 56$ at $\beta = 6$. I will finish with some conclusions and remarks about our future plans.

2. APPROXIMATE BLOCK DIAGONALIZATION OF THE FERMION MATRIX

In the framework of lattice gauge theories the light hadron masses are extracted from the exponential decay of hadron hadron correlation functions at large time separations. Therefore, it is the infrared behaviour of the quark propagator, which fixes the lowest masses in the hadronic channels. The approximate block diagonalization scheme starts from the observation that the infrared behaviour of the quark propagator is governed by the small eigenvalues modes of the fermion matrix Δ^{-1}. It has been shown in ref. 1) how to isolate these small eigenvalue modes:

In the first step, the fermion matrix for one fermion species on the original lattice with spacing a is written in terms of a fermion matrix for 16 fermion species on a coarser "blocked" lattice with spacing 2a. In a second step the mass matrix of these 16 fermionic modes is diagonalized. It turns out that there are 15 heavy modes with masses:

$$M_\ell = m_q + a^{-1} \cdot \rho_\ell \quad , \rho_\ell = 2,4,6,8 \quad , \ell = 1 \cdots 15 \tag{2.1}$$

and one light mode with mass

$$M_0 = m_q \tag{2.2}$$

m_q is the bare quark mass entering in the fermionic part of the Wilson action on the original lattice. In a third step the interaction between the 16 fermionic modes is treated perturbatively. It turns out, that the perturbation expansion is a power series in the lattice spacing a. The result of these three steps can be phrased as follows:

The fermion matrix for Wilson fermions on the original lattice with spacing a:

$$\Delta^{-1}(x,x') = \frac{1}{2a}\sum_\mu \gamma_\mu S_{1\mu}(x,x') + \frac{1}{2a}\sum_\mu C_{1\mu}(x,x') - m_q\,\delta_{x,x'} \tag{2.3}$$

where

$$S_{1\mu}(x,x') = U(x,x+\mu)\,\delta_{x',x+\mu} - U(x,x-\mu)\,\delta_{x',x-\mu} = O(a) \tag{2.4}$$

$$C_{1\mu}(x,x') = U(x,x+\mu)\,\delta_{x',x+\mu} + U(x,x-\mu)\,\delta_{x',x-\mu} - 2\,\delta_{x',x} = O(a^2) \tag{2.5}$$

can be brought into a approximate block diagonal form by means of an appropriate unitary transformation:

$$V\Delta^{-1}V^+ = \left(\Delta^{-1}(\ell,\ell')\right) \qquad \ell,\ell' = 0,1,\dots,15 \tag{2.6}$$

The 16×16 block matrices $\Delta^{-1}(\ell,\ell')$ with matrix elements $\Delta^{-1}(\tilde{x}\ell,\tilde{x}'\ell')$ are defined on a coarser "blocked" lattice with spacing 2a:

$$\tilde{x} = 2a(\tilde{n}_0, \tilde{n}_1, \tilde{n}_2, \tilde{n}_3) \tag{2.7}$$

In the limit $a \to 0$, the non-diagonal blocks vanish

$$\Delta^{-1}(\ell,\ell') = O(a) \qquad \ell \neq \ell' \tag{2.8}$$

whereas the diagonal blocks $\Delta^{-1}(\ell,\ell)$ look like fermion matrices on the blocked lattice with masses as given in eqs.(2.1) and (2.2).

The large distance behaviour of the quark propagator is governed by the light fermionic mode with mass $M_o = m_q$ and the effective fermion matrix:

$$\Delta^{-1}(\tilde{x}0,\tilde{x}'0) = \frac{1}{4a}\sum_\mu \gamma_\mu s_{2\mu}(\tilde{x},\tilde{x}') + \frac{1}{4a}\sum_\mu C_{2\mu}(\tilde{x},\tilde{x}') - m_q\,\delta_{\tilde{x}\tilde{x}'}$$

$$+ \frac{1}{4a}\sum_\mu \gamma_\mu W_\mu(\tilde{x})\,\delta_{\tilde{x},\tilde{x}'}$$

$$- \frac{1}{32a}\sum_\mu C_{4\mu}(\tilde{x},\tilde{x}') + O(a^2) \quad (2.9)$$

on the blocked lattice. The effective gluonic fields $W(\tilde{x},\tilde{x}+2\mu)$ entering in the combinations $s_{2\mu}$, $C_{2\mu}$, $C_{4\mu}$ (analogous to eqs.(2.4) - (2.5)):

$$s_{2\mu}(\tilde{x},\tilde{x}') = W(\tilde{x},\tilde{x}+2\mu)\,\delta_{\tilde{x}',\tilde{x}+2\mu} - W(\tilde{x},\tilde{x}-2\mu)\,\delta_{\tilde{x}',\tilde{x}-2\mu} \quad (2.10)$$

$$C_{n\mu}(\tilde{x},\tilde{x}') = W(\tilde{x},\tilde{x}+n\mu)\,\delta_{\tilde{x}',\tilde{x}+n\mu} + W(\tilde{x},\tilde{x}-n\mu)\,\delta_{\tilde{x}',\tilde{x}-n\mu} - 2\,\delta_{\tilde{x}',\tilde{x}} \quad (2.11)$$

$$n = 2, 4$$

are averages over 8 unitary SU(3) matrices $U^{(\tau)}(\tilde{x},\tilde{x}+2\mu)$:

$$W(\tilde{x},\tilde{x}+2\mu) = \frac{1}{8}\sum_\tau U^{(\tau)}(\tilde{x},\tilde{x}+2\mu) \quad (2.12)$$

The latter are paralleltransporters along a standard path $\mathcal{C}^{(\tau)}(\tilde{x},\tilde{x}+2\mu)$ on the original lattice connecting two neighbouring sites on the coarser blocked lattice with spacing 2a

$$\mathcal{U}^{(\tau)}(\tilde{x}, \tilde{x}+2\mu) = \prod_{\ell \in \mathcal{C}^{(\tau)}(\tilde{x},\tilde{x}+2\mu)} \mathcal{U}(\ell) \tag{2.13}$$

Similarly, one finds for the effective fields $W_\mu(\tilde{x})$ in the "contact" term on the right hand side of eq. (2.9)

$$W_\mu(\tilde{x}) = \frac{1}{8} \sum_\tau \{ \mathcal{U}_\mu^{(\tau)}(\tilde{x},\tilde{x}) - (\mathcal{U}_\mu^{(\tau)}(\tilde{x},\tilde{x}))^+ \} \tag{2.14}$$

where the $U_\mu^{(\tau)}(\tilde{x},\tilde{x})$ are Wilson loops along 8 standard paths $\mathcal{C}^{(\tau)}(\tilde{x},\tilde{x})$ starting and ending at $\tilde{x}$. The new gauge fields $W(\tilde{x},\tilde{x}+2\mu)$ are not exactly unitary. However, one can show, that the deviation from unitarity is of order a^4:

$$W(\tilde{x},\tilde{x}+2\mu)\, W(\tilde{x}+2\mu,\tilde{x}) - 1 = O(a^4) \tag{2.15}$$

In order to repeat the blocking procedure a second time, it is convenient to unitarize the W's

$$W \to H W \qquad H = (W W^+)^{-\frac{1}{2}} \tag{2.16}$$

We have tested first the block diagonalization scheme in the free case, where the gluon field is switched off and the quark propagator can be computed by Fourier transformation. In figure 1 you see the difference of the propagators on a $16^3 \times 28$ lattice and (after two blocking steps) on a $4^3 \times 7$ lattice. For very small quark masses, i.e. $\kappa = (8+2ma)^{-1}$ close to 1/8, you observe no deviations. For larger quark masses you find deviations at smaller distances. For larger distances however these deviations become again very small. This demonstrates numerically, that the block diagonalization scheme is a good approximation for the large distance behaviour of the quark propagator.

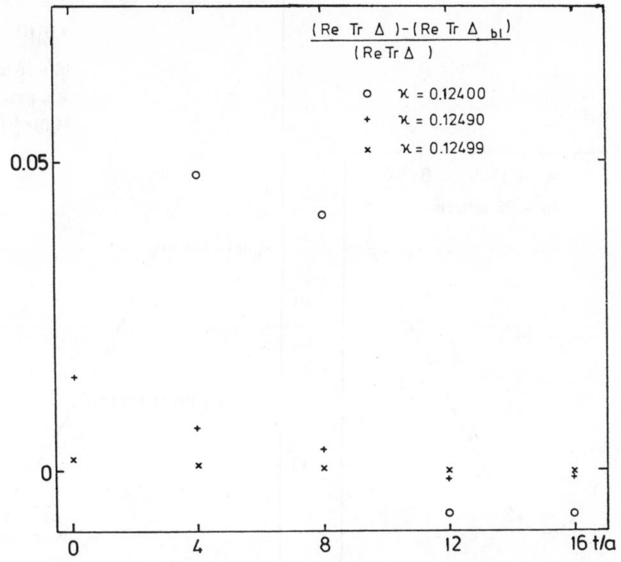

FIG. 1: The normalized difference of the free propagator on a $16^3 \times 28$ lattice and the free propagator after two blocking steps. Traces in spin have been taken and summation over the time slice t was performed. Three different κ-values were chosen.

3. HADRON MASS COMPUTATION ON A $16^3 \times 28$ LATTICE

We have produced - on the CYBER 205 in Bochum - 48 gluonic background fields at $\beta = 6$. For thermalization we needed 1200 sweeps. Two successive background fields are separated by 50 sweeps. We have checked that the configurations are statistically independent by measuring phase and size of Polyakov-loops as proposed by Martinelli et al[4). The approximate block diagonalization procedure was applied twice. Thus we have reduced the rank of the matrix to be inverted by a factor $16^2 = 256$. For each background field configuration the quark propagator was computed by means of conjugate gradient algorithm at three different source points and 5 values of the hopping parameter (κ = .133, .1335, .134, .1345, .135). In total we performed 72o inversions.

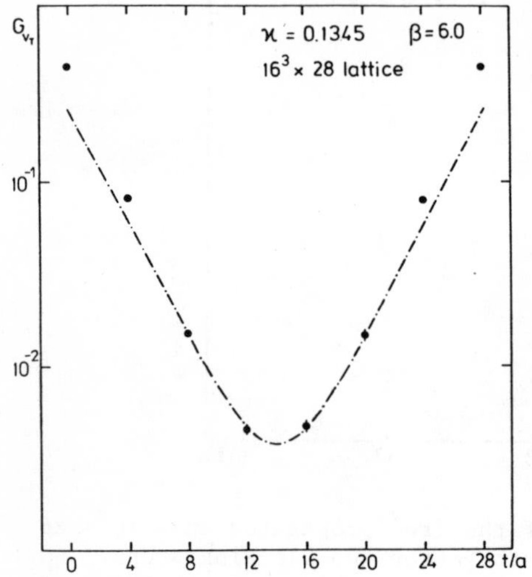

FIG. 2: Propagator of transversal polarized vector meson at κ = .1345 vs time distance. The dashed dotted line represents a fit to the large distance points t/a = 8,12, 16,2o.

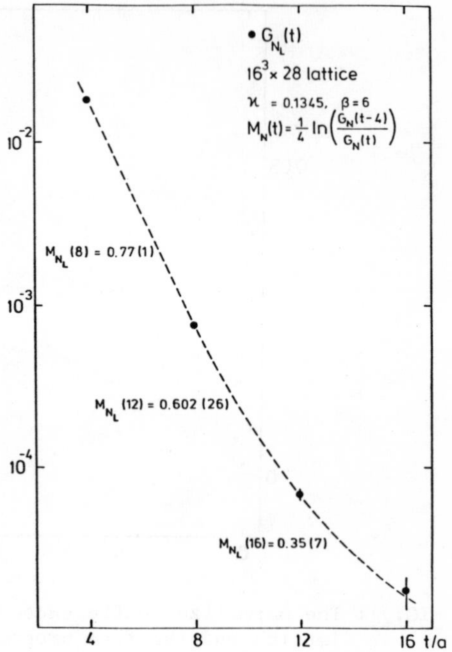

FIG. 3: Nucleon propagator, traced over lower spinor components at κ = .1345.

Figure 2 shows the ρ propagator for transversally polarized ρ's and κ = .1345. The dotted dashed line is a fit with parametrization $C \cosh m(t - \frac{T}{2})$ to the data points at large time separations $\frac{t}{a}$= 8, 12, 16, 2o.

Figure 3 shows the nucleon propagator for κ = .1345. The numbers M(t) denote the local masses extracted from the slope of two successive data points. The apparent deviation from a pure exponential behaviour and the decrease of the local masses indicates, that we have not yet achieved a clean signal for the large distance behaviour of the correlation function. A. Billoire et al[5)] met similar problems with the nucleon propagator. (They computed the hadron mass spectrum with Wilson fermions on a $10^3 \times 20$ lattice at $\beta = 6$.)

We have extracted the hadron masses by means of a least square fit to the propagators at large time separations t/a = 8, 12, 16, 24 and using the following parametrizations:

$$C \cosh m\left(t - \frac{T}{2}\right) \qquad \text{for mesons} \qquad (3.1)$$

$$C_+ e^{-m_+ t} + C_- e^{-m_-(T-t)} \qquad \text{for baryons} \qquad (3.2)$$

In the baryonic channels $m_\pm$ refer to the masses of the parity partners. Then the standard ansatz was made for the dependence of the hadron masses on the hopping parameter κ:

$$m_i^2(\kappa) = A_i \cdot \kappa^{-1} + B_i \qquad \text{for mesons} \qquad (3.3)$$

$$m_i(\kappa) = A_i \cdot \kappa^{-1} + B_i \qquad \text{for baryons} \qquad (3.4)$$

Extrapolation to zero pion mass yields a critical hopping parameter

$$\kappa_c = .13544(9) \qquad (3.5)$$

Figure 4 shows the hadron masses as function of the quark mass

$$m_q = \frac{1}{2}\left(\kappa^{-1} - \kappa_c^{-1}\right) \qquad (3.6)$$

To check for rotational symmetry we have determined the ρ mass from correlation functions with transverse and longitudinal polarization of the ρ. Similarly, the proton and Δ mass have been extracted from two different correlation functions, corresponding to the upper and lower components in the Dirac spinors. Our results are summarized in table 1.

Table 1: Hadron masses in lattice units

Mesons		Baryons	
π	= 0.0568 (13)	$N(939)_u$	= 0.532 (20)
ρ_T	= 0.314 (7)	$N(939)_\ell$	= 0.549 (44)
ρ_L	= 0.325 (18)		
K	= 0.197 (3)	$\Delta(1232)_u$	= 0.654 (28)
K^*_T	= 0.353 (5)	$\Delta(1232)_\ell$	= 0.635 (63)
K^*_L	= 0.362 (10)		
ϕ_T	= 0.389 (3)	Ω^-_u	= 0.684 (15)
ϕ_L	= 0.394 (6)	Ω^-_ℓ	= 0.700 (62)

Massparameters from 4 point fit

κ_c	= .13544 (9)	κ_{s+u}	= .13434 (8)
κ_u	= .13535 (9)	κ_s	= .13335 (8)

Errors were estimated from the variance of fitting independently to six subsamples.

If we identify the masses m_{V_T} and m_{ps} in the vectormeson- and pseudoscalar channel with the experimental value of the ρ and π mass,

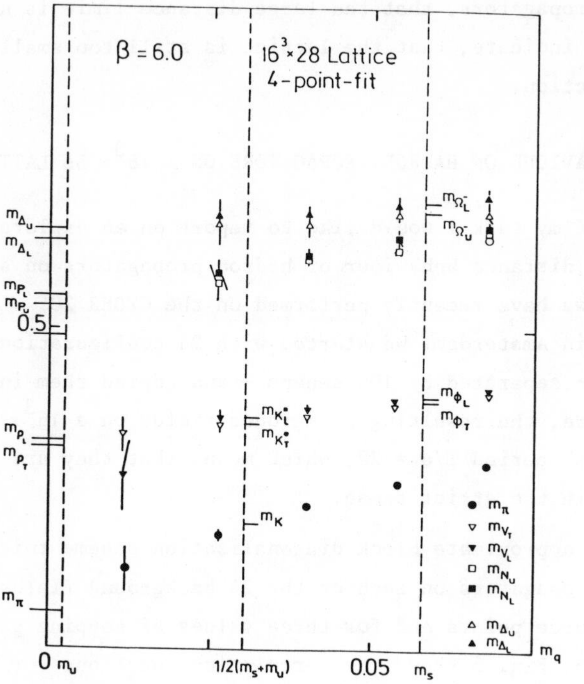

FIG. 4:

Hadron masses vs quark mass m_q at $\beta = 6$ as obtained from the fits to the large distance points t/a = 8, 12, 16, 20. The predicted spectrum of physical hadrons is given by the horizontal lines.

we obtain for the lattice spacing at $\beta = 6$

$$a(\beta = 6) = .0803(18)\ \text{fm} \approx (2.47\ \text{GeV})^{-1} \quad .$$

Thus the spatial extension of our lattice is about 1.3 fm, not very much to accommodate a nucleon with diameter 1 fm. It turns out, that the mass ratio

$$\frac{m_N}{m_\rho} = 1.7 \text{ is too large,} \quad \frac{m_\Delta}{m_N} = 1.2 \text{ is too small}$$

in comparison with the experimental number:

$$\frac{m_N}{m_\rho} = 1.25 \quad , \qquad \frac{m_\Delta}{m_N} = 1.31 \quad .$$

There are many sources for systematical errors, which might be responsible for this failure. For example, I mentioned already in the discussion of the hadron propagators, that the large distance limit is not yet reached. This would indicate, that the lattice is still too small - at least in time direction.

4. LARGE DISTANCE BEHAVIOUR OF HADRON PROPAGATORS ON A $16^3 \times 56$ LATTICE

In the last part of my talk I would like to report on an exploratory study of the large distance behaviour of hadron propagators on a $16^3 \times 56$ lattice, which we have recently performed on the CYBER 205 machines in Bochum and in Amsterdam. We started with 24 configurations on the $16^3 \times 28$ lattice - separated by 100 sweeps - and copied them in time direction. Therefore, the resulting background fields on a $16^3 \times 56$ lattice are periodic with period $T/a = 28$, which means that they are not in thermal equilibrium in the strict sense.

Applying again the approximate block diagonalization scheme twice, we computed the quark propagators on each of the 24 background fields with three different source points and for three values of hopping parameter $\kappa = .133, .134, .1375$. Fig. 5 shows the correlation functions for transversally and longitudinally polarized ρ's for $\kappa = .134$. The solid and dashed curves are fits to a parametrization of the type (3.1). In the interval $8 \leq \frac{t}{a} \leq 48$ the data points nicely follow these curves, which indicates that the large distance limit is now realized.

The nucleon and the Δ propagators for $\kappa = .134$ are plotted in Figs. 6 and 7. Again, in the interval $8 \leq \frac{t}{a} \leq 48$ the data points signal a clean exponential decay over 6 orders of magnitude. For $\kappa = .133$ and .13475 all propagators are shown together in Figs. 8 and 9. Notice the apparent ordering of the slopes, which manifestly demonstrates the correct sequence of the hadron masses $m_\pi < m_\rho < m_N < m_\Delta$.

Finally, in Fig. 10 the hadron masses are plotted as functions of the quark mass (eq.(3.6)). The critical κ-value turns out to be $\kappa_c = .13549$. The extrapolation of the hadron masses to κ_c yields within the statistical error the same results as we obtained in our calculation on the $16^3 \times 28$ lattice. In particular, the mass ratio m_π/m_ρ stays at 1.7.

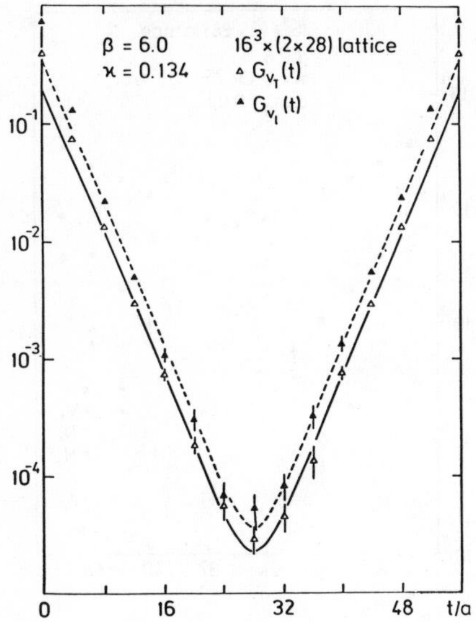

FIG. 5: ρ propagator plotted against time separation t at κ = .134 and β = 6.0. Symbols ▲ (Δ) refer to longitudinal (transverse) ρ's with respect to 3-direction. The transverse propagator is shifted by a factor two. The dashed and solid lines are cosh-fits with a mass $am_{\rho_L} = .353$ $am_{\rho_T} = .349$.

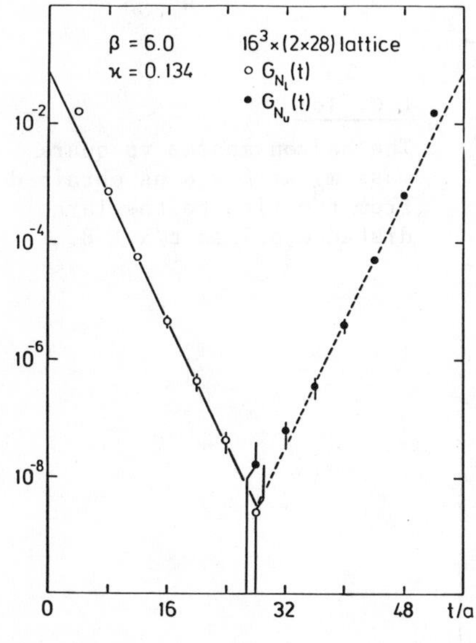

FIG. 6: The nucleon propagators, traced over upper and lower spin components, G_{N_u} (●) and G_{N_L} (o) at κ = .134 (am_q = .041) and β = 6.0. The fits to an exponential yield masses $am_{N_u} = .609$ and $am_{N_L} = .614$.

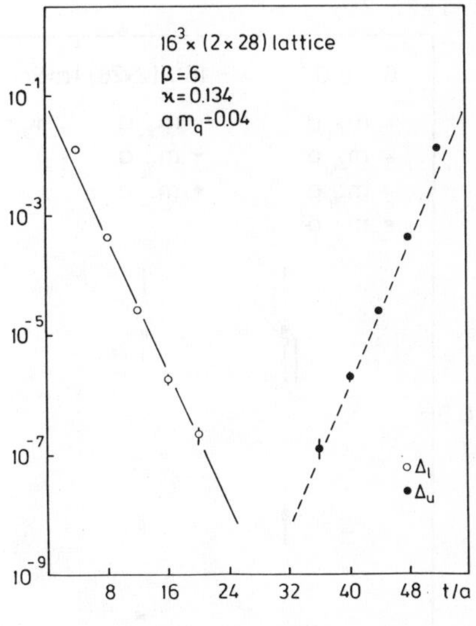

FIG. 7: Same as fig. 6, but for the Δ propagators. The fits to an exponential yield masses $am_{\Delta_u} = .659$ and $am_{\Delta_L} = .649$.

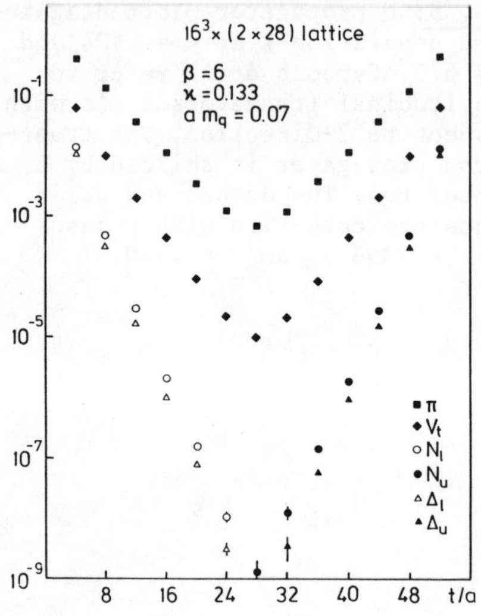

FIG. 8: A synopsis of propagators at κ = .133 (am_q = .069) in the notation of ref. 2.

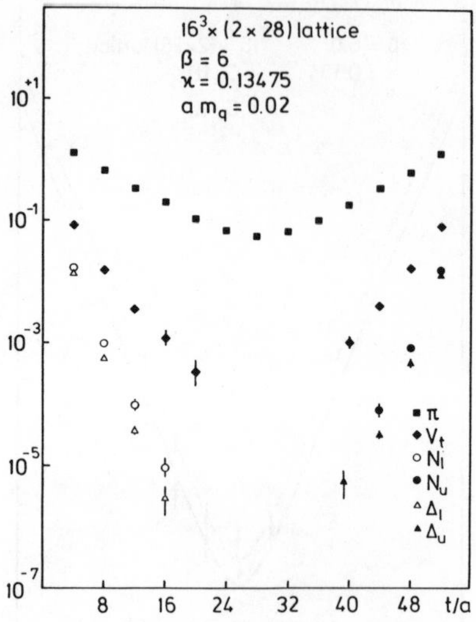

FIG. 9: Same as fig. 8, but for κ = .13475 (am_q = .02).

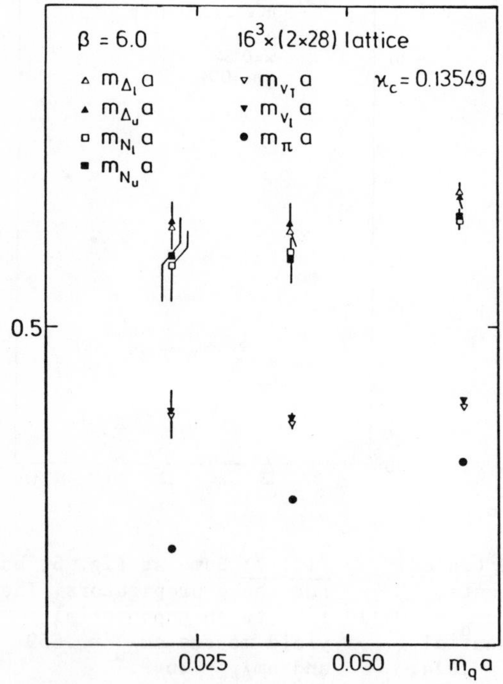

FIG. 1o:

The hadron masses vs quark mass m_q at β = 6 as obtained from the fits to the large distance points t/a ≥ 8.

5. DISCUSSION AND CONCLUSIONS

Recent Monte-Carlo simulations on large lattices - our results on a $16^3 \times 56$ lattice with Wilson fermions and the results of D. Barkai, K.J.M. Moriarty and C. Rebbi[6)] on a $16^3 \times 32$ lattice with Kogut-Susskind fermions at $\beta = 6$ - look very promising:

To our knowledge for the first time a neat exponential decay of hadron propagators could be observed over a wide range of time separations and for rather small quark masses. The resulting meson masses are in reasonable agreement with the experimental number. The mass ratio $m_N/m_\rho = 1.7$ - obtained in our calculation with Wilson fermions - still deviates considerably from the experimental value. In this respect, the prediction of ref. 6) ($m_N/m_\rho = 1.4$) with Kogut-Susskind fermions is in better shape.

In both calculations the lattice spacing a was fixed by the physical values of the π and ρ masses. As in previous calculations on smaller lattices[5)], the values

a^{-1} ($\beta = 6$) = 2.47 GeV (Wilson fermions)

a^{-1} ($\beta = 6$) = 1.66 GeV (Kogut-Susskind fermions)

differ considerably. A possible explanation could be, that the "sick" particle in our calculation with Wilson fermions is not the nucleon, but the ρ. If we fix the lattice spacing by the experimental values of the π and nucleon masses, the lattice spacing turns to be a^{-1} ($\beta = 6$) = 1.77 GeV in fair agreement with the result of ref. 6).

There are various sources for systematical errors, which might be responsible for the wrong mass ratios m_N/m_ρ:

- finite size effects in particular in space directions
- the quenched approximation in the updating of the gluon fields
- the approximate block diagonalization scheme for the computation of the large distance behaviour of quark propagators.

We hope to get control over the systematic errors, which are introduced by the approximate block diagonalization scheme in the near future. Ph. de Forcrand has produced 31 background fields at $\beta = 6.3$ on a $24^3 \times 48$ lattice. He intends to compute the hadron masses "exactly", i.e. without making any approximation in the inversion of the fermion matrix. Using his configurations, we are also going to compute the hadron spectrum with our block diagonalization method. A comparison of the results will tell us how large the systematic errors introduced by this method really are.

6. REFERENCES

1) K.H. Mütter and K. Schilling, Nucl. Phys. B23o [FS10], 275 (1984)

2) A. König, K.H. Mütter and K. Schilling, Phys. Lett. 147B, 145 (1984) and Wuppertal-preprint WU B 84-25, to be published in Nucl. Phys.

3) A. König, K.H. Mütter, K. Schilling and J. Smit, Wuppertal-preprint WU B 85-11, to be published in Phys. Lett. B.

4) G. Martinelli, G. Parisi, R. Petronzio and F. Rapuano, Phys. Lett. 122B, 283 (1983)

5) A. Billoire, E. Marinari and R. Petronzio, Nucl. Phys. B251 [FS 13], 141 (1985)

6) D. Barkai, K. J. M. Moriaty and C. Rebbi, CERN-preprint Ref.-TH 4155 (1985)

FERMION MONTE CARLO ALGORITHMS

Don Weingarten

IBM T. J. Watson Research Center
Yorktown Heights, NY 10598

Abstract

Rough estimates are given for the asymptotic rate of growth of the number of arithmetic operations required by the fermion Monte Carlo algorithm of Petcher and the present author and by the algorithm of Fucito, Marinari, Parisi, and Rebbi. For the first algorithm we find a rate of N^{11}/m_q and for the second we obtain N^8/m_q^2 on an N^4 lattice with renormalized quark mass m_q. Numerical experiments are presented comparing the two algorithms on a 2^4 lattice. With a hopping constant K of 0.15 and β of 4.0, the number of operations for the second algorithm is about 2.7 times larger than for the first and about 13000 times larger than for corresponding Monte Carlo calculations with a pure gauge theory. An estimate is made of the number of arithmetic operations required by each algorithm for a full calculation of hadron masses.

A detailed treatment will appear in the Proceedings of the American Physical Society Division of Particles and Fields Meeting, Santa Fe, November, 1984, and in Nuclear Physcis B.

Wave Functions and Finite Size Effects in a Two-Dimensional Lattice Field Theory†

H. B. Thacker
Fermi National Accelerator Laboratory
P.O. Box 500
Batavia, IL 60510

Abstract

A study of finite size corrections to the masses of fermions and bound states in the Baxter/ massive Thirring/ sine Gordon lattice field theory is discussed. It is shown that information on bound state wave functions may be used to extrapolate Monte Carlo mass calculations to infinite volume.

---

†Talk delivered at the Conference on Advances in Lattice Gauge Theory, Tallahassee, FL, April, 1985

Numerical lattice QCD calculations are often carried out inside a box whose physical size is, at best, only slightly larger than a hadron.[1] In this situation, the question of how large the box must be to have a negligible (or calculable) effect on the physics is an important consideration in selecting the parameters of the calculation. Recent work by Weingarten and Velikson[2] and by Gottlieb[3] has shown that the data generated in Monte Carlo investigations of the hadron spectrum can also be used to study hadron wave functions, which are extracted from correlation functions involving spatially extended operators. The results are quite reasonable and indicate, for example, that at $\beta=5.7$ ($a\approx0.2$ fermi) hadron wave functions fall off substantially with quark separations of 3 or 4 sites. In addition to the intrinsic interest of studying hadron wave functions, these results may also provide a very economical way of estimating finite size effects on spectrum calculations. The important practical point here is that, once the gauge configurations and quark propagators are generated for the spectrum calculations, the study of hadron wave functions requires a negligible amount of additional computer time. In contrast, a direct study of finite size effects by varying the size of the box is a very expensive proposition requiring independent simulations with very good statistics at several different box sizes. It would clearly be advantageous to avoid such a calculation by using the wave functions on a fixed size lattice to estimate finite size effects. To study this possibility David Hochberg and I carried out a Monte Carlo investigation[4] of the two-dimensional Baxter/ massive Thirring/ sine Gordon model.[5-9] This is a very convenient toy model for several reasons. It has a nontrivial spectrum consisting of fermions and an adjustable number of fermion-antifermion bound states (sine Gordon mesons). For an infinite volume lattice, the spectrum is exactly known for all lattice spacings (i.e. not just in the continuum limit).[8,9] One may also carry out Monte Carlo calculations of the spectrum and of the bound state wave functions which are quite analogous to the calculations in QCD. By exploiting the fact that the Baxter model is equivalent to the lattice massive Thirring model, we were able to carry out the Monte Carlo calculations in terms of the Ising-like spins $\sigma_{ij}=\pm 1$ of the Baxter model with an action consisting of local 2-spin and 4-spin couplings. The simplicity of the action combined with the fact that the model is two-dimensional enabled us to obtain very accurate Monte Carlo results (±3% for masses) with only a modest investment of computer time. It should be noted that the Baxter model constitutes an exact treatment of massive Thirring fermions[9] including closed loops, so there is no need for a "quenched" approximation.

To study the dynamics of finite size effects, we varied three different parameters in the calculation: (1) The size of the box in lattice units, i.e. the number of lattice sites in the space direction; (2) The infinite volume mass of the fermion in lattice units (this is determined by an elliptic modulus in the standard Baxter parametrization of the spin couplings[8]); (3) The coupling constant g which controls the strength of the Thirring interaction (and also determines the bound state masses). The spatial size of the lattice varied from 6 to 100 sites. The number of sites in the time direction was always taken to be ≥30, long enough to see pure exponential time-dependence for Fourier transformed correlation functions over a large range of time separations. For an infinite volume lattice, the fermion and bound state masses are given in terms of the Baxter couplings by analytic formulas involving elliptic functions. These were first derived in Ref. 8 and are summarized in Ref. 4. In the Monte Carlo calculation, the mass of the fermion m_F was extracted from the spin-spin correlation function (for $T>T_c$), while the mass of the lowest lying bound state m_B was obtained from the correlation function for composite operators constructed from nearest-neighbor spin pairs $\sigma_{i,j}\sigma_{i+1,j}$ separated by one site in the spatial direction. The latter operator corresponds to $\bar{\psi}\gamma_5\psi$ in the fermion representation. Thus, we looked at the long range behavior of the functions

$$F(\tau) = \sum_x \langle\sigma_{0,0}\sigma_{x,\tau}\rangle \sim Z_F(e^{-m_F\tau} + e^{-m_F(N_t-\tau)}) \tag{1}$$

$$B(\tau) = \sum_x \langle\sigma_{0,0}\sigma_{1,0}\sigma_{x,\tau}\sigma_{x+1,\tau}\rangle \sim Z_B(e^{-m_B\tau} + e^{-m_B(N_t-\tau)}) \tag{2}$$

where N_t is the length of the lattice in the time direction. The bound state wave function $\Psi(y)$ was extracted from the correlation function between a nearst neighbor spin pair and a spatially extended spin pair,

$$B(\tau;y) = \sum_x \langle\sigma_{0,0}\sigma_{1,0}\sigma_{x,\tau}\sigma_{x+y,\tau}\rangle \sim Z_B\Psi(y)(e^{-m_B\tau} + e^{-m_B(N_t-\tau)}) \tag{3}$$

with y odd. It is a nontrivial test of this method that the exponent m_B in (3) is the same as that in (2). We found this to be true to very high accuracy. The coefficient $Z_B\Psi(y)=Z_B^{\frac{1}{2}}\times Z_B^{\frac{1}{2}}\Psi(y)$ represents the factorized residue of the bound-state pole, so that $Z_B^{\frac{1}{2}}\Psi(y)$ is the vacuum-to-bound-state matrix element of the spatially extended operator.

We began by considering the Ising/free-fermion case of the Baxter model ($g=0$ or in conventional Baxter model notation $\mu=\pi/2$ where $g=-2\cot\mu$). In this case, the spectrum consists of only the fermion. The Monte Carlo results for the spin-spin correlation function on a 30×30 lattice are shown in Fig. 1 along with the exponential fit defined in Eq. (1). Note the extremely high quality of the exponential fit over a large number of sites. This was typical of all the correlation functions we measured, including those for nonlocal operators, which made the task of extracting masses and wave functions quite straightforward and unambiguous. The mass of the fermion in the free field case is shown for various box sizes in Fig. 2. The infinite volume fermion mass was chosen to be $m_{F\infty}=0.1178$, which corresponds to a Compton wavelength (spin-spin correlation length) of about $8\frac{1}{2}$ sites. We see that the shift of the mass due to finite size effects is positive and monotonically increasing as the box size is decreased. The shift becomes large when the box is approximately twice the Compton wavelength of the fermion. The solid curve in Fig. 2 is an empirical fit given by

$$m_F = \left(m_{F\infty}^2 + \frac{0.93}{L^2}\right)^{1/2} \tag{4}$$

It should be noted that the Jordan-Wigner transformation between spins and fermions in a finite volume involves boundary terms (c.f. Ref. 10) which shift the allowed momenta by $O(1/L)$. This might explain the form of Eq. (4). [Our lattice is periodic in the spin variables, i.e. in the language of Ref. 10 we treat the "a-cyclic" problem rather than the "c-cyclic" problem.]

Next we considered the more interesting case of the interacting theory. We took $\mu=0.65\pi$ corresponding to the weakly attractive coupling region in which there is a single fermion-antifermion bound state. [In general, there are n bound states for $(\frac{n}{n+1})\pi<\mu<(\frac{n+1}{n+2})\pi$.] Taking the fermion mass $m_{F\infty}=0.1178$ as before, the infinite volume bound state mass is $m_{B\infty}=0.1762$. The Monte Carlo results for the fermion and bound state masses on lattices of spatial dimension 100, 50, 30, 20, 10, and 6 sites are shown in Fig. 3. The fermion is close to its infinite volume value for $L\approx30$, dips slightly at $L=20$ and then increases rapidly for smaller lattices in a manner similar to the free fermion case. In contrast, the bound state mass is noticeably below $m_{B\infty}$ even for $L=50$ and is more than 20% low for $L=20$. For very small lattices, the bound state mass turns around and begins to increase rapidly, apparently in unison with the fermion mass.

The rest of our Monte Carlo calculations were devoted to understanding the bound state mass curve in Fig. 3. We found that this curve is the result of two competing effects. The large positive mass shift of the bound state for very small lattices is a result of the increasing mass of its fermion constituents. The size scale at which this effect becomes relevant is roughly $2m_F^{-1}\approx17$ sites. To verify this interpretation, we increased the fermion mass to $m_{F\infty}=0.2043$ or $2m_F^{-1}\approx10$ sites. For lattices down to $L=10$, the fermion mass was essentially equal to its infinite volume value while the bound state mass was monotonically decreasing with no tendency to turn upward (see Ref. 4 for details), thus confirming the idea that the relevant length scale for the positive component of the mass shift is the fermion Compton wavelength.

The other finite size effect exhibited by the bound state mass curve in Fig. 3 is a negative mass shift for lattices of moderate size. We found that the box size at which this effect becomes relevant is determined by the spatial extent of the bound state wave function. We looked at wave functions on a 30×30 lattice for three different values of the coupling constant: $\mu = 0.65\pi$, 0.72π, and 0.83π ($g = 1.02$, 1.76, and 3.15 respectively). The Monte Carlo results for the square of the wave function $\Psi(y)$, which were extracted from the correlation functions (3), are shown in Fig. 4. The theoretical infinite volume bound state mass was held fixed at $m_{B\infty}=0.1762$. The measured values of the mass on a 30×30 lattice were $m_B=.1348\pm.0024$, $.1548\pm.0042$, and $.1751\pm.0050$ for $\mu=.65\pi$, $.73\pi$, and $.82\pi$ respectively. Note that as we increase the coupling and pull in the wave function to a smaller size, the mass approaches the correct infinite volume value. For the strongest of the three couplings, $.82\pi$, the wave function is quite well contained in the 30 site box, and correspondingly, we see no finite size correction to the mass ($m_B \approx m_{B\infty}$ within errors). In fact, the mass correction in all cases appears to be roughly proportional to $|\Psi(L/2)|^2$, i.e. the squared wave function at maximum separation on the periodic lattice. Returning to the bound state mass curve in Fig. 3, we find that a very good description of the finite size effects for L>30 is given by

$$m_B = m_{B\infty} + c\ |\Psi(L/2)|^2 \qquad (5)$$

This gives the solid curve in Fig. 3. In Eq. (5) we measured the wave function for each value of L directly, using a lattice of length L. However, these values could have been quite accurately estimated using only the data from the 30 site lattice by simply fitting the tail of the wave function to a periodic exponential (analogous to the right hand side of (1)) and extrapolating. Similarly, we might use Eq. (5) to define a procedure for extrapolating to infinite volume by measuring the mass and wave functions for two values of L and solving for $m_{B\infty}$ and c. For example, if we use the data for L=30 and 50 we obtain the extrapolated value $m_{B\infty}$=.1771±.005 which is actually somewhat closer to the correct infinite volume value (.1762) than the result we obtained from a direct calculation on a 100×100 lattice.

Some of our results may have implications for QCD spectrum calculations (particularly those including closed loops, which is presently only feasible on rather small lattices). The formula (5) might have an analog for QCD bound states. [Presumably, in three space dimensions, $|\Psi(L/2)|^2$ should be replaced by an integral of $|\Psi|^2$ over the surface of a cube.] It would then be possible to do spectrum calculations on moderate size lattices and correct for finite size effects by calculating hadron wave functions.

REFERENCES

1. For discussion of finite size effects in lattice QCD, see P. Hasenfratz and I. Montvay, Phys. Rev. Lett. 50, 309 (1983); M. Luscher, Proceedings of the Nato Advanced Study Institute, Cargese (1983).
2. D. Weingarten and B. Velikson, Nuc. Phys. B249, 433 (1985).
3. S. Gottlieb, Proceedings of this Conference.
4. D. Hochberg and H. Thacker, Nuc. Phys. B (to be published).
5. W. Thirring, Ann. Phys. (N.Y.) 3, 91 (1958).
6. S. Coleman, Phys. Rev. D11, 2088 (1975).
7. R. J. Baxter, Ann. Phys. (N.Y.) 70, 193 (1972).
8. J. D. Johnson, S. Krinsky, and B. M. McCoy, Phys. Rev. A8, 2526 (1973).
9. A. Luther, Phys. Rev. B14, 2153 (1976).
10. E. Lieb, T. Schultz, and D. Mattis, Ann. Phys. (N.Y.) 16, 407 (1961); T. Schultz, D. Mattis, and E. Lieb, Rev. Mod. Phys. 36, 856 (1964).

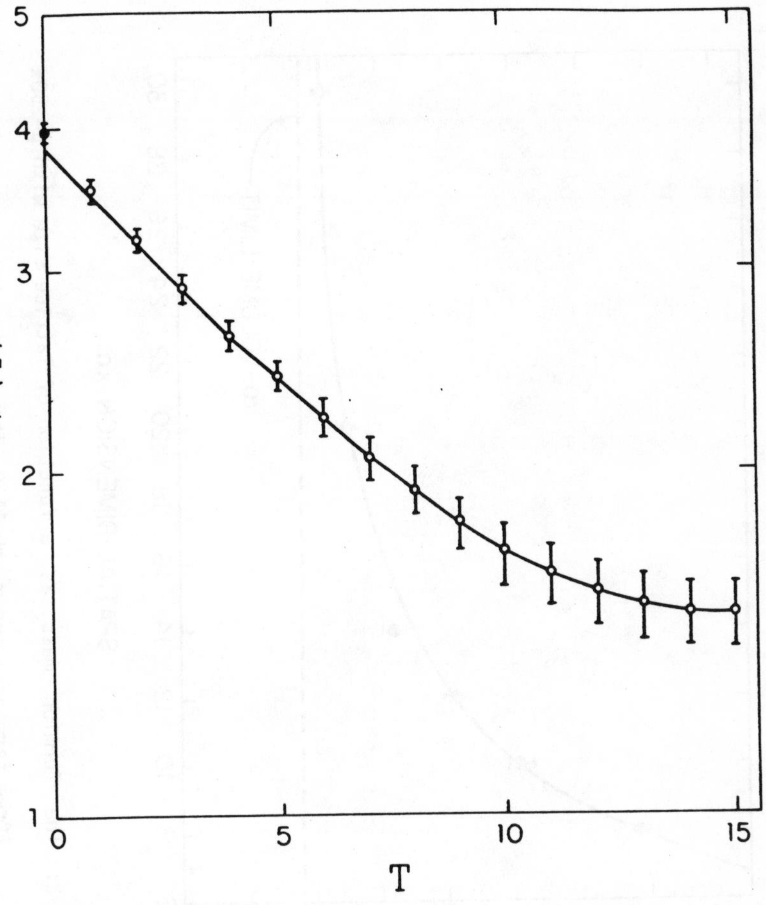

Fig. 1: Spin-spin correlation function on a 30×30 lattice with $\mu=.50\pi$ (free fermions) and $m_{F\infty}=0.1178$. The solid line is the fit defined in Eq. (1).

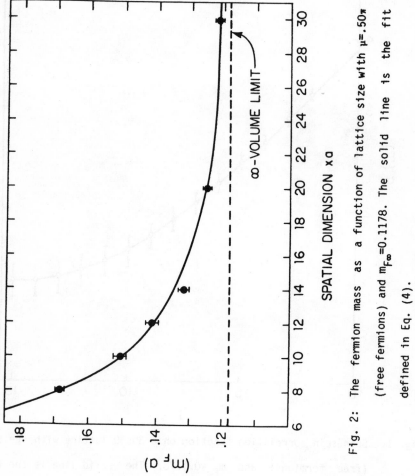

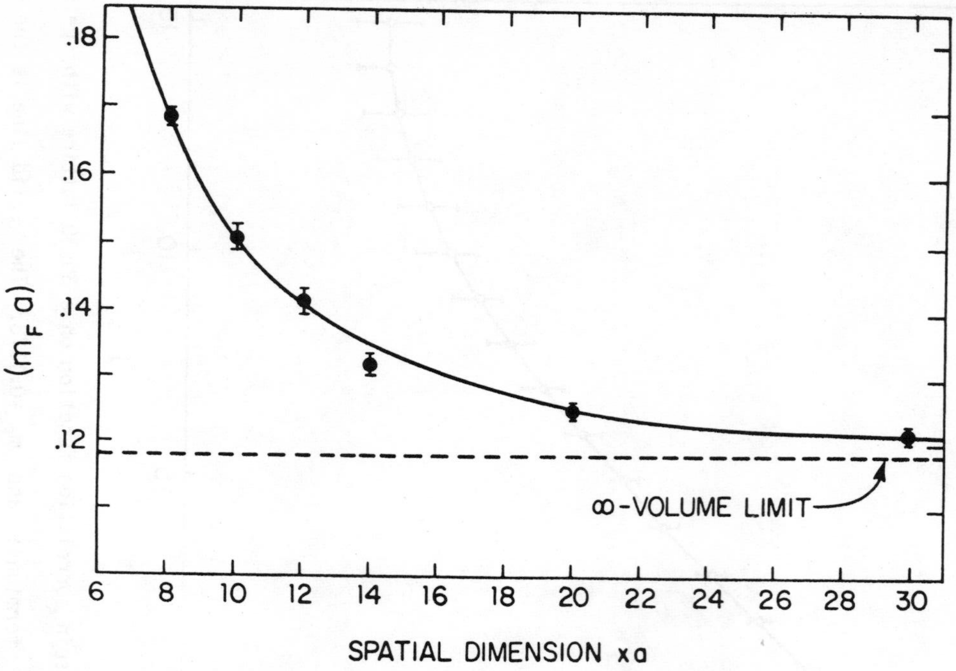

Fig. 2: The fermion mass as a function of lattice size with $\mu = .50\pi$ (free fermions) and $m_{F\infty} = 0.1178$. The solid line is the fit defined in Eq. (4).

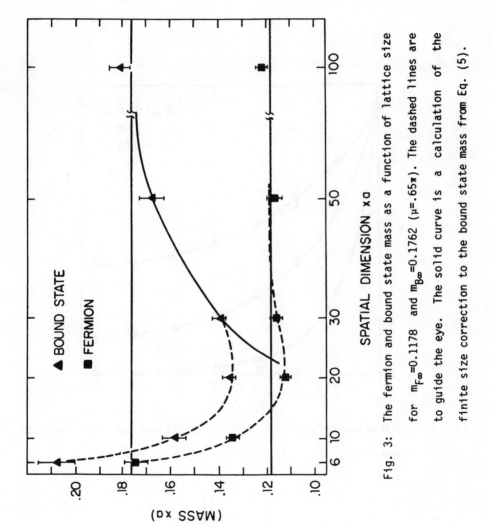

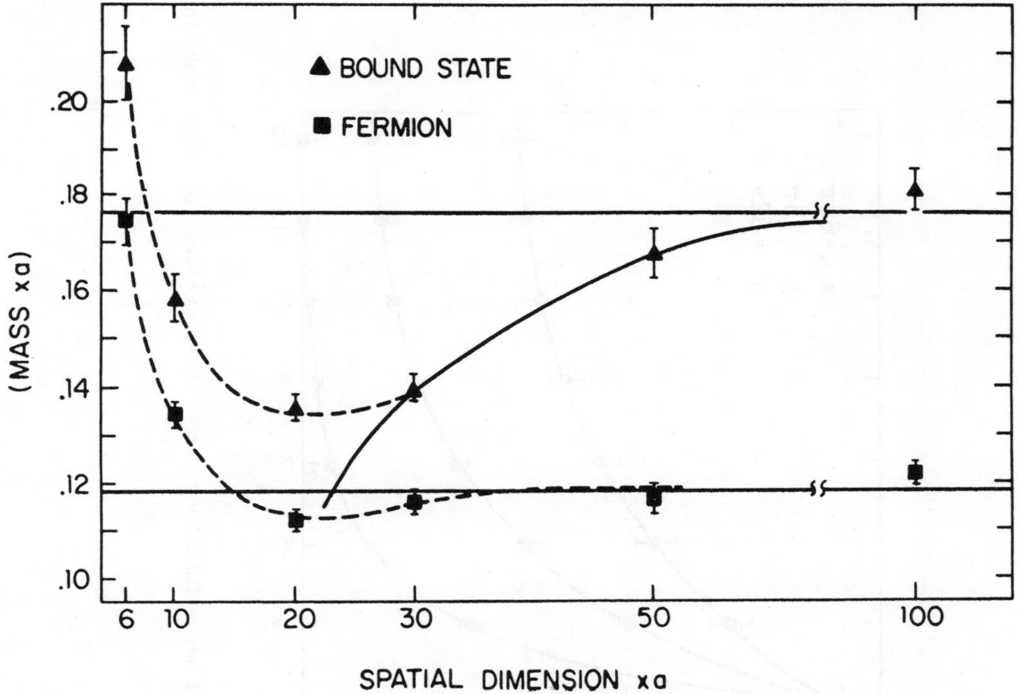

Fig. 3: The fermion and bound state mass as a function of lattice size for $m_{F\infty}=0.1178$ and $m_{B\infty}=0.1762$ ($\mu=.65\pi$). The dashed lines are to guide the eye. The solid curve is a calculation of the finite size correction to the bound state mass from Eq. (5).

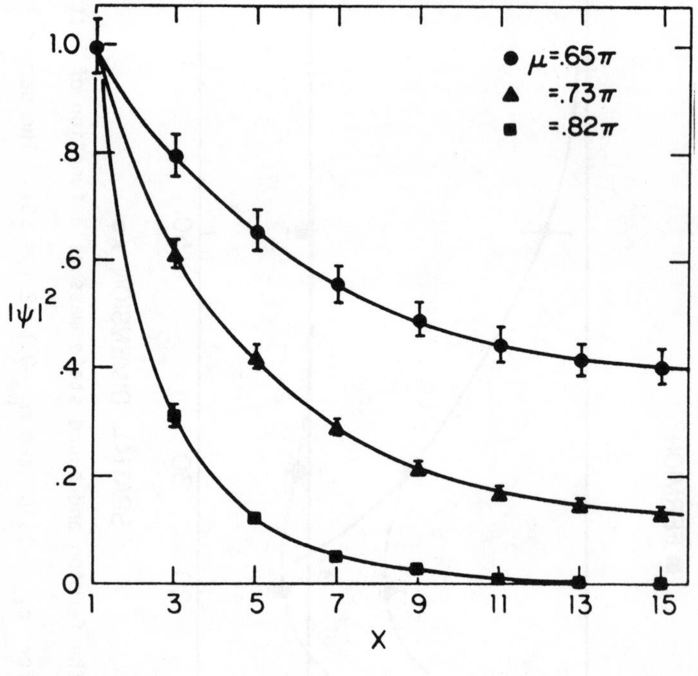

Fig. 4: Squared wave functions for three values of the coupling constant.

PORTRAIT OF A PROTON

Steven Gottlieb
University of California, San Diego
La Jolla, CA 92093
U. S. A.

ABSTRACT

I present the results of a calculation of hadron wave functions in SU(3) lattice gauge theory. The proton, delta, pion and rho meson are all studied. The first three particles have wave functions which fall off more quickly than that for the rho. This suggests that the rho is much more subject to finite size effects, and may explain the difficulty so far in calculating the pi-rho mass splitting.

1. INTRODUCTION

Finite size effects have presented a serious challenge to hadronic spectrum calculations[1] in lattice gauge theory from the very beginning. The need to reduce computing costs leads one to choose the smallest acceptable box size. However, the most straightforward way of demonstrating the adequacy of the box size, to increase it and to see that the spectrum doesn't change, is very costly. It would be much better to have some sort of test which could be carried out without changing the box size. With recent calculations showing that scaling does not set in by β=6.0,[2] where several spectral calculations have been done,[3] it becomes even more crucial to pick a box size neither too small nor too large in future calculations at weaker coupling.

I present here the results of a calculation of hadronic wave functions for pion, rho, nucleon, and delta in SU(3) gauge theory in the valence approximation. The falloff of the wave functions as the distance between the quarks grows indicates the size of the bound state, and, at least qualitatively, whether the boundary conditions significantly affect the state. It is hoped that further study may yield a quantitative basis for estimating finite size effects on particle masses from knowledge of the wave function.[4]

This study has its roots in finding a purer source for the pion[5] than that afforded by the local operator $\overline{\psi}(x)\gamma^5\psi(x)$. It may be possible to use knowledge of the wave function to improve spectral calculations by reducing the contributions of excitations,[6] as had been done in the case of glueballs.[7] Further, it would be very interesting to compare wave functions in the valence approximation with those obtained from a proper treatment of dynamical quark effects.

2. OPERATORS AND WAVE FUNCTIONS

In an earlier calculation of the $\rho\pi\pi$ coupling constant,[5] non-local meson operators were considered as sources for pion and rho. These gauge invariant operators are defined as :

$$\pi_{NL}(\vec{\Delta}) = \sum_{\vec{x}} \overline{\psi}(x)\gamma^5\psi(x+\vec{\Delta})U(x,x+\vec{\Delta}) \tag{1}$$

$$\rho^i_{NL}(\vec{\Delta}) = \sum_{\vec{x}} \overline{\psi}(x)\gamma^i\psi(x+\vec{\Delta})U(x,x+\vec{\Delta}) \tag{2}$$

The factors $U(x,x+\Delta)$ are products of gauge matrices along the shortest path between x and $x+\Delta$ (or the average if there isn't a unique path). The operators have the advantage of being gauge invariant; however, their proper renormalization in the continuum limit is nontrivial,[8] especially when the path is not straight. It is quite interesting to consider, instead, Bethe-Salpeter wave functions defined in lattice Coulomb gauge:

$$\phi_\pi(\vec{\Delta}) = \langle \Omega \mid \sum_{\vec{x}} \overline{\psi}(x)\gamma^5\psi(x+\vec{\Delta}) \mid \pi \rangle \tag{3}$$

$$\phi^i_\rho(\vec{\Delta}) = \langle \Omega \mid \sum_{\vec{x}} \overline{\psi}(x)\gamma^i\psi(x+\vec{\Delta}) \mid \rho \rangle \tag{4}$$

The operators used to define these wave functions are the natural generalizations of the gauge invariant operators defined above, but without reference to any particular path. To calculate the wave functions we must fix the gauge, as we discuss below.

We also wish to study the wave functions for baryons. We will consider here only operators with two quarks at one site and the third at a separate site. This restriction is one which should be relaxed in future studies in order to get a fuller picture of the baryon wave functions. Thus we define

$$\phi_P(\vec{\Delta}) = \langle \Omega \mid \sum_{\vec{x}} \psi^\alpha_i(x+\vec{\Delta})\psi^\beta_j(x)\psi^\gamma_k(x)\epsilon^{ijk}\chi^P_{\alpha\beta\gamma} \mid P \rangle , \tag{5}$$

where i, j and k are color indices and α, β and γ are spinor indices, with χ^P being the spinor wave function for the proton.[9] We have suppressed an index indicating the spin state of the proton which should be appended to χ. A similar wave function is defined for the Δ with a different spinor wave function χ^Δ.

In order to evaluate the wave functions on the lattice, we calculate the long time behavior of two point functions of the corresponding operators. A local operator is always used at the source end in accord with the dictates of the Gauss-Seidel matrix inversion method used to calculate the quark propagators.[1] Denoting the baryon and meson operators as $B(\vec{\Delta},t)$ and $M(\vec{\Delta},t)$, respectively, for large values of t and N_t-t, we find

$$\langle B(\vec{\Delta},t)B^\dagger(0,0)\rangle \rightarrow e^{-m_B t}\,\phi_B(\vec{\Delta})\phi_B^*(0) , \tag{6}$$

and

$$\langle M(\vec{\Delta},t)M^\dagger(0,0)\rangle \rightarrow [e^{-m_M t} + e^{-m_M(N_t-t)}]\phi_M(\vec{\Delta})\phi_M^*(0) . \tag{7}$$

Normalizing to one at the origin, we will plot

$$\frac{\langle B(\vec{\Delta},t)B^{\dagger}(0,0)\rangle}{\langle B(\vec{0},t)B^{\dagger}(0,0)\rangle} = \frac{\phi_B(\vec{\Delta})}{\phi_B(\vec{0})} . \tag{8}$$

Any t dependence of these ratios is indicative of contamination from excited states.

3. GAUGE FIXING

In order to carry out our calculation, it is necessary to fix a gauge; however, it should be clear that this must only be done on the few configurations for which fermion propagators will be calculated. Typically, one stores these gauge configurations on disk or magnetic tape. Thus, we see that the analysis can be done "offline," or after the gauge Monte Carlo calculation has been completed. Given a gauge configuration and the quark propagator on that configuration, three things must be done: determine the gauge transformation to transform to Coulomb gauge, transform the quark propagators, and calculate the two point functions for the various meson and baryon operators.

The lattice Coulomb gauge is a simple generalization of the lattice Landau gauge employed earlier by Wilson in a renormalization group study of SU(2) gauge theory.[10] The gauge condition is that

$$\sum_{i \text{ spatial}} \text{Re Tr}\Big(U(x,x+i)+U(x,x-i) \Big) \tag{9}$$

is to be maximized at each site x. In Landau gauge, i runs over all unit vectors, not just spatial ones. Ideally, one would perform this maximization by simultaneously varying the gauge transformation matrix at each site of the lattice. Unfortunately, simultaneous maximization in $8N_x N_y N_z$ variables is beyond our ability. The transformation is carried out by an iterative scheme, just as Wilson used for Landau gauge. At a given site, the links are gauge transformed to maximize Eq. (9). One then goes to another site and performs the maximization at that site. If a link joins the two sites, after the second maximization, the links at the first site will no longer obey condition (9) because the link joining the two sites has changed. One continues on to the next site until all have been maximized once. This constitutes one gauge fixing sweep. After many sweeps through the lattice the procedure converges.

The calculation was done using 18 gauge configurations on a $6^2 \times 12 \times 18$ lattice with $\beta = 5.7$. It was found that 400 gauge fixing sweeps were sufficient to reduce the maximum change in the gauge transformation matrix at any site from one sweep to the next to $< 10^{-6}$.

4. RESULTS

The hadron wave functions were calculated for only one value of hopping parameter since it had been found in SU(2) that meson wave functions[11] are relatively independent of the quark mass. The hopping parameter was set to 0.34 which corresponds to $m_N = 1.59$, $m_\Delta = 1.78$, $m_\pi = 0.84$, and $m_\rho = 0.94$ in lattice spacing units. (Note that the chiral symmetry breaking parameter has been set to 0.5.)

In the figures, the wave functions are plotted as a function of the distance between the quark and antiquark for mesons, or between one quark and the other two for baryons. Figure 1 shows a typical wave function. It shows the pion wave function six time planes from the local pion source. In this plot, all displacement vectors are shown, e. g., at $r=1$, there are three points plotted corresponding to the vectors (1,0,0), (0,1,0) and (0,0,1). They cannot be resolved in this figure, however. The most striking aspect of Fig. 1 is that there seem to be three curves, not just one. The points on the two higher curves come from displacement vectors where at least one coordinate is halfway across the box. This finite size effect had been seen in the SU(2) meson wave functions.[11] In subsequent curves we throw out all displacement vectors which touch the boundary halfway across the box. Next, we remark that at short distance, there is a small but noticeable departure from Euclidean symmetry. The points at $r=\sqrt{2}$, and $r=\sqrt{3}$ [corresponding to displacements (1,1,0) and (1,1,1)] fall below a smooth curve joining the point $r=1$ and points with $r\geq 2$. Careful examination of subsequent figures reveals even at longer distances a sawtooth structure of the curves which decreases as t increases, corresponding to an improvement in rotational symmetry. Also note that although our box size is rectangular the non-boundary points with largest distance, which come from points with $z=5$ and $x,y<3$, fall on a smooth extrapolation of the curve formed by closer points. (This will be easier to see in subsequent figures where the points on the boundary are not plotted.) It appears that the noncubic box size does not have any appreciable effect upon the wave function at long distance.

Next we note that the wave functions fall off by a large factor at a distance of 6 lattice spacings. For the nucleon, plotted in Fig. 2, the falloff is about a factor of 20. This is very different from a free quark confined in a box by periodic boundary conditions. In that case the wave functions hardly fall off at all. The nucleon wave function shows very little dependence upon the time plane from which we calculate it. This, of course, is a good sign, and indicates that excited states are not very important.

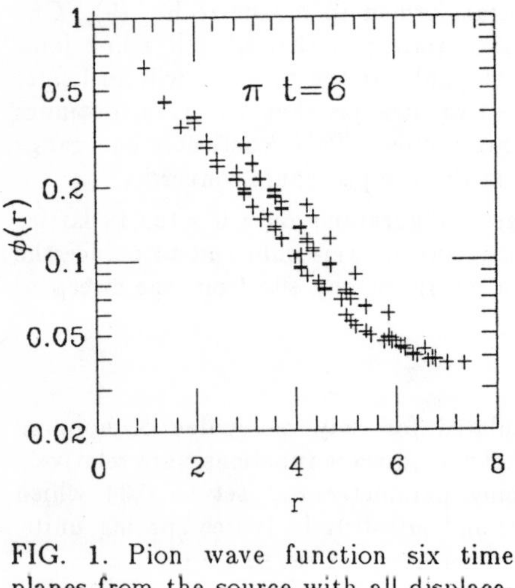

FIG. 1. Pion wave function six time planes from the source with all displacements plotted.

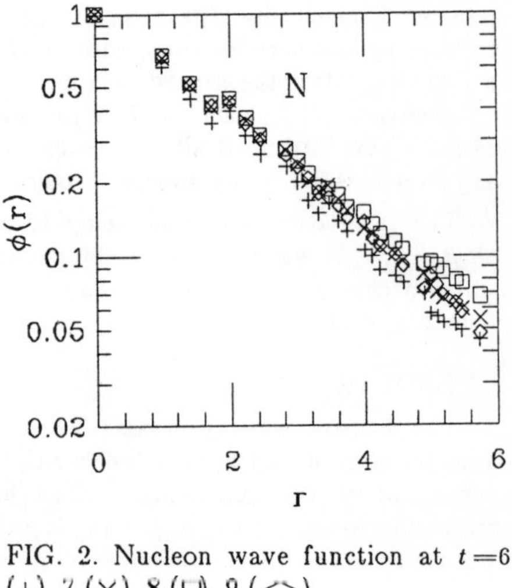

FIG. 2. Nucleon wave function at $t=6$ (+), 7 (×), 8 (□), 9 (◇).

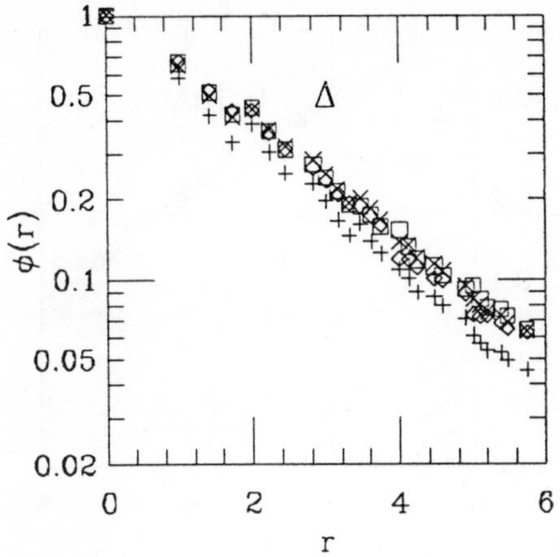

FIG. 3. Same as Fig. 2, but for Δ.

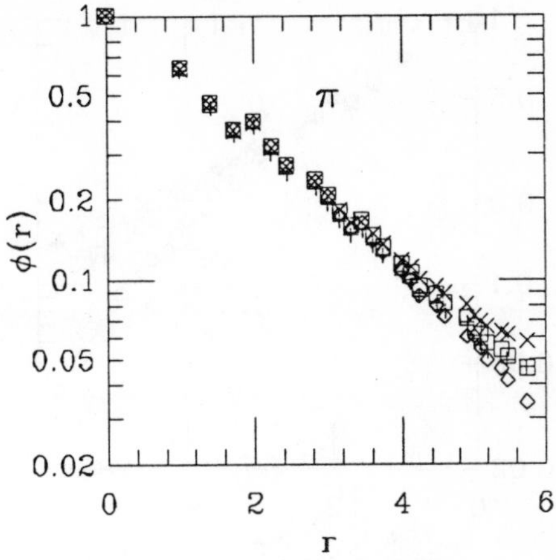

FIG. 4. Same as Fig. 2, but for π.

The picture of the delta which emerges in Fig. 3 is very similar to that of the nucleon. The shape of the delta is almost identical to that of the nucleon, and its t dependence is even smaller. Recalling the mass difference between the two states, we see that it cannot be attributed to a different spatial arrangement of the quarks. Rather, it must come from the difference in the spins.

Now we turn to the mesons. As seen in Fig. 4, the pion falls off a bit more quickly than either the nucleon or delta. Comparing the pion here to the result for SU(2) gauge theory,[11] we see that the pion wave function falls off much more quickly in SU(3). In SU(2), where the coupling was chosen to give the same value of the string tension, the pion wave function fell to only 0.25 by a distance of 5. For SU(3), the value is more than three times smaller.

The picture of the rho meson that emerges from Fig. 5 is a bit confusing because of the abrupt change in the wave function for $t=9$. From smaller t we surmise that the rho is a larger state than any of the others. It's wave function seems to fall to only 0.10 or 0.15 at distance of 6. However, the result for $t=9$ seems to be just a statistical effect. In Fig. 6, I show the rho wave function from the first nine lattices only. This looks more like the curves for smaller t in Fig. 5. Wave functions at $t=9$ are more subject to statistical error because, for smaller t, measurements come from two different time planes symmetrically displaced from the source point. On a lattice with 18 time planes, there is a unique plane displaced by 9 from the source. The most likely conclusion is that the rho really is a larger state than the other hadrons examined here.

Finally, it is interesting to look at a plot of $r^2 \mid \phi \mid^2$ which is the probability to find a quark at a distance r. This is plotted in Figs. 7 and 8. Only the proton is shown in Fig. 7 since the delta is so similar. In Fig. 8, the difference between pion and rho meson is quite striking.

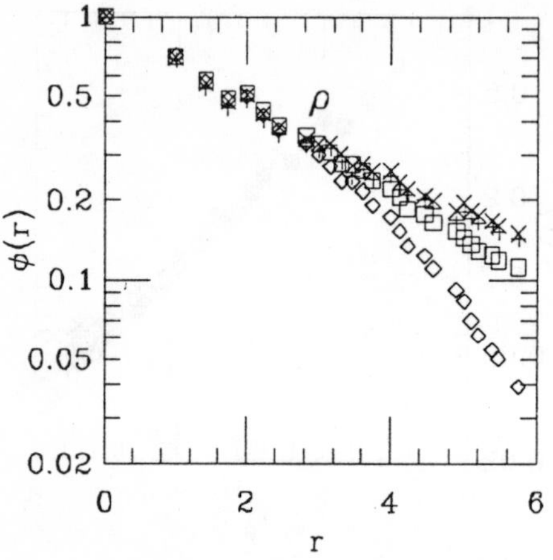

FIG. 5. Same as Fig. 2, but for ρ.

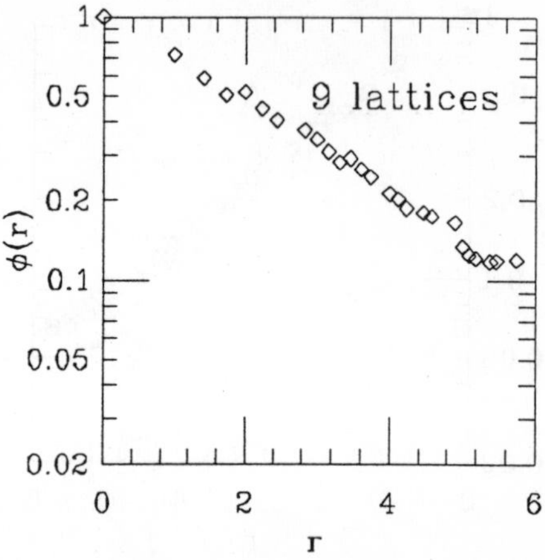

FIG. 6. Rho wave function at $t = 9$ from first nine lattices only.

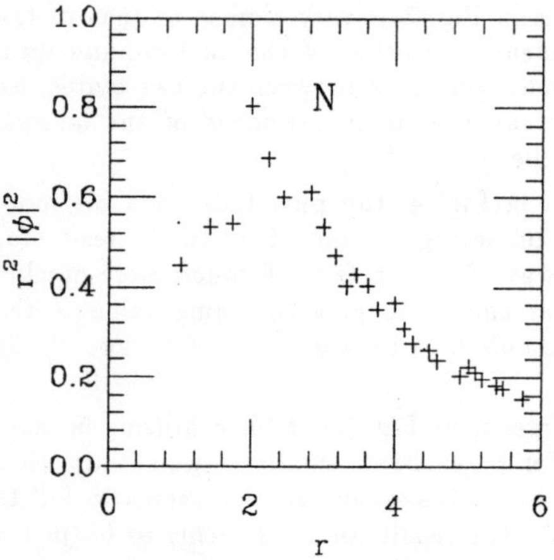

FIG. 7. Radial probability for nucleon.

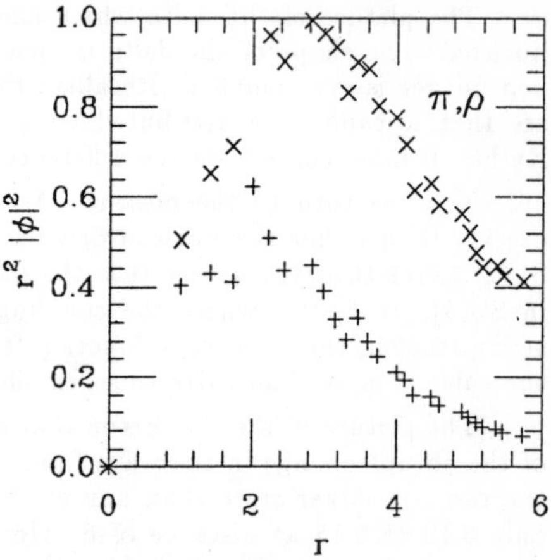

FIG. 8. Radial probability for π (+), and ρ (×).

5. CONCLUSIONS

The portrait of the proton and other states presented here is a very reasonable one. If we determine the lattice spacing from the value of the string tension, it is about 0.2 F. The radial distribution for the proton then peaks at a distance of $\approx$0.4 F. The spatial box size should clearly be increased from $6^2 \times 12$ if the proton is to be well contained inside. For the rho, the situation is even more critical since its wave function falls off more slowly. The rho distribution peaks at $\approx$0.6 F and is much more subject to finite size effects than the other particles. This helps to explain why spectral calculations have had so much trouble getting the pi-rho mass splitting correct.

The analysis done here can easily be done for other simulations if gauge configurations and fermion propagators are stored. In this way, the adequacy of the spatial size of the lattice may be judged without merely increasing the size and repeating the calculation. It would be interesting to develop a quantitative theory of finite size effects based on some characteristic of the wave function, such as its value at the boundary.

It would also be very interesting to repeat the current calculation with dynamical fermions and see what effects they have on the wave functions. The baryon wave functions could be calculated for all three quarks on different sites to get a fuller picture of the nucleon and delta.

ACKNOWLEDGMENTS

It is a pleasure to thank Hank Thacker and Don Weingarten for many discussions throughout the course of this work. In addition, I am indebted to Apoorva Patel for valuable input at one point. The hospitality of the Fermilab Theory Group and Computing Department to myself and my programs, respectively, is gratefully acknowledged. This work was supported by the Department of Energy, Contract No. DE-AT03-81ER-40029. Finally, I would like to thank Dennis Duke and Jeff Owens for a very stimulating and well organized conference.

REFERENCES

1. E. Marinari, G. Parisi, and C. Rebbi, Phys. Rev. Lett. 47, 1795 (1981); H. Hamber and G. Parisi, *ibid*. 47, 1792 (1981); Phys. Rev. D 27, 208 (1983); D. Weingarten, Phys. Lett. 109B, 57 (1982); Nucl. Phys. B215, 1 (1983); P. Hasenfratz and I. Montvay, Phys. Rev. Lett. 50, 309 (1983).

2. K. Bowler *et al*., in *Gauge Theory on a Lattice: 1984*, edited by C. Zachos, W. Celmaster, E. Kovacs and D. Sivers (Argonne National Laboratory, Argonne, 1984); R. Gupta, G. Guralnik, A. Patel, T. Warnok, and C. Zemach, Phys. Rev. Lett. 53, 1721 (1984); A. Hasenfratz, P. Hasenfratz, U. Heller, and F. Karsch, Phys. Lett. 143B, 193 (1984); A. D. Kennedy, J. Kuti, S. Meyer, and B. J. Pendleton, Phys. Rev. Lett. 54, 87 (1985); Phys. Lett. 155B, 414 (1985); D. Barkai, K. J. M. Moriarty, and C. Rebbi, Phys. Rev. D 30, 1293 (1984); for more recent results, consult these proceedings.

3. C. Bernard, T.Draper, and K. Olynyk, Phys. Rev. D 27, 227 (1982); R. Gupta and A. Patel, Phys. Lett. 124B, 94 (1983); K. Bowler, G. Pawley, D. Wallace, E. Marinari, and F. Rapuano, Nucl. Phys. B220, 137 (1983); H. Lipps, G. Martinelli, R. Petronzio, and F. Rapuano, Phys. Lett. 126B, 250 (1983); A. Hasenfratz, P. Hasenfratz, Z. Kunszt, and C. B. Lang, Phys. Lett. 110B, 289 (1982); J. P. Gilchrist, H. Schneider, G. Schierholz, and M. Teper, Phys. Lett. 136B, 87 (1984).

4. For a discussion of this possibility in a two dimensional model see: D. Hochberg and H. B. Thacker, Report No. FERMILAB-PUB-84/123-T; H. B. Thacker, these proceedings.

5. S. Gottlieb, P. Mackenzie, H. Thacker, and D. Weingarten, Phys. Lett. 134B, 346 (1984); S. Gottlieb, in *Gauge Theory on a Lattice: 1984*, edited by C. Zachos, W. Celmaster, E. Kovacs and D. Sivers (Argonne National Laboratory, Argonne, 1984).

6. This is under investigation by P. Mackenzie and myself.

7. See for example K. Ishikawa, G. Schierholz, and M. Teper, Z. Phys. C 19, 329 (1983).

8. S. Aoyama, Nucl. Phys. B194, 513 (1982); K. Dietz and T. Filk, Phys. Rev. D 27, 2944 (1983); D. Knauss and K. Scharnhorst, Report No. PHE 84-06 (unpublished).

9. The spinor wave functions may be found in D. Weingarten, Ref. 1.

10. K. G. Wilson, in *Recent Developments in Gauge Theories, (Cargese 1979)*, edited by G. 't Hooft *et al.* (Plenum, New York, 1980)

11. B. Velikson and D. Weingarten, Nucl. Phys. B249, 433 (1985).

LATTICE CALCULATION OF WEAK MATRIX ELEMENTS: A PROGRESS REPORT*

Claude Bernard, George Hockney, and A. Soni
University of California, Los Angeles, CA 90024, USA
and
Terrence Draper
University of California, Irvine, CA 92717, USA

ABSTRACT

This is a progress report of a project that is designated to calculate non-leptonic weak matrix elements on a space-time lattice. The emphasis here is on presentation of the preliminary results from a pilot Monte-Carlo done with a "small" (i.e. $6^3 \times 17$ fermion) lattice at $\beta = 5.7$ with 16 gauge configurations. Because of its large magnitude the origin of the $\Delta I = 1/2$ enhancement is investigated as a test case. For a lattice meson of mass ≃ kaon mass we find the $\Delta I = 3/2$ amplitude compatible with zero within our statistics and a clear and significantly enhanced signal for the $\Delta I = 1/2$ amplitude. The dominance of the $\Delta I = 1/2$ amplitude appears to be a manifestation of the dominance of a class of graphs called the eye graphs. A preliminary measurement of the $(\Delta S = 2)$ LL,LR operators is also discussed.

*Presented by A. Soni.

This project was initiated over two years ago.[1,2] As it has progressed we have collaborated with various people on specific issues. The Monte Carlo (MC) calculation on a small lattice was done in collaboration with Minick Rushton;[3] application of chiral perturbation theory was worked out with David Politzer and Mark Wise;[4] some aspect of the lattice weak coupling perturbative calculations are being examined with James Jennings; the possibility of using algebraic routines is being explored with Karl Schilcher and the second generation MC on a bigger lattice has been initiated in collaboration with Steve Otto. Following is the outline of the rest of this talk.

1. Motivation. We will briefly discuss what processes we are currently examining and what their physical significance is.

2. Key Theoretical Ingredients. There are three important theoretical ingredients. These are (a) The Operator Product Expansion (OPE) and the Renormalization Group (RG), (b) lattice weak coupling (w.c.) perturbation theory, and (c) chiral perturbation theory (CPTh). Their role will be briefly discussed.

3. A Computational Difficulty in Implementing the MC.

4. Preliminary Results from the Pilot MC.

5. Work in Progress. (a) for completion of the first generation MC; (b) towards the second generation MC.

6. Summary.

The emphasis here will be on results that we have obtained so far.

1. MOTIVATION

Lattice Monte Carlo (MC) techniques offer a unique opportunity for calculating from the fundamental theory quantities that govern low energy hadronic phenomena. The efforts of the past few years in this direction suggest that practical difficulties often seriously limit attainable accuracies with these methods. Thus although they are providing very interesting qualitative results, considerable refinements must still be made before precise calculations can be done. At this stage, one may therefore want to apply MC techniques to study effects for which even qualitative results can have some physical interest and significance. One such effect that comes to mind is the $\Delta I = 1/2$ rule that governs non-leptonic weak decays of strange particles. We recall that what has come to be known as the $\Delta I = 1/2$ puzzle is the empirical statement that $\Delta I = 1/2$ amplitudes are enhanced over $\Delta I = 3/2$ ones by as much as a factor of 20 to 25. Thus, for example, the life-times of neutral and charged kaons are very different:

$$\frac{\Gamma(K^+\to\pi^+\pi^o)}{\Gamma(K_S^o\to\pi^+\pi^-)} \simeq \frac{1}{450}$$

which is easily understood on the basis of $\Delta I = 1/2$ dominance. This is such a large effect that even an inaccuracy of order 50%, which is often the case for some current MC calculations, need not mask the origin of this phenomenon. The same theoretical and computational machinery could of course also be used in calculation of many other important hadronic matrix elements that characterize non-leptonic weak decays. Here are some examples:

As is well known the $K^o\bar{K}^o$ mass difference and the CP violation parameter ϵ in neutral kaon complex have provided very useful constraints on electroweak theories. In the standard model these are governed by the matrix elements of the $\Delta S = 2$ LL operator: $\langle K^o|\bar{s}\gamma_\mu(1-\gamma_5)d\,\bar{s}\gamma^\mu(1-\gamma_5)d|\bar{K}^o\rangle$. Indeed in the standard KM phase convention the CP violation parameter ϵ is proportional to this matrix element. It has become customary to parametrize this matrix element in terms of the B parameter which is the ratio

$$B = \frac{\langle K^o|\bar{s}\gamma_\mu(1-\gamma_5)d\,\bar{s}\gamma^\mu(1-\gamma_5)d|\bar{K}^o\rangle}{\langle K^o|\bar{s}\gamma_\mu(1-\gamma_5)d\,\bar{s}\gamma^\mu(1-\gamma_5)d|\bar{K}^o\rangle|_{\text{vacuum saturation}}}$$

Thus Donoghue et al.[5)] have made the useful observation that B can be related to charged kaon decay ($K^+ \to \pi^+\pi^o$) (assuming SU(3) and PCAC) thereby finding $B \simeq .33$. The recent measurements of B meson lifetime has constrained KM parameters well enough that a knowledge of the B parameter can be used to put a nontrivial bound on the mass of the top quark. Thus, for example, with B = .33 the standard model with three generations requires $m_{top} \gtrsim 40$ GeV for it to successfully account for the experimentally observed value of ε.[6)]

Left-right symmetric theories constitute a popular means for extending the standard model. The $K^o\bar{K}^o$ mass difference can be used to learn about the lower bound on the mass scale of such theories. These bounds depend linearly on $M_{LR} \equiv \langle K^o|\bar{s}\gamma_\mu(1-\gamma_5)d\ \bar{s}\gamma^\mu(1+\gamma_5)d|\bar{K}^o\rangle$, and the current bound (1.6 TeV for the mass of the right handed gauge bosons) was deduced by assuming vacuum saturation.[7)] It will therefore be useful to calculate M_{LR} on the lattice.

The matrix elements of some penguin operators control in the standard model another CP violation parameter, namely ε'/ε.[6,8)] Indeed efforts are now underway for an improved measurement of this important parameter.[10)] In the absence of a reliable calculation for these parameters, the experimental measurements, often achieved at tremendous effort, cannot be used effectively for constraining the theory. It is therefore clearly important to see how far one can go with MC techniques in alleviating this old but very difficult problem of non-leptonic weak decays.

2. KEY THEORETICAL INGREDIENTS.

In the following we briefly review the important theoretical ingredients that relate the $K \to \pi\pi$ amplitudes to matrix elements that are amenable to a MC lattice calculation.[1,2)]

(a) Operator Product Expansion and the Renormalization Group.

The use of the operator product expansion and the renormalization group becomes necessary because the characteristic scale for weak interactions is the mass, M_W, of the W-boson (i.e. about 80 GeV) whereas the lattice ultra-violet cutoff (dictated by low energy hadronic physics and computer time) is π/lattice spacing $\simeq$ 3 GeV. Thus the W boson field has to be integrated out from the weak Hamiltonian. Such a procedure is very similar to that in the continuum,[11] the multiplicatively renormalizable four quark operators $O_\pm$ emerge:

$$O_\pm = [\bar{s}\gamma_\mu(1-\gamma_5)u\ \bar{u}\gamma^\mu(1-\gamma_5)d \pm \bar{s}\gamma_\mu(1-\gamma_5)d\ \bar{u}\gamma^\mu(1-\gamma_5)u] - [u \to c]\ . \quad (1)$$

These operators have different anomalous dimensions ($\gamma_+ = -g^2/4\pi^2$, $\gamma_- = g^2/2\pi^2$). Since O_- is a pure $\Delta I = 1/2$ operator and O_+ is a mixture of $\Delta I = 1/2$ and 3/2 operators the difference in their anomalous dimensions leads to a mild enhancement of the $\Delta I = 1/2$ amplitude by about a factor of two which originates purely from the difference in their Wilson coefficients. The remaining observed enhancement ought to arise as a result of the difference in the matrix elements of these operators. So far this has been very difficult to prove because of the nonperturbative nature of the problem; lattice MC techniques offer the promise of progress in this area.

We next recall that there is a second step in the operator product approximation which becomes relevant in the so-called penguin approach.[12-14] In that case one makes the further assumption that charm quark mass can be treated as large and integrates it out from the weak Hamiltonian. As a result one arrives at the following set of operators:*

$$\begin{aligned}
O_1 &= \bar{s}_L\gamma_\mu d_L\bar{u}_L\gamma_\mu u_L - \bar{s}_L\gamma_\mu u_L\bar{u}_L\gamma_\mu d_L \\
O_2 &= \bar{s}_L\gamma_\mu d_L\bar{u}_L\gamma_\mu u_L + \bar{s}_L\gamma_\mu u_L\bar{u}_L\gamma_\mu d_L + 2\bar{s}_L\gamma_\mu d_L\bar{d}_L\gamma_\mu d_L + 2\bar{s}_L\gamma_\mu d_L\bar{s}_L\gamma_\mu s_L \\
O_3 &= \bar{s}_L\gamma_\mu d_L\bar{u}_L\gamma_\mu u_L + \bar{s}_L\gamma_\mu u_L\bar{u}_L\gamma_\mu d_L + 2\bar{s}_L\gamma_\mu d_L\bar{d}_L\gamma_\mu d_L - 3\bar{s}_L\gamma_\mu d_L\bar{s}_L\gamma_\mu s_L \\
O_4 &= \bar{s}_L\gamma_\mu d_L\bar{u}_L\gamma_\mu u_L + \bar{s}_L\gamma_\mu u_L\bar{u}_L\gamma_\mu d_L - \bar{s}_L\gamma_\mu d_L\bar{d}_L\gamma_\mu d_L \\
O_5 &= \bar{s}_L\gamma_\mu t^a d_L[\bar{u}_R\gamma_\mu t^a u_R + \bar{d}_R\gamma_\mu t^a d_R + \bar{s}_R\gamma_\mu t^a s_R] \\
O_6 &= \bar{s}_L\gamma_\mu d_L[\bar{u}_R\gamma_\mu u_R + \bar{d}_R\gamma_\mu d_R + \bar{s}_R\gamma_\mu s_R]\ . \qquad (2)
\end{aligned}$$

---

*O_1 to O_6 are defined according to the conventions of Ref. 12.

Except for O_4 all of these are $\Delta I = 1/2$ operators. Under $SU(3)_L \times SU_R(3)$ O_3 and O_4 transform as (27,1) whereas the others transform as (8,1). Since the charm mass is not that large we do not regard this as a particularly safe approximation. However, we plan to study this on the lattice as well, so we will be able to compare the results with and without the approximation.

As is well known, in the standard model with the usual phase convention, the first row and the first column of the KM matrix is purely real. This implies that CP violation in the kaon system can arise only through virtual emission and absorption of charm and/or top quark in the intermediate states. So if we want to calculate matrix elements that are relevant to CP violation in the kaon system (say the ε' parameter) then since $m_{top} >> a^{-1}$ we necessarily need to integrate out the top quark from the $\Delta S = 1$ weak Hamiltonian. Thus for studying CP violation amplitudes a penguin-like approximation becomes unavoidable.

(b) Operator Mixing.

As mentioned in the preceding discussion the procedure for OPE on the lattice is very similar to that in the continuum. One serious complication is due to the lack of chiral invariance of lattice fermions. In particular the standard Wilson action that we are using violates chiral invariance even for massless quarks when they are off their mass shell. This can lead to mixing of operators of different chiral structure. So one needs to do lattice weak coupling perturbation theory which is very clumsy. As a result of highly tedious calculations involving one loop Feynman graphs one finds the relation (to order g^2) between the continuum operators and their lattice counterparts which has the following generic appearance:[15,16]

$$O_i^{cont} = \left[1 + \frac{g^2}{16\pi^2} Z_{1i}(r) \right] O_i^{latt} + \frac{g^2}{16\pi^2} r^2 Z_{2ij}(r)\, O_j^{latt} . \tag{3}$$

Here Z's are finite renormalization constants and r is the parameter in the Wilson action that controls the mass of extra fermionic fields. So, as an example, the usual (V-A) × (V-A) four quark operator such as

$\bar{\Psi}_1\gamma_\mu(1-\gamma_5)\Psi_2\bar{\Psi}_3\gamma_\mu(1-\gamma_5)\Psi_4$ gets corrected by terms proportional to g^2 which contain operators of mixed chirality e.g. $\bar{\Psi}_1\Psi_2\bar{\Psi}_3\Psi_4$, $\bar{\Psi}_1\gamma_5\Psi_2\bar{\Psi}_3\gamma_5\Psi_4$, etc. Thus the lattice Monte Carlo must have the flexibility to deal with all Dirac operators and not just the usual LL,LR that enter in the continuum $\Delta S=1$ effective Hamiltonian.

We have finished all the necessary weak coupling calculations for the M_W large case in the OPE. For the m_{charm} or m_{top} large case the calculations are now in progress.

Our numerical calculations show that for the case of LL operators the individual matrix elements of the wrong chirality operators are very large but there is some cancellation of the contributions among themselves. In the end they often change the matrix elements of the naive (i.e. order g^0 operators) by as much as a few hundred percent. For the case of LR operators the w.c. corrections were found to be much smaller. For the LL case these corrections are large because the matrix elements of the naive LL operators on the lattice are quite small so that the relative contribution of the $O(g^2)$ SS,PP ... operators can be very significant. One can think of the $O(g^2)$ calculation as determining the correct "O^{th} order" operators, much in the same sense as in degenerate perturbation theory. Once these "O^{th} order" operators are found, the perturbation theory is expected to be well behaved.

(c) Chiral Perturbation Theory.

We thus wish to calculate matrix elements of the type $\langle\pi\pi|O|K\rangle$ where O is a generic four-quark operator, for instance, $\bar{s}\Gamma u\bar{u}\Gamma d$, $\bar{s}\Gamma c\bar{c}\Gamma d$ etc. and Γ is any Dirac operator. Calculation of such a four point function on the lattice is difficult, and one has to resort to some approximation scheme for relating it to a three point function. That is where chiral perturbation theory becomes useful. It allows us to reduce one of the pion fields and relate the $K \to \pi\pi$ matrix element to a suitable linear combination of $K \to \pi$ and $K \to 0$ matrix elements.[4)]

Thus:

$$\langle \pi^+\pi^- | O | K^o \rangle = \frac{i(m^2_{K^o} - m^2_{\pi^+})}{m^2_M f_M} \langle \pi^+ | O | K^+ \rangle + \frac{m^2_{K^o} - m^2_{\pi^+}}{f^2_M (m'^2_{K^+} - m'^2_{\pi^+})} \langle 0 | O | K^o \rangle \ . \quad (4)$$

Here m_M is the common mass of the K^+ or π^+ meson on the lattice and f_M is its decay constant (normalized to $f_\pi \simeq 132$MeV), $m_{K'}$ and $m_{\pi'}$ are masses of lattice K^+ and π^+, and O is any one of the four quark operators that occurs in the $\Delta S=1$ effective Hamiltonian; it transforms under $SU_L(3) \times SU_R(3)$ as (8,1) or (27,1). This equation is valid for both $\Delta I = 1/2$ or $\Delta I = 3/2$.

In passing we like to mention two other useful implications of CPTh (1) Matrix elements of (8,1) or (27,1) operators must $\propto m^2_M$. (2) Meson decay constants can be evaluated on the lattice by calculating the matrix elements of a two quark operator: $\bar{s}(1-\gamma_5)d$, which transforms as $(3,\bar{3})$, between K and π and K and vacuum.* Although this method requires the ability to calculate a three point function on the lattice it has the advantage that (unlike some usual procedures) it does not require a knowledge of the quark masses.

3. A COMPUTATIONAL DIFFICULTY: THE EYE GRAPHS.

Let us now consider a typical three point function that one needs to evaluate on the lattice e.g. $\langle \pi^+ | \bar{s}\Gamma u \bar{u}\Gamma d | K^+ \rangle$. For MC calculations the mesons are replaced by their interpolating diquark operators i.e. $\pi^+ \sim \bar{u}\gamma_5 d$ and $K^+ \sim \bar{s}\gamma_5 u$. We thus arrive at the Green's functions $\langle 0 | \bar{d}(x)\gamma_5 u(x)\ \bar{s}(0)\Gamma u(0)\ \bar{u}(0)\ \Gamma d(0)\ \bar{u}(y)\ \gamma_5 s(y) | 0 \rangle$. On Wick contraction we immediately see that there are two ways to contract the two u's and the two $\bar{u}$'s as shown in Fig. 1 and Fig. 2. Fig. 1 is referred to as "figure eight" graph[17)] and Fig.2 is referred to as the "eye graph". In the eye graph the u quark loop (or the charm quark loop that enters when one considers $\bar{s}\Gamma c\bar{c}\Gamma d$) is not higher order in a 1/N expansion, N being the numbers of colors. So the eye graph cannot be eliminated on the basis of the quenched approximation.

---

*For further details, see Ref. 4.

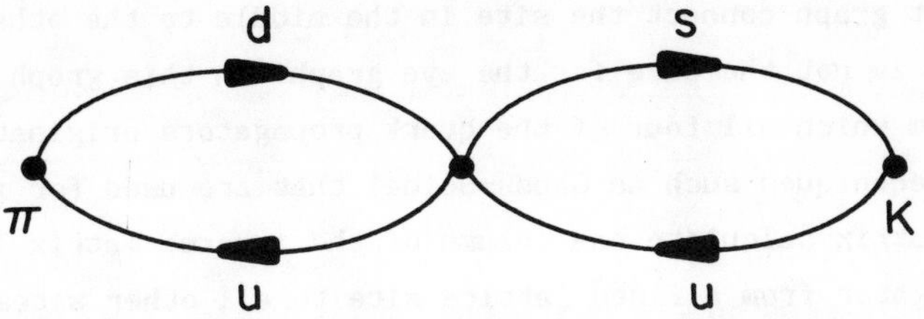

Fig. 1. The "figure eight" graph. All propagators begin or end at the location of the 4-quark operator which sits in the middle.

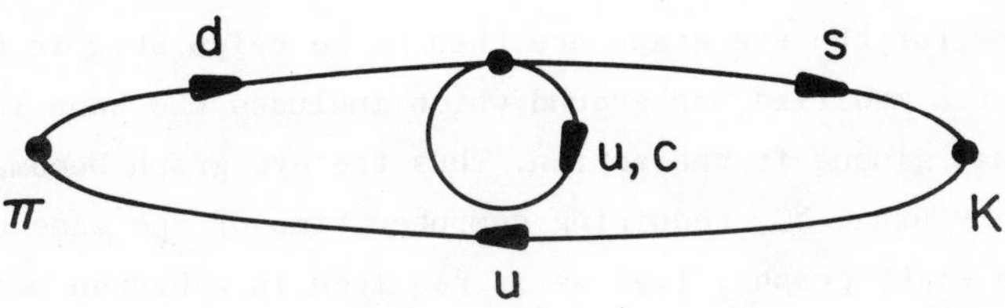

Fig. 2. The "eye" graph. The u and $\bar{u}$ bar or the c and $\bar{c}$ fields in the 4-quark operator are contracted with each other. Unlike the "figure eight" graph (Fig. 1), there is no single point from which all propagators begin or end.

Furthermore, the propagators in these figures are quark propagators on the lattice i.e. they are calculated in a gluon background and therefore the quark loop in Fig. 2 is not a vacuum expectation value. Thus the eye graph cannot be excluded even when the operator is normal ordered.

The inclusion of the eye graphs introduces a technical complication. Notice that all four of the quark propagators in the figure eight graph connect the site in the middle to the other two sites. Such is not the case for the eye graph: In this graph there is no site from which all four of the quark propagators originate. Now the numerical techniques such as Gauss-Seidel that are used for inverting the quark matrix calculate one column of the inverse matrix i.e. the quark propagator from a fixed lattice site to all other sites in the lattice. So, for example, for the figure eight graph one can simply use the propagators from a hadron mass calculation. The eye graph on the other hand requires a totally different approach.

To deal with the eye graph we exponentiated the kaon; i.e. we put it into the action.[18)] Then the eye graph effectively becomes a two point function in which all the quark propagators originate from a single lattice site, namely the site of the operator. However these propagators for the eye graph are then to be calculated in the presence of a modified background which includes the kaon in addition to the usual gluons in the action. Thus the eye graph becomes tractable by usual MC, requiring computer time of the same order as the figure eight graphs, i.e. as is required in a hadron mass calculation. Notice that the technique then allows us to study separately the contribution of each class of graphs. We are also able to compare the contribution from each Dirac operator, and indeed the up quark loop versus the charm quark loop in the eye graph can be compared.

4. PRELIMINARY RESULTS FROM THE PILOT MC.[3)]

The background gauge fields were generated by the standard Metropolis MC technique on a lattice of size $6^3 \times 10$ with periodic boundary conditions at a lattice gauge coupling such that $\beta=6/g^2=5.7$. Two independent sets of eight configurations were used which evolved into equilibrium from different starting points. 1000 passes of 12 Metropolis hits/site were used for thermalization; 500 passes separated each configuration used. Each of the configurations was copied in the time direction to form the background in which $6^3 \times 17$ quark lattices were embedded. For the quarks we used periodic bounadry conditions in the spatial direction and "free" boundary conditions in the time direction. This is chosen to allow the entire length of the lattice in the time direction to be used and thereby it permits clearer separation of the contamination from radial excitations. Furthermore, spectator quarks cannot propagate from the K to the π by going "backward" in time. The price one pays is in the form of boundary contamination close to the edge and a slight violation of the CPT invariance of the lattice action. Thus the usual equality: $\gamma_5 Q(y,x) \gamma_5 = Q^{\dagger}(x,y)$ where Q(x,y) is a quark propagator from site y to x, is no longer exactly true. We found however that the discrepancy was extremely small ~ 1%.

All numerical calculations were done by setting r = 1 in the Wilson action. The quark propagators were calculated for six values of the hopping constant k = .094,.150,.155,.162,.164 and .165. The critical value for k was found to be .171±.001.* Table I shows the corresponding pseudoscalar meson mass and the decay constant f_M (normalized to f_π = 132 MeV). We see that the meson mass varies from am ≃ 3.0 to am ≃ .5. Our estimate for a^{-1} ~ 1 GeV.[19] Thus the meson with k = .162,.164 or .165 has roughly the mass of a kaon. However the lattice meson is made of degenerate quarks, unlike the physical kaon. Table 1 shows that af_M and (for k>.155) m_M^2/m_q are fairly constant, indicating compatibility with the chiral behavior. However

---

*Our Gauss-Seidel did not converge for k ≥ .166. Indeed for k = .165, one of the 16 gauge-configurations did not converge. Where it did converge, this configuration produced a very small meson mass (am_M < 0.1 at k = .164) for two time slices near the location of the weak operator. (The mass took on normal values elsewhere in the configuration). This would imply very large finite volume effects on results obtained from the configuration, and we therefore believe it should be discarded. The data shown here are therefore all based on the remaining 15 configurations; however, if we include all 16 configurations (and disregard the k = .165 data) the qualitative conclusions reported here remain unchanged, though the error bars for k = .162 and .164 are significantly increased.

the magnitude of the decay constant is at least a factor of two larger than that of the physical kaon. The quark masses shown are deduced by use of the formula $am_q = \ln[1 + 0.5(k^{-1}-k_c^{-1})]$ and are > .10.

Fig. 3 shows the isospin 1/2 and the isospin 3/2 amplitudes for K→π as a function of k. The charm quark was held fixed at k = .094, the corresponding $c\bar{c}$ pseudoscalar (i.e. η_c) has am ≃ 3. We see that as

K	am_q	am_M	af_M	am_M^2/m_q
.094		2.95±.05		
.150	.34	1.09±.05	.36±.01	3.4± .3
.155	.26	.90±.06	.36±.01	3.1± .5
.162	.15	.64±.09	.38±.03	2.7± .8
.164	.12	.57±.09	.38±.04	2.8±1.0
.165	.10	.54±.10	.38±.04	2.9±1.1

Table 1. Masses and decay constants as a function of k. $k_c = .171 \pm .001$, $a^{-1} \sim 1$ GeV (see Ref. 19). Note that f_M is normalized to correspond to $f_\pi = 135$ MeV in the continuum.

the meson mass is reduced the 3/2 amplitude decreases and for k = .162 to .165 (i.e. $m_M \simeq M_K$) it is compatible with zero within our accuracy. On the other hand the isospin 1/2 amplitude is seen to increase for lighter mesons. For $m_M \sim m_K$ the ΔI=1/2 amplitude appears much larger than the 3/2 amplitude.

Table 2 shows a ratio of the eye-graph contribution to the figure-eight graph contribution for the matrix elements of some four quark operators between K and π. We see that the ratio eyes/eights rises sharply as meson mass gets lighter. This happens because the eye graphs are intrinsically much larger than the figure eights but for heavier K,π mass the up quark becomes closer to the charm quark mass and there is a greater GIM cancellation between the up and charm loops in the eyes. By examination of the numbers shown in Table 2 and by recalling that the eye graph is purely $\Delta I=1/2$ whereas the figure eight is a mixture of $\Delta I=1/2$ and $\Delta I=3/2$, one is led to conclude that the dominance of $\Delta I=1/2$ amplitude is a manifestation of the dominance of the eye graphs over figure eights. Thus any results obtained by ignoring the eye graphs can be very misleading.

To relate the $K\to\pi$ amplitude to the $K\to\pi\pi$ amplitude via Eq. (4) we have also measured the $K\to 0$ amplitudes. We found that the $K\to 0$

		EYES/EIGHTS	
K	$aA^{latt}(K_S\to\pi\pi)$	$\langle\pi\|O_+\|K\rangle$	$\langle\pi\|O_-\|K\rangle$
.150	.52±.03	−1.7	−2.4
.155	.89±.04	−3.5	−5.2
.162	2.20±.10	366	22
.164	3.53±.10	41	10
.165	4.79±.60	29	8

Table 2. $A^{latt}(K_S\to\pi\pi)$ shown in the table is the $K_S\to\pi\pi$ amplitude calculated from the $K^+\to\pi^+$ ($\Delta I=1/2$) amplitude obtained by Monte Carlo (see Eqs. (5-7). The corresponding experimental result is $A^{expt}(K_S\to\pi\pi) \simeq 2.25$ GeV. Ratio of the contributions of the eye graph to figure eight graph is also shown for O_+ and O_-.

amplitudes were much smaller (~5%) in comparison to the $K\to\pi$ amplitudes. Thus for all practical purposes $K\to 0$ can be ignored and the enhancement seen in Fig. 3 for the isospin 1/2 amplitude for $K\to\pi$ implies an enhancement for $K\to\pi\pi$ amplitudes.

As Fig. 3 shows the $\Delta I=3/2$ amplitude is compatible with zero within our accuracy so that a ratio of the two amplitudes is too unreliable at this stage for comparison with experiment. But we can compare the absolute magnitude of the $\Delta I=1/2$ amplitude that we calculated to the experimentally measured $K\to\pi\pi$ amplitude. For this purpose we deduce a reduced amplitude from the measured partial width for $K_S \to \pi\pi$:

$$A^{expt}(K_S \to \pi\pi) = |M_{expt}| 2\sqrt{2}/G_F \sin\theta_c (m_K^2 - m_\pi^2) \tag{5}$$

where

$$|M_{expt}|^2 = 16\pi\, m_K^2\, \Gamma(K_S \to \pi\pi)/\sqrt{m_K^2 - 4m_\pi^2}\ . \tag{6}$$

The corresponding amplitude measured on the lattice through $K \to \pi$ is related via CPTh (Eq. (4)) and the rest of the stated theoretical machinery resulting in:

$$A^{latt}(K_S \to \pi\pi) = \sqrt{3}\, f_M m_M^2 \langle \pi^+ | C_+ O_+^{1/2} + C_- O_-^{1/2} | K^+ \rangle_{latt} / 4m_q^2 \tag{7}$$

where C_+, C_- are the Wilson Coefficients ($C_+ \simeq 1$, $C_- \simeq -1$),* and where we define

$$\langle \pi | O | K \rangle_{latt} \equiv \sum_{xy} \frac{\langle 0 | \chi_\pi(\vec{x}, t_x)\ O(\vec{0}, 0)\ \chi_K^+(\vec{y}, t_y) | 0 \rangle}{\sum_x \langle 0 | \chi_\pi(\vec{x}, t_x)\ \chi_\pi^+(\vec{0}, 0) | 0 \rangle} \times \sum_y \langle 0 | \chi_K(\vec{0}, 0)\ \chi_K^+(\vec{y}, t_y) | 0 \rangle$$

with χ_π, χ_K the lattice interpolating field for the π and K, $\langle \pi | O | K \rangle$ is measured directly in our lattice calculation.

---

*In order to make the "best" (i.e., most "convergent") correspondence between lattice and continuum operators, we argue that one should choose $\mu_{\overline{MS}}^{continuum} = \Lambda_{\overline{MS}} / a\Lambda_{latt}$. This implies $g_{latt}(a) = g_{cont}(\mu)$. At our current qualitative stage the choice of μ is not particularly significant; but the proper choice would be important in future, more quantitative calculations. Of course, if one wishes to avoid guesswork, such questions can only be answered to a *two*-loop calculation.

Table 2 shows A^{latt} for various valuess of m_M. We see that for $m_M \sim m_K$ i.e. k = .162 to .165 A^{latt} is in the right ball park of $A^{expt} \simeq$ 2.25 GeV, given that $a \sim 1\ GeV^{-1}$ as is suggested by studies of the static quark potential and the string tension.[19)]

There is however one feature of Fig. 3 that is disturbing. From CPTh one expects both the amplitudes to $\to 0$ as $m_M \to 0$. The I=1/2 amplitude in the figure shows no such behavior. At this point we are unable to say whether this is due to our inability to go to lighter meson masses or whether there is some other reason for this discrepancy, such as the bad chiral behavior of Wilson fermions at the not-very-weak coupling of $\beta = 5.7$.

We have also measured $K^o\bar{K}^o$ matrix elements for the $\Delta S=2$ LL,LR operators, namely the quantities: $\langle K^o|\bar{s}\gamma_\mu(1-\gamma_5)d\ \bar{s}\gamma^\mu(1+\gamma_5)d|\bar{K}^o\rangle$. Note that only figure eights contribute to this class of matrix elements so that they are simpler to evaluate than $K\to\pi$ transitions. In addition, one should note that here CPTh is not being used to reduce a meson as was done for $K\to\pi\pi$. However to the extent that invidual masses of the d and s quarks on the lattice differ from the physical masses one does have to rely on CPTh for relating the results of the lattice meson to that of the experimental one. Rather than comparing the absolute magnitude of this matrix element to that calculated in the continuum literature it is perhaps better to merely test the idea of vacuum saturation. For that purpose we define dimensionless quantities:

$$B_{LL}^{latt} = -\langle K^o|\bar{s}\gamma_\mu(1-\gamma_5)d\ \bar{s}\gamma_\mu(1-\gamma_5)d|\bar{K}^o\rangle \Big/ \frac{8}{3} f_M^2\, m_M^2 \tag{8}$$

$$B_{LR}^{latt} = \langle K^o|\bar{s}\gamma_\mu(1-\gamma_5)d\ \bar{s}\gamma_\mu(1+\gamma_5)d|\bar{K}^o\rangle \Big/ 2f_M^2\, m_M^2 \left(\frac{m_M^2}{(m_s+m_d)^2} + \frac{1}{6} \right) \tag{9}$$

This has the advantage that we do not need to know a^{-1} in GeV (the continuum counterpart of these matrix elements goes as GeV^4). We see (Fig. 4) that the LR case is fairly independent of meson mass and seems to agree quite well with vacuum saturation. For LL case (see

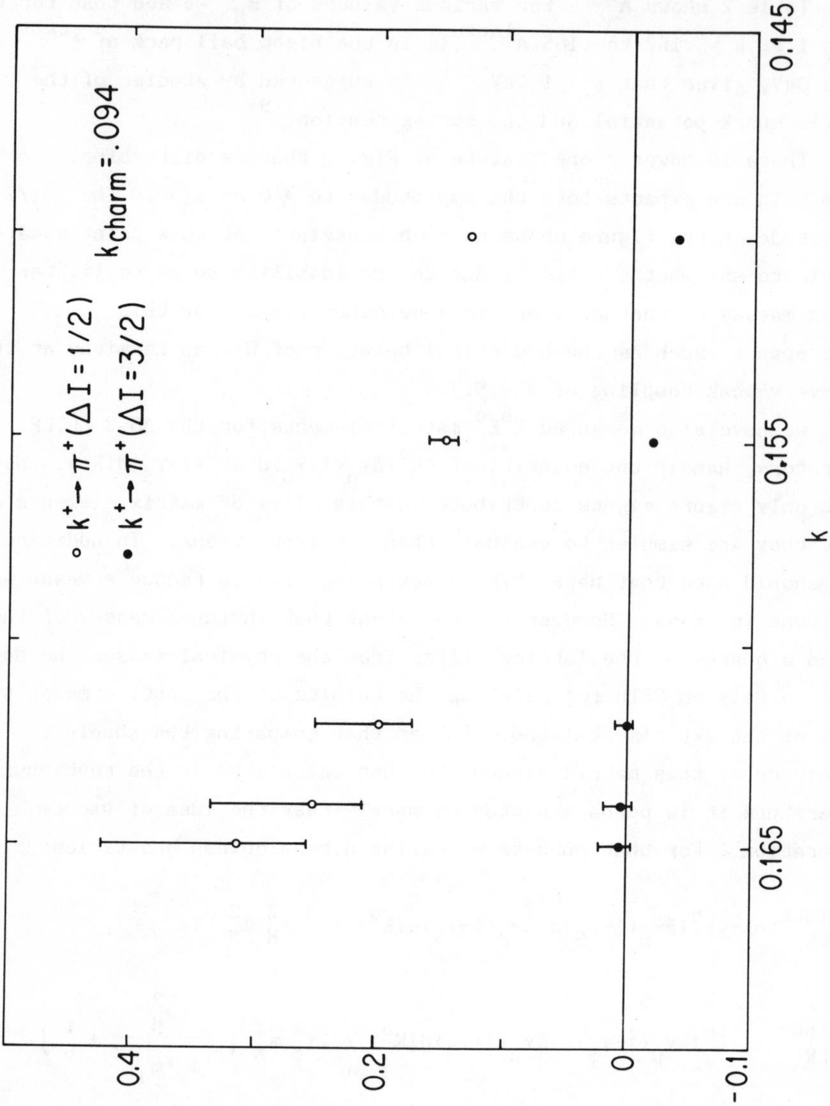

Fig. 3. The $\Delta I = 1/2$ and $\Delta I = 2/1$ amplitudes for $K^+ \to \pi^+$ transition as a function of the hopping constant K which controls the common mass of the kaon and pion. See Eq. (4).

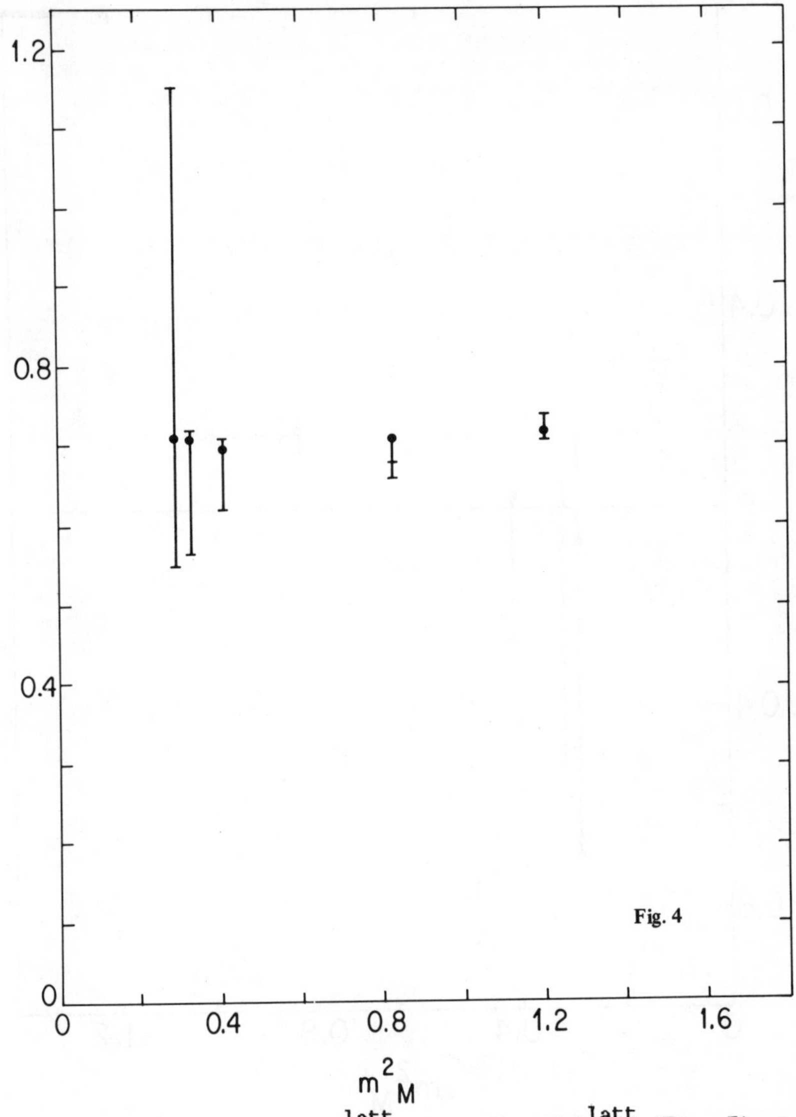

Figs. 4 and 5. Shown are B^{latt}_{LR} (Fig. 4) and B^{latt}_{LL} (Fig. 5) as a function of (meson mass)^2. These are the ratio of the matrix elements $\langle K^o|(\Delta S=2)_{LR,LL}|\bar{K}^o\rangle$ calculated directly and divided by its value implied by vacuum saturation. See Eqs. (8) and (9). Note that these matrix elements are renormalization point (μ) dependent and μ is "guessed"

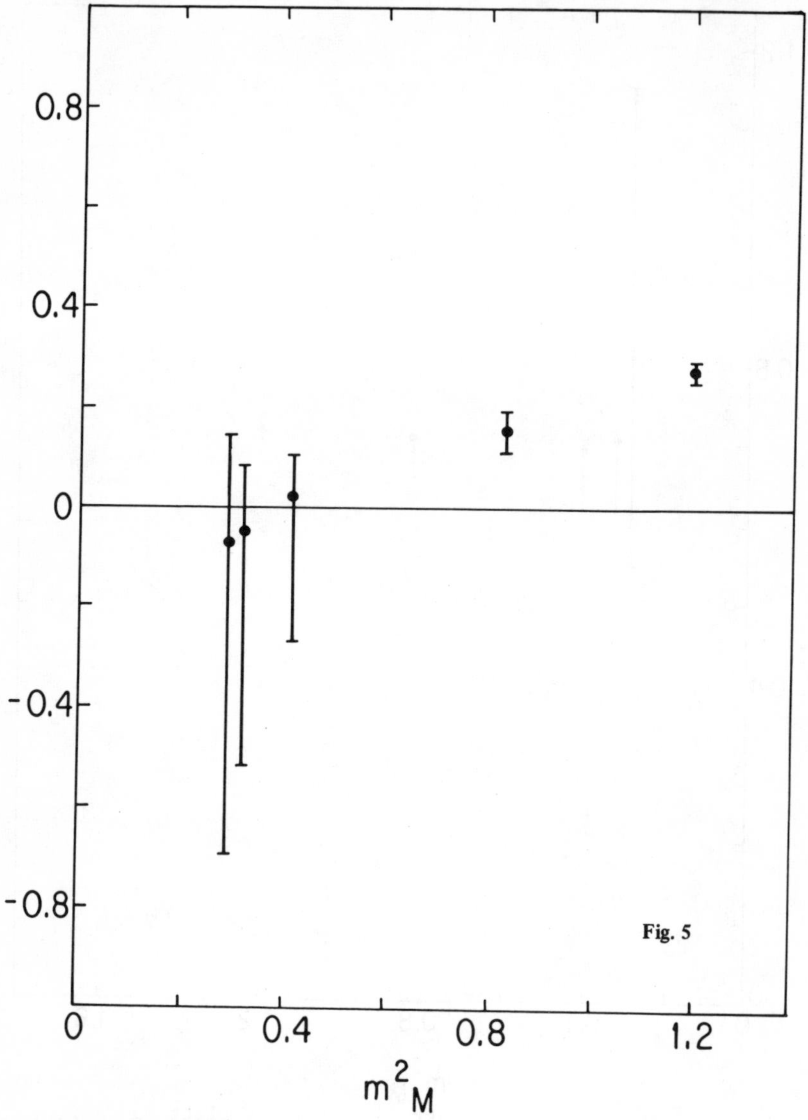

(see footnote, p. 14) to be ~ 28.8 GeV. To obtain the matrix elements at $\alpha_s(\mu^2) = 1$ (which is often needed) the values given in the figures should be multiplied by ~ 1.65. Note also that the uncertainty in m_M^2 is omitted from these figures for clarity (see Table 1).

Fig. 5) vacuum saturation seems to hold well only for $m_M >> m_K$. In the domain of the physical kaons i.e. $m_M \sim m_K$ (k=0.162 to 0.165) our result for B_{LL}^{latt} is compatible with zero with substantial error bars.

5. WORK IN PROGRESS.

There are a number of things that still need to be done for the completion of the 1st. generation of our project. So far all our numerical calculations are done with r = 1 in the Wilson action. We thus need to study the r dependence of our numerical results. In addition our current MC is done by exponentiating the kaon at site 5 i.e. three time slices away from the site at which the weak operator sits. We will shift the site of the exponentiation and study the dependence of our numerical results.

We have to apply $O(g^2)$ perturbative corrections to the decay constants and to the quark masses. We also need to complete the weak coupling perturbative calculations of the penguin like graphs which are needed when m_{charm} or m_{top} are removed by the OPE and which would enter calculations of ϵ'.

For the second generation MC for which we are collaborating with Steve Otto we are using a $12^3 \times 16$ gauge (with perhaps $12^3 \times 25$ fermion) lattice at β=6.1. The total physical dimension of the lattice will thereby be not much larger than in our first MC i.e. $\simeq$ 1.2 fm but now the $(\text{lattice spacing})^{-1} \sim 2$ GeV so the resolution of the lattice will be improved. We may do 50 gauge configurations rather than just 16 that were done in the pilot MC, also we may do Kogut-Susskind fermions in addition to Wilson fermions. Since our current $6^3 \times 10$ gauge ($6^3 \times 17$ fermion) lattice with 16 gauge configurations took 75 hours of Cray time we estimate that we will need : $(12/6)^3 \times (25/17) \times (50/16) \times (75) > 2000$ Cray hours perhaps spanned over a couple of years.

6. SUMMARY.

To summarize, our first generation (quenched approximation) MC calculations of non leptonic weak matrix elements shows unmistakable signal but with appreciable errors. For the M_W large case in the OPE the calculation is essentially complete. Relevant to the $\Delta I=1/2$ rule our findings are: $\langle\pi|O_-|K\rangle \sim \langle\pi|O_+|K\rangle$ but $\langle\pi|O_+^{1/2}|K\rangle \gg \langle\pi|O_+^{3/2}|K\rangle$. The $\Delta I=1/2$ amplitude, which is proportional to $\langle\pi|O_+^{1/2}-O_-^{1/2}|K\rangle$, is much larger than the $\Delta I=3/2$ amplitude for $m_M \sim m_K$. Our data shows that for $m_M \sim m_K$ the contribution of the eye graphs (which are pure $\Delta I=1/2$) is much larger than the contribution of figure eight graphs(which are mixtures of $\Delta I=1/2$ and $\Delta I=3/2$ amplitudes).This suggests that the $\Delta I=1/2$ rule seen in kaon decays is a manifestation of the dominance of the eye graphs. This also implies that the vacuum saturation approximation used by dropping the eye graphs is numerically completely misleading besides being technically incorrect (since it does not have the correct chiral behavior).[20] The $K\to\pi$ amplitude calculated on the lattice implies a $K_S\to\pi\pi$ amplitude that appears to be in ball park agreement with experiment. However, in the range of meson masses that are numerically accessible to us the $\Delta I=1/2$ amplitude does not seem to scale with m_M^2 as is suggested by CPTh.

We have also calculated ($\Delta S=2$) LL,LR $K^o\bar{K}^o$ matrix elements. Our preliminary results show that the LR matrix element is compatible with vacuum saturation for all meson masses. The LL matrix element however appears to agree with vacuum saturation only when the meson mass is $\gg$ mass of the kaon. For meson mass ~ kaon mass the LL matrix element is compatible with zero with large errors.

In conclusion, this work illustrates that lattice MC techniques can be very useful for attacking the old but very difficult problem of the nonleptonic decays of strange partcles. Even at this exploratory stage one is able to obtain information that should be useful in understanding the origin of the $\Delta I=1/2$ rule. The numerical values of some other quantities such as B^{latt} appear interesting and could even prove to be important if they are confirmed by refined measurements now underway. The theoretical machinery that we have set up will now be used on a bigger (i.e. $12^3 \times 16$ gauge) lattice requiring > 2000 Cray X-MP hours. Hopefully that will enable us to attain a reasonable level of accuracy (say ≈ 30%) in our calculations.

We are grateful to Bill Celmaster, Fred Gilman, Steve Otto, David Politzer and Mark Wise for useful discussions. The research of C.B., T.D. and G.H. was supported by the NSF and the research of A.S. was supported by the DOE Outstanding Junior Investigator Program under Grant No. DE-AM03-7GSF0034/PA-AE-AT03-81ER40024/Task J. The computing was done at the DOE supported Livermore MFE Computing Network.

REFERENCES.

1. For a previous report see C. Bernard in Gauge Theory on a Lattice (edited by C. Zachos, W. Celmaster, E. Kovacs, and D. Sivers), p. 85.
2. See also T. Draper, "Lattice Evaluation of Strong Corrections to Weak Matrix Elements--The Delta I Equals One-Half Rule," Ph.D. Thesis (UCLA), August, 1984.
3. Bernard, C., Draper, T., Hockney, G., Rushton, M., and Soni, A., in preparation.
4. Bernard, C., Draper, T., Soni, A., Politzer, D., and Wise, M., UCLA/84/TEP/14, CALT 68-1211, to be published in Physical Review D.
5. Donoghue, J. F., Golowich, E., and Holstein, B. R., Phys. Lett. 119B, 412 (1982).
6. See, e.g., Wise, M., CALT Preprint-68-1179 (1984).
7. Beall, G., Bander, M., and Soni, A., Phys. Rev. Lett. 48, 848 (1982); see also Gilman, F. J. and Reno, M. H., Phys. Lett. 127B, 426 (1983) and references therein.
8. Gilman, F. J. and Hagelin, J. S., Phys. Lett. 133B, 443 (1983)

9. Black, J. K., et al., Phys. Rev. Lett. 54, 1628 (1985); Bernstein, R. H., et al., Phys. Rev. Lett. 54, 1631 (1985).
10. Winstein, B. (private communication).
11. Gaillard, M. K. and Lee, B. W., Phys. Rev. Lett. 33, 108 (1974).
12. Shifman, M. A., Vainshtein, A. I., and Zakharov, V. I., Nucl. Phys. B120, 316 (1977).
13. Wise, M., SLAC-PUB-227 (1980).
14. Gilman, F. J. and Wise, M. B., Phys. Rev. D20, 2392 (1979); Guberina, B. and Peccei, R. D., Nucl. Phys. B163, 289 (1980); Hill, C. T., and Ross, G., Phys. Lett. 94B, 234 (1980); Donoghue, J., Golowich, E., Holstein, B., and Ponce, N., Phys. Rev. D23, 1213 (1981).
15. Martinelli, G., Phys. Lett. 141B, 395 (1984).
16. Because of its importance to the current program and because Ref. 15 does not include the (particularly important) eye graphs in its weak coupling calculations, we have undertaken our own calculations of this mixing. See Bernard, C., Draper, T., and Soni, A. (to be published).
17. During the time that this project has been in progress there have been two groups [Cabibbo, N., Martinelli, G., and Petronzio, R., Nucl. Phys. B244, 381 (1984); and Brower, R. C., Gavela, M. B., Gupta, R., and Maturana, G., Phys. Rev. Lett. 83, 1318 (1984)] who have reported lattice calculations of weak matrix elements. Both of these groups have evaluated only the figure eight graphs and have not included the eye graphs in their calculations. As we will discuss later in this work our calculations show that for meson mass ≃ kaon mass the eye graph contributions dominate $\Delta I = 1/2$ amplitudes.
18. This method has previously been used by Gottlieb, S., Mackenzie, P. B., Thacker, H. B., and Weingarten, D. H., Phys. Lett. 134B, 346 (1984). See also Bernard, C., Draper, T., Olynyk, K., and Rushton, M., Phys. Rev. Lett. 49, 1076 (1982). There are some important differences between the "exponentiation" used in these works and ours. These

differences are detailed in Ref. 2.

19. Otto, S. (private communication); Barkai, D., Moriarty, K. J. M., and Rebbi, C., Phys. Rev. D30, 1293 (1984). Adjusting a so that the β = 5.7 results scale with the weaker coupling results gives a (β = 5.7) $\simeq$ 1.0 GeV^{-1}; assuming asymptotic scaling at β = 5.7 would give a $\simeq$ 1.25 GeV^{-1}.
20. Dupont, Y. and Pham, T. N., Phys. Rev. D29, 1368 (1984); Pham, T. N., Phys. Lett. 145B, 113 (1984).

LATTICE HADRON STRUCTURE AND THE ELECTRIC FORM FACTOR*

W. Wilcox and R.M. Woloshyn

TRIUMF, 4004 Wesbrook Mall, Vancouver, B.C. V6T 2A3

ABSTRACT

After a brief introduction and review of the present status of hadronic structural investigations in lattice quantum chromodynamics, we emphasize the need for clean and experimentally verifiable tests of hadronic size. Such a test is provided by the lattice electric form factor. We present our latest results and comment on directions of future applications.

1. INTRODUCTION AND REVIEW

We would like to briefly review here the approaches that several groups have taken to the study of quenched hadronic structure, as well as to report on recent results on the staggered pion electric form factor.

As numerical simulations of QCD on the lattice improve, it is necessary to devise calculations which provide more detailed tests and insights into the theory. One of the exciting new developments in this

---

* Work supported in part by the Natural Sciences and Engineering Research Council of Canada.

direction is the study of the internal structure of hadrons. There are of course many ways to investigate this subject and we will begin by reviewing some of these.

Possibly the most direct method of investigating the size and structure of hadronic states is through the formation of lattice matrix elements of the form

$$\frac{\langle G(\vec{x},t_2)|P(\vec{y},t_1)|G(\vec{x},t)\rangle}{\langle G(\vec{x},t_2)|G(\vec{x},t)\rangle\langle 0|P(\vec{y},t_1)|0\rangle} \tag{1}$$

for $\langle 0|P|0\rangle \neq 0$. Movement of the probe point, $\vec{y}$, in the vicinity of the creation-annihilation point, $\vec{x}$, then yields information on the spatial correlation length, an indicator of hadron size. Such studies have been carried out for SU(2) glueballs[1)] as well as for mesonic states.[2)] The most effective probe in these types of studies also appears to be the simplest, the elementary plaquette. The continuum quantity being probed is just the $(\vec{B}^a)^2$ or $(\vec{E}^a)^2$ content of such particles. Although these studies have indeed yielded qualitative information on the states investigated, this technique also has several drawbacks. Since the strongest statistical correlations are found when $t_2=t_1=t$, there will of course be significant admixtures of radially excited states as well as higher momentum states - not the situation of direct experimental relevance. One may try to partially damp out such higher energy states by separating the creation and annihilation time points t_2 and t; however, one is then investigating the structural signal from an object with an intrinsically more uncertain position in the lattice. Perhaps the most effective place to apply such a technique is in the investigation of the gluonic structure of heavy, static $Q\bar{Q}$ systems[3)] where these drawbacks do not arise.

An improved method of measuring spatial structure in lattice hadronic states has been proposed recently by Barad, Ogilvie and Rebbi.[4)] They suggest measurement of the four-point function

$$\tilde{\rho}(r,t';x,t) = \sum_{y} \rho(y,t';y+r,t';x,t), \tag{2}$$

where

$$\rho(x',t';x'',t'';x,t) = \langle(\bar{\psi}^{u}\Gamma\psi^{d})(x,t)\ j_4^{u}\ (x',t')$$
$$j_4^{d}(x'',t'')\ (\bar{\psi}^{d}\Gamma\psi^{u})(0)\rangle. \tag{3}$$

This type of measurement, which measures the current-current overlap based on the relative distance between the quark and antiquark, has so far been carried out only for staggered fermions in two dimensional QCD, but the results are encouraging. By increasing the creation-annihilation time interval and/or by a partial summation over initial or final spatial position one is projecting out low momentum states of known quantum numbers. Thus, because of the relative nature of the four-point function (2), there is only an advantage and no disadvantage as in the previous situation, to a more time-stretched measurement. It will be extremely interesting to extend these results to the realistic case of four dimensions. It is crucial in this approach to adopt a current density $j_\mu(x)$ which results from a continuous exact vector symmetry on the lattice. Fortunately, there exists such a symmetry for both the Wilson and Kogut-Susskind formulations of fermions.

We shall mention only briefly here the structural technique, due to Velikson and Weingarten,[5)] of measuring a particle's Coulomb gauge wave-function since this method has been discussed and further results presented at this conference by Gottlieb. Suffice to say that these studies support the notion that the states simulated seem indeed to be rather localized objects on the lattice, as one would expect for the theory near to the continuum limit.

Through methods of the type described above, one hopes to arrive at a better understanding of the size and structure of lattice hadrons, and from this information to develop more physical insight and intuition. However, there is also a need at this stage, we believe, to more directly confront the experimental situation. One needs to be able to calculate in a clean and controlled manner, physical observables which may be used to compare directly to experimental determinations of hadronic structure. In the laboratory, measurement of hadronic size and structure involves the interaction of the particle in question with an electromagnetic probe through a well understood vertex. It is through the measurement of the strength of this vertex as a function of momentum transfer that the experimental determination of hadron size comes about. This strength function, the electric form factor, may be calculated on the lattice; we now describe such a calculation.

2. THE LATTICE ELECTRIC FORM FACTOR

The definition of an electric form factor on the lattice makes sense only if a conserved vector current exists there. Fortunately, as pointed out earlier, such a quantity exists both for Wilson and Kogut-Susskind fermions. We will now describe the calculation carried out[6]), which involves the KS fermion generic pion and SU(2) color.

It is necessary to group lattice points in hypercubes in order to construct interpolating fields of known quantum numbers when using single component KS fermion fields. In spite of these non-local interpolating fields, it is known that the mass for some continuum states may be determined from completely local combinations of staggered fermion fields. We have shown elsewhere[7]) that the quantity of primary interest in this study, a pionic three-point function, can also be calculated in a form that includes only local fermion field operators as interpolating fields. Therefore, we will quote here only some of the necessary formulae and the final results in our numerical study.

We are modeling here a non flavor-singlet meson with a net non-dynamic electric charge. Specifically, we will be considering the pseudo-goldstone meson (the generic pion) associated with the remnant continuous axial symmetry present in the staggered scheme of lattice fermions.[8]) The usual flavor structure associated with the staggered fermions is not useful for constructing charged states since an "electric charge" defined within these flavors is not conserved. Such a conserved current within this formulation is available; however, it has the effect of assigning identical charges to all four staggered fermion flavors. We therefore introduce two sets of the four flavors (this has no repercussions in the quenched approximation) labelled by u and d, with charges q_u and q_d, $q_u - q_d = 1$. The staggered fermion action with SU(2) gauge fields and a two component χ field is

$$S_F(U) = 1/2 \sum_{x,\mu,f} \alpha_\mu(x)\bar{\chi}^f(x)[U_\mu(x)\, \chi^f(x+a_\mu) - U^\dagger_\mu(x-a_\mu)\chi^f(x-a_\mu)] + ma \sum_{x,f} \bar{\chi}^f(x)\chi^f(x), \tag{4}$$

where

$$\alpha_\mu(x) = \alpha_\mu(x \pm a_\mu) = \begin{cases} 1 & , \quad \mu=1 \\ (-1)^{x_1} & , \quad \mu=2 \\ (-1)^{x_1+x_2} & , \quad \mu=3 \\ (-1)^{x_1+x_2+x_3} & , \quad \mu=4. \end{cases}$$

Local phase transformation of the fermion field

$$\chi(x) \to e^{iw(x)}\chi(x) , \tag{5a}$$

$$\bar{\chi}(x) \to \bar{\chi}(x)e^{-iw(x)}, \tag{5b}$$

yields a vector current

$$j_\mu(x) = - \frac{\delta S_F(U)}{\delta \Delta_\mu w(x)} \tag{6a}$$

$$= - \frac{i}{2} \alpha_\mu(x)[\bar{\chi}(x)U_\mu(x)\chi(x+a_\mu) + \bar{\chi}(x+a_\mu)U^\dagger_\mu(x)\chi(x)], \tag{6b}$$

where $\Delta_\mu w(x) = w(x+a_\mu) - w(x)$. Using the notation

$$\langle \ldots \rangle \equiv \int dU d\bar{\chi} d\chi e^{-S_g - S_F} (\ldots), \tag{7}$$

the local three-point function we calculate is

$$A(\vec{p},\vec{q};t_2,t_1) = -\frac{1}{4} Z^{-1} (-1)^{t_2} \sum_{\vec{x}_2} e^{-i\vec{p}\cdot\vec{x}_2} (-1)^{\vec{x}_2}$$
$$\langle \bar{\chi}^d(\vec{x}_2,t_2)\chi^u(\vec{x}_2,t_2) \sum_{\vec{x}_1} e^{i\vec{q}\cdot\vec{x}} \rho(\vec{x}_1,t_1)\bar{\chi}^u(0)\chi^d(0)\rangle \tag{8}$$

with $\rho(x) = q_u j_4^u(x) + q_d j_4^d(x)$, $(-1)^{\vec{x}} = (-1)^{x_1+x_2+x_3}$ and where color indices have been suppressed. Z above is the normalization integral. We also need the two-point function

$$G(\vec{p},t) = -\frac{1}{4} Z^{-1} (-1)^t \langle \sum_{\vec{x}} e^{-i\vec{p}\cdot\vec{x}} (-1)^{\vec{x}} \bar{\chi}^d(\vec{x},t) \chi^u(\vec{x},t) \bar{\chi}^u(0)\chi^d(0)\rangle. \tag{9}$$

Using the relation

$$G(\vec{p},t) \underset{t>>1}{\rightarrow} Z(p) e^{-E_p t}, \tag{10}$$

we find (at least for low momentum states) that

$$A(\vec{p},\vec{q};t_2,t_1) \xrightarrow[t_1,(t_2-t_1)>>1]{} \left\{ Z(p)Z(p') \frac{(1+e^{E_p a})}{(1+e^{-E_p a})} \frac{(1+e^{-E_{p'} a})}{(1+e^{E_{p'} a})} \right\}^{1/2}$$
$$e^{-E_{p'} t_1} e^{-E_p (t_2-t_1)} \frac{(E_p+E_{p'})}{2\sqrt{E_p E_{p'}}} F(q) \tag{11}$$

with $E_p = \{\vec{p}^2 + m_\pi^2\}^{1/2}$ and $\vec{p}' = \vec{p} - \vec{q}$. In order to achieve this result it is necessary to assume the continuum limit is closely approximated in order to relate lattice matrix elements to continuum ones. The

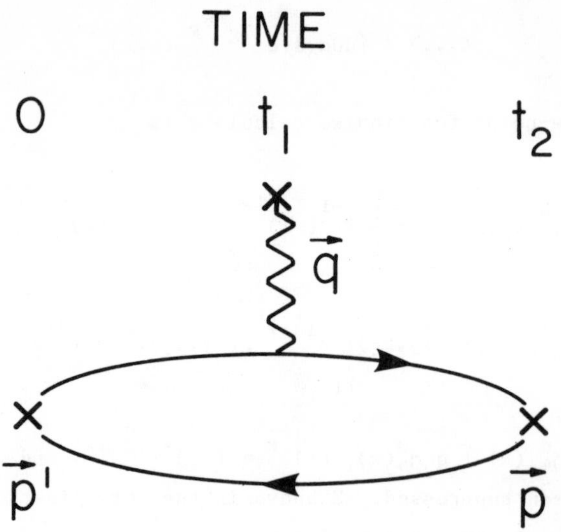

Fig. 1 Schematic representation of the mesonic measurement being carried out.

physical process involved is sketched in Fig. 1. The two spatial sums present in (8) mean that in general one needs quark propagators starting from all intermediate spatial points - an essentially impossible task done the brute force way. However, using the source method[9,10] one may elegantly avoid this problem. The application of this method to our problem will be described shortly.

In our actual numerical simulation we find it convenient to use antiperiodic boundary conditions on the quark fields in the space directions but to set the fermion coupling to zero across the time boundaries. Coupling the time boundaries together would lead to difficulties when introducing the charge operator since the amount of charge flowing in the forward time direction would no longer be a constant in time. When calculating the form factor, the problem of non-vacuum contamination can be corrected by taking a geometric mean of correlation functions calculated for appropriate momenta. This combination is:

$$\left\{\frac{A(0,\vec{q};t_2,t_1)\ A(\vec{q},\vec{q};t_2,t_1)}{G(0,t_2)\ G(\vec{q},t_2)}\right\}^{1/2} \xrightarrow[t_1,(t_2-t_1)\gg 1]{} \frac{E_q + m_\pi}{2\sqrt{E_q m_\pi}} F(q) \qquad (12)$$

yielding a quantity independent of time. Notice that the Z(p), Z(p') factors have dropped out, leaving the one matrix element of interest. Since it is the Z factors which contain contribution form states other than the vacuum in our calculations with nonperiodic time boundary conditions[11)], we expect the ratio to be independent of time boundary effects. We note that it is important in calculating the statistical error in F(q) from Monte Carlo data to include the covariances between the various factors in (12).

Use of the source method in summing the effect of the charge operator is crucial in this calculation. Let us define a new action:

$$S_F(U,\alpha_t^{u,d},\vec{q}) = S_F(U) - \sum_{\vec{x},f} \alpha_t^f\, e^{i\vec{q}\cdot\vec{x}} j_4^f(\vec{x},t). \qquad (13)$$

This gives a new fermion matrix ($S_F = \sum_{\alpha,\beta} \bar{\chi}_\alpha\, 1/2\, M_{\alpha\beta}\chi_\beta$)

$$M(\alpha_t,\vec{q})_{xA;yB} = M_{xA;yB} + i/2\, \alpha_t\, e^{i\vec{q}\cdot\vec{x}}\, \alpha_4(\vec{x})$$
$$\delta_{\vec{x},\vec{y}} \left[\delta_{t_x,t}\, \delta_{t_y,t+1}\, (U_4(x))_{AB} + \delta_{t_x,t+1}\delta_{t_y,t}\, (U_4^\dagger(x-a_4))_{AB}\right] \qquad (14)$$

for each flavor, f, where $M_{xA;yB}$ is the standard staggered fermion matrix. This new fermion matrix has the useful properties

$$M^*(\vec{\alpha}_t,\vec{q})_{xA;yB} = (-1)^{x+y}\, M(\alpha_t,-\vec{q})_{yB;xA} \qquad (15)$$

and

$$\sigma_2 M(\alpha_t,\vec{q})\sigma_2 = M^*(-\alpha_t,-\vec{q}) \qquad (16)$$

for σ_2 in color space. The relation (16) is an extremely handy one.

It is a consequence of the fact that 2 and $\bar{2}$ are equivalent representations of SU(2). It shows that there are only four fermionic degrees of freedom per flavor per hypercube, i.e. the other color is just the antiparticle of the first. The lack of such a relation for SU(3) will be one of the major differences in generalizing the present SU(2) considerations.

The three-point function (8) can now be written:

$$A(\vec{p},\vec{q};t_2,t_1) = -1/4\, Z^{-1}(-1)^{t_2}\sum_f q_f \frac{\partial}{\partial\alpha^f_{t_1}} \sum_{\vec{x}_2} e^{-i\vec{p}\cdot\vec{x}_2}(-1)^{\vec{x}_2}$$

$$\langle\langle \bar{\chi}^d(\vec{x}_2,t_2)\chi^u(\vec{x}_2,t_2)\bar{\chi}^u(0)\chi^d(0)\rangle\rangle, \tag{17}$$

where the double bracket notation signifies integration with respect to the modified fermion action. In the quenched approximation using (15) and (16) one may show that the three-point function (17) becomes

$$A(\vec{p},\vec{q};t_2,t_1) = \frac{1}{N_c}(q_u-q_d)\sum_{\vec{x}_2,c} e^{-i\vec{p}\cdot\vec{x}_2}\mathrm{Tr}$$
$$[M^{-1}(\vec{x}_2,t_2;0)]^\dagger \frac{\delta}{\delta\alpha_{t_1}} M(\alpha_{t_1},\vec{q}\mid\vec{x}_2,t_2;0). \tag{18}$$

Notice the physically necessary factor of $(q_u-q_d)=1$ which has emerged. Thus, the necessary three-point function can be calculated as the numerical derivative of a two-point function. In our simulations, this derivative is obtained by calculating two-point functions at $\alpha=0$ and $\alpha=0.05$.

Our numerical results obtained so far are shown in Figs. 2 and 3. The lattice size was $10^2\cdot 20\cdot 16$ with current carrying momentum in the three-direction. Thirty-two gauge field configurations were prepared using a heat bath Monte Carlo in quenched approximation and the Wilson gauge field action at $\beta=2.3$. The gauge fields were constructed on a $10^3\cdot 16$ lattice and then doubled in the three-direction. Propagators with and without the source were calculated for three different spatial

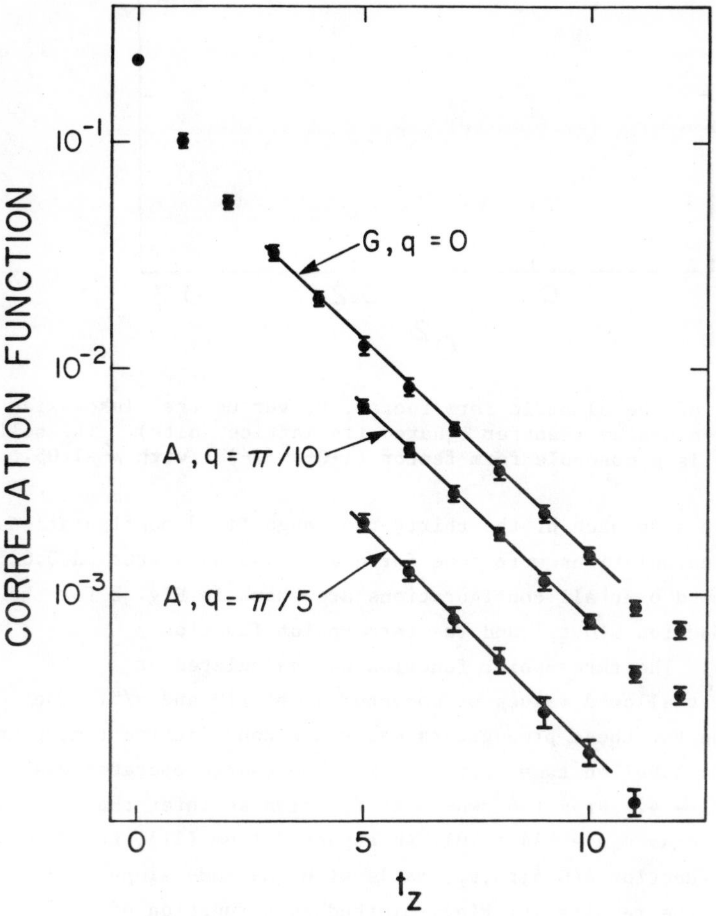

Fig. 2 Plot of the two-point function $G(0,t_z)$ and the three-point function $A(0,q;t_x,t_z)$ (for $q=\pi/10$ and $\pi/5$) as a function of t_z. The solid lines are single exponential fits.

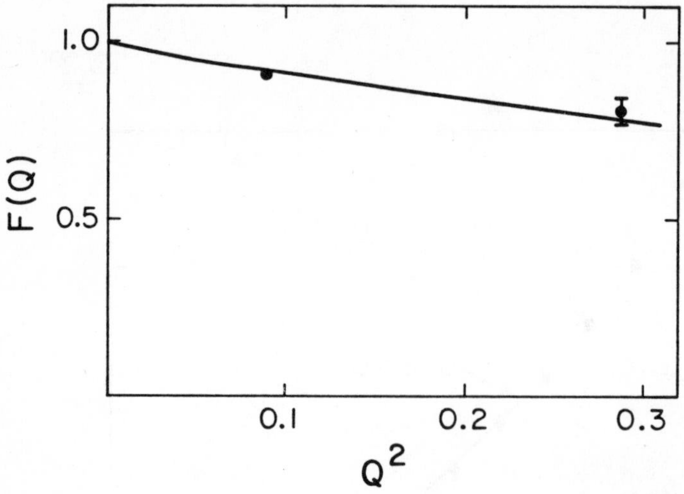

Fig. 3 Plot of the electric form factor, F, versus the minkowskian four-momentum transfer squared (in lattice units). The solid line is a monopole form factor $(1+Q^2a^2/\lambda^2)^{-1}$ with $\lambda^2=1.05$.

starting points in each of the thirty-two gauge field configurations. The present calculations were done for quark mass parameter ma=0.025. The data summed over all configurations are shown in Fig. 2 for the two-point function $G(0,t_z)$ and the three-point function $A(0,\vec{q};t_z,t_x)$. The three-point function was calculated at the two lowest allowed values of momentum $q_3 = \pi/10$ and $\pi/5$. The starting time for these propagators was the second lattice time point; this point is labelled $t_z = 0$ in Fig. 2. The charge operator was located at $t_x = 4$. From the two-point function we infer the pseudoscalar mass $m_\pi a = .44 \pm .01$. As expected from (11), the three-point function $A(0,\vec{q};t_z,t_x)$ falls with the same slope as $G(0,t_z)$. The results for F(q), plotted as a function of minkowskian four-momentum transfer squared $Q^2 = \vec{q}^2 - (E - m_\pi)^2$, are shown in Fig. 3. These results were obtained by averaging (12) over time steps five to nine. The errors are statistical only. The solid line in Fig. 3 is a monopole form factor $(1+Q^2a^2/\lambda^2)^{-1}$ with $\lambda^2 = 1.05$. The conversion to physical units depends on the calculation of the lattice spacing.

Using the value $a \simeq 0.16$ fm ($\beta=2.3$) obtained by Gutbrod and Montvay,[12] we infer $\sqrt{\langle r^2 \rangle} \approx .38$ fm. The older value of $a \approx .23$ fm based on the Cruetz ratio would give .55 fm.

3. SUMMARY

We have emphasized here the need for experimentally verifiable tests of hadronic size. The natural candidate for such a measurement is the electric form factor. Using the source method in quenched approximation, such a measurement can be made with an amount of computer time approximately the same as that required for a hadron mass calculation. Of course, the results here should be repeated for different quark masses as well as for SU(3). The extension of these results to the proton and the eventual removal of the quenched approximation should also be tackled. Our SU(2) numerical results are gratifying in that they provide further evidence that the quark and antiquark in a lattice meson are indeed localized in a compact object significantly smaller than the lattice volume.

REFERENCES

1. K. Ishikawa, G. Schierholz, H. Schneider and M. Teper, Nucl. Phys. B227, 221 (1983).
2. W. Wilcox and R.M. Woloshyn, TRIUMF Report TRI-PP-85-33.
3. M. Fukugita and T. Niuya, Phys. Lett. 132B, 374 (1982).
4. K. Barad, M. Ogilvie and C. Rebbi, Phys. Lett. 143B, 222 (1984).
5. B. Velikson and D. Weingarten, Nucl. Phys., B249, 433 (1985).
6. W. Wilcox and R.M. Woloshyn, Phys. Rev. Lett. (to be published).
7. W. Wilcox and R.M. Woloshyn, TRIUMF Report TRI-PP-85-32.
8. H. Kluberg-Stern, A. Morel and O. Napoly, Nucl. Phys. B220, 447 (1983).
9. S. Gottlieb, P.B. Mackenzie, H.B. Thacker and D. Weingarten, Phys. Lett. 134B, 346 (1984).
10. C. Bernard, in Gauge Theory on a Lattice: 1984, edited by C. Zachos, W. Celmaster, E. Kovacs and D. Sivers (National Technical Information Service, Springfield, VA, 1984).
11. C. Bernard, T. Draper, K. Olynyk and M. Rushton, Nucl. Phys. B220, 508 (1983).
12. F. Gutbrod and I. Montvay, Phys. Lett. 128B, 411 (1984).

PSEUDOFERMIONIC SIMULATION OF QCD THERMODYNAMICS

F.Karsch
Department of Physics, University of Illinois at Urbana-Champaign,
1110 West Green Street, Urbana, IL 61801, USA
and
R.V.Gavai
Department of Physics, Brookhaven National Laboratory,
Upton, Long Island, NY 11973,USA

ABSTRACT

We discuss recent results obtained for the thermodynamics of QCD in the presence of three light quark flavors. The influence of dynamical fermions has been simulated utilizing the pseudofermion algorithm. Our MC results indicate a rapid crossover behavior of all thermodynamic observables in the presence of light fermions. In the zero mass limit we find evidence for a chiral phase transitions at $T_C/\Lambda_L = 183 \pm 10$.

While the thermodynamics of pure SU(N) gauge theories has been analyzed in great detail by now, the simulation of QCD in the presence of dynamical fermions on Euclidean lattices is still in an exploratory stage. Various approximation schemes have been investigated to simulate the effect of dynamical fermions among which the pseudofermion [1] and microcanonical [2] methods seem to be very promising. These methods have recently been used to explore the effect of dynamical fermions on the thermodynamics of QCD [3-6].

Much evidence has by now been collected for the nce of a first order deconfining phase transition in pure SU(3) gauge theory [7]. The nature of this transitions is well understood and can be related to the breaking of a global Z(3) symmetry of the pure gauge action [8]. However, in the presence of dynamical fermions this symmetry is broken explicily. This lead to the speculation that dynamical fermions may give rise to qualitative changes in the phase structure of QCD at finite temperature. Indeed considerations based on the analysis of effective action models suggest that the influence of dynamical fermions is similar to the effect of an external magnetic field in spin models [8]. This indicates that for sufficiently light fermions the first order deconfinement transitions present in the pure gauge sector might disappear for sufficiently light fermions. However, in the zero mass limit one still expects a chiral phase transition [9,10]. In the quenched limit (pure gauge theory) this transition has been found to be first order and coincide with the deconfinement transition [11]. An analysis of effective chiral models with $SU(n_f) \times SU(n_f)$ flavor symmetry suggests that in general the chiral transition is a fluctuation induced first order transition for $n_f \geqslant 3$ [10].

The chiral transition has been studied in MC simulations including dynamical fermions within different approximation schemes [3-6,12]. These calculations show that physical observables undergo a rapid change from a low temperature hadronic regime to a high temperature quark-gluon plasma in a very narrow temperature range. However, it is difficult to obtain decisive results on the order of the chiral transition and indeed different groups came to somewhat different conclusions [3-6, 12].

In the following we will discuss our results [6] of a MC simulation of QCD with 3 light flavors on a lattice of size $8^3 \times 4$ and point out the problems arising when one tries to determine the order of a phase transition in the presence of dynamical fermions. The thermodynamics of QCD with dynamical fermions is described by the partition function

$$Z(\beta,V) = \int \prod_{x,\mu} dU_{x,\mu} \prod_x d\chi_x d\bar{\chi}_x e^{-S(U,\bar{\chi},\chi)} \tag{1}$$

with the action

$$S(U,\bar{\chi},\chi) = S_G(U) + S_F(U,\bar{\chi},\chi) \tag{2}$$

where the gluonic part S_G is the standard Wilson action with coupling $\beta = 6/g^2$

$$S_G = \beta \sum_{x,\mu<\nu} [1 - \frac{1}{3} \mathrm{ReTr}\, (U_{x,\mu} U_{x+\mu,\nu} U^\dagger_{x+\nu,\mu} U^\dagger_{x,\nu})] \tag{3}$$

and S_F denotes the fermion action which for staggered fermions of mass m reads

$$S_F = m \sum_{x,\mu<\nu} \bar{\chi}_\chi \chi_\chi + \frac{1}{2} \sum_{x,\mu} \bar{\chi}_\chi \eta_\mu(x) [U_{x\mu} \chi_{x+\mu} - U^\dagger_{x-\mu,\mu} \chi_{x-\mu}] \tag{4}$$

In eq. (4) small $\eta_\mu(x) = (-1)^{x_1+\ldots x_{\mu-1}}$, x labels the sites of a hypercubic lattice of size $N_\sigma^3 \times N_\tau$. Temperature and volume of the thermodynamic system are then given as

$$1/T = N_\tau a \quad ; \quad V = (N_\sigma a)^3 \tag{5}$$

for a lattice with lattice spacing a.

The fermionic fields $\chi, \bar{\chi}$ appear only quadratically in the action, eq. (4), and can be integrated out explicitly. This yields a partition function in terms of gluonic variables only. For n_f fermion flavors we get

$$Z = \int \prod_{x,\mu} dU_{x\mu} [\det(m^2 + D^2)]^{n_f/8}\, e^{-S_G} \tag{6}$$

with $Q = m + \sum_\mu D^\mu$

$$D^\mu_{xy} = \frac{1}{2} \eta_\mu(x)\, [U_{x\mu} \delta_{y,x+\mu} - U^\dagger_{y\mu} \delta_{y,x-\mu}]. \tag{7}$$

Eq. (6) is the starting point for a pseudofermionic simulation of the partition function. The determinant appearing in eq. (6) can be thought of as resulting from an integration over scalar fields [1]. Actually all one needs to generate an ensemble of equilibrated gauge field configurations is the knowledge of the change of the determinant under a change of a set of gauge field variables. For small changes, δU, in the gauge fields this is given by

$$\frac{\det\,(Q + \delta Q)}{\det Q} = \det\,(1 + Q^{-1}\delta Q) \qquad (8)$$

$$= 1 + \mathrm{Tr}\,Q^{-1}\delta Q + O(\delta U^2)$$

Of course this procedure introduces systematic errors and is thus expected to be valid only for small changes δU. This, however, leads to slow convergence of the MC-simulations as subsequent configurations differ only little and are thus highly correlated. In fig. 1 we illustrate this problem. Varying the step size δU from a rather large value (53% acceptance rate) to a small value (79% acceptance rate) clearly leads to a large increase in the equilibration time.

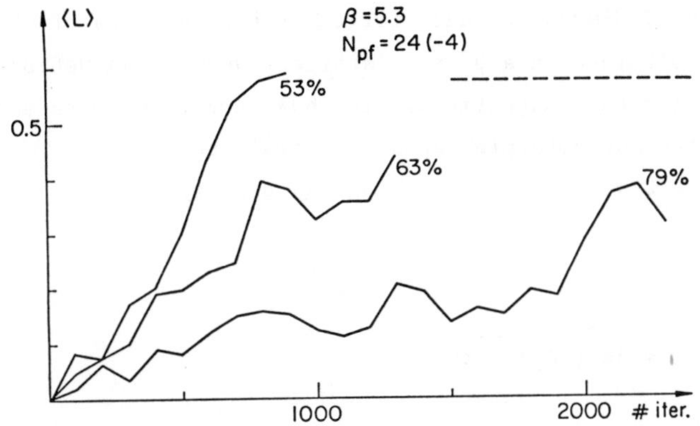

Figure 1: The Polyakov line versus number of MC iterations for various acceptance rates at $\beta = 5.3$. The dashed line shows the equilibrium value obtained from a long run with the intermediate value for the acceptance rate. The data have been averaged over 50 subsequent iterations.

Thus although it is desirable to use a large acceptance rate in order to reduce systematic errors, it is an advantage of intermediate values that the approach to equilibrium is conciderably faster. This allows to distinguish slow convergence from the possible appearance of metastable states in the transition region from the low to the high temperature regime. Thus a moderate value for the acceptance rate seems to be adequate to study the dynamics in the crossover region. Of course one has to check that the acceptance rate can be lowered without introducing too large systematic errors. Comparing our results presented below [6] obtained with an acceptance rate of 63% with those of ref. [4] where the acceptance rate was 80%, we find that local observables like the plaquette expectation value and the Polyakov loop agree within errors.

Another source of errors in the pseudofermion algorithm is the use of a finite, in general small, number of pseudofermionic configurations to evaluate the matrix elements of Q^{-1}. Too small a number again slows down the convergence to equilibrium [6]. In the following we will discuss MC-data for the thermodynamics of QCD with 3 light flavors obtained on a $8^3 \times 4$ lattice. We used an Metropolis update scheme for the gauge fields with 63% acceptance rate and 50 iteration in the pseudofermion update to evaluate Q^{-1}.

In figure 2 we show the total energy density

$$\varepsilon = T^2 V^{-1} \delta \ln Z / \delta T \equiv \varepsilon_G + \varepsilon_F \tag{9}$$

with

$$\varepsilon_G = 3\beta \left(\langle P_\sigma \rangle - \langle P_\tau \rangle \right) \tag{10}$$

$P_{\sigma(\tau)} = 1 - 1/3 \ \mathrm{ReTr} U_{\sigma(\tau)}$ denoting space-space (space-time) like plaquettes, and

$$\varepsilon_F = \frac{n_f}{4} \langle \mathrm{tr}\, D^4 (D + m)^{-1} \rangle - \left\{ \frac{3n_f}{16} - \frac{m}{4} \langle \bar{\chi}\chi \rangle_{T=0} \right\} \tag{11}$$

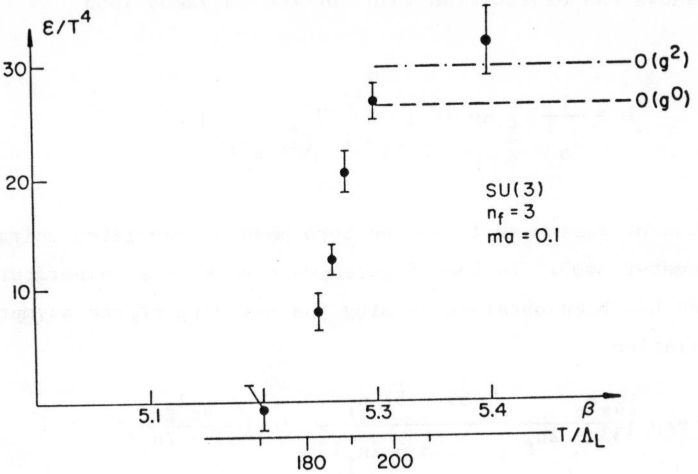

Figure 2: Energy density versus coupling β for SU(3) with 3 flavors of mass ma = 0.1 on a $8^3 \times 4$ lattice. Also shown are the lowest order (- -) and $O(g^2)$ (-•-) weak coupling perturbative results of ref. [13]. The temperature scale has been obtained by assuming the validity of the asymptotic scaling relation eq. (13).

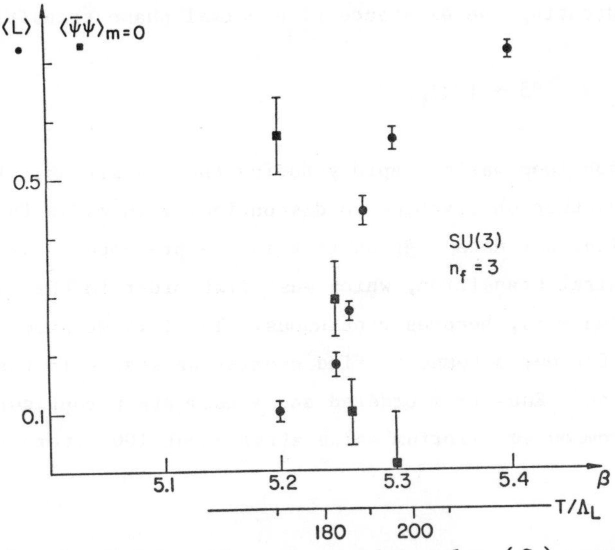

Figure 3: The Polykov line expectation value (●) versus β and the zero mass extrapolated chiral order parameter (■). $\langle\bar{\psi}\psi\rangle_{m=0}$ has been obtained from a linear extrapolation of date at ma = 0.075 and 0.1.

Figure 3 shows the expectation value of the Polyakov loop

$$L = \frac{1}{N_\sigma^3} \sum_{\vec{x}} \mathrm{Re}\,\mathrm{Tr}\,\Big(\prod_{x_4=1}^{N_\tau} U_{(\vec{x},x_4),4} \Big) \tag{12}$$

for fermions of mass ma=0.1 and the zero mass extrapolated chiral order parameter $\langle\bar{\psi}\psi\rangle$. In both figures we also show a temperature scale which has been obtained asuming the validity of the asymptotic scaling relation

$$a\Lambda_L = \exp\Big\{\frac{4\pi^2\beta}{33-2n_f} - \frac{459-57\,n_f}{(33-2n_f)^2}\,\ln\Big(\frac{8\pi^2\beta}{33-2n_f}\Big)\Big\}. \tag{13}$$

All 3 quantities plotted in fig. 2 and 3 clearly show a rapid variation in a narrow temperature interval around T = 180 MeV. Above T ≈ 200 MeV the energy density agrees well with the behaviour of a free lattice gas with perturbative corrections due to one gluon exchange [13]. The chiral order parameter seems to vanish above β_c = 5.25 ± 0.05 indicating the existence of a chiral phase transition at

$$T_C = (183 \pm 10)\Lambda_L. \tag{14}$$

Also the Polyakov loop varies rapidly during the transition. However, like in the two other observables no discontinuity shows up in the transition region. Our data suggest that in the presence of light fermions the chiral transition, which was first order in the quenched approximation (ma = ∞), becomes continuous. In Fig.4 we show a typical result for our attempt to find coexisting states in the transition region. Runs from ordered and random start configurations converge to a common equilibrium value after about 1000 iterations.

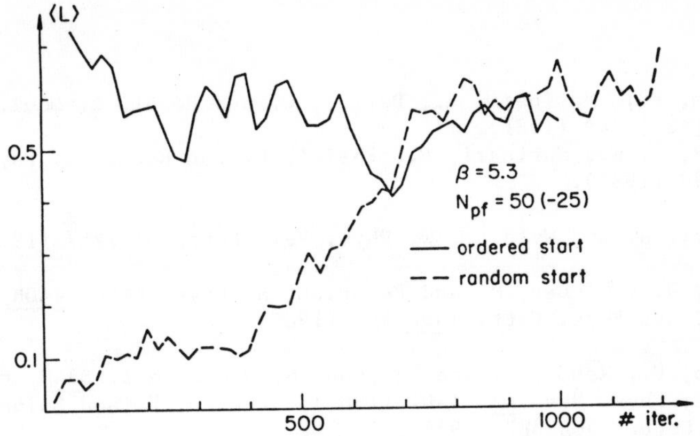

Figure 4: The Polyakov line versus number of MC iterations. Shown is the evolution of ⟨L⟩ from random (- -) and ordered (——) start configurations at β = 5.3. The date have been averaged over 20 subsequent iterations.

To summarize, we find evidence for a chiral phase transition at $T_c/\Lambda_L = 183 \pm 10$. At least on the size of lattices ($8^3 \times 4$) used by us there is no indication for a strong first order phase transition in all observables considered. Of course with the amount of statistics gathered by us we still cannot exclude the existence of a weak first order phase transition in the crossover region. It is certainly of interest to study the behavior in the crossover region in more detail. In particular it will be interesting to see whether on larger lattices which allow to go deeper into the continuum regime, this behavior persists or whether indications for a fluctuation induced first order transition can be found as advoctated in ref.[10].

REFERENCES

1) Fucito, F., Marinari, E., Parisi, G. and Rebbi, C. Nucl. Phys. B180 [FS3] 369 (1981).
Hamber, H. W., Marinari, E., Parisi, G. and Rebbi, C. Phys. Lett. 124B 99 (1983).

2) Polonyi, J. and Wyld, H. W. Phys. Rev. Lett. 51 2257 (1983).

3) Gavai, R. V., Lev, M. and Peterson, B. Phys. Lett. 140B 397 (1984) and Phys. Lett. 149B 492 (1984).

4) Fucito, F., Rebbi, C. and Solomon, S. Nucl. Phys. B248 615 (1984); Phys. Rev. D31 1461 (1985); Fucito, F. and Solomon, S. Phys. Lett. 140B 387 (1984).

5) Polonyi, J., Wyld, H. W., Kogut, J. B., Shigemitsu, J. and Sinclair, D. K. Phys. Rev. Lett. 53 644 (1984).

6) Gavai, R. V. and Karsch, F. Nucl. Phys. B. to appear, Illinois Preprint, ILL-(TH)-85-#19, April, 1985.

7) For recent reviews and further references see, Karsch, F., "The Deconfinement Transition in Finite Temperature Lattice Gauge Theory", To appear in The Proceedings of the Enrico Fermi International School of Physics 1984;

Cleymens, J., Gavai, R. V. and Suhonen, E., to appear in Phys. Rep.

8) For a review see Svetitsky, B., "Symmetry Aspects of Finite Temperature Confinement Transitions", Phys. Rep. to appear.

9) Tomboulis, E. T. and Yaffe, L. G., Phys. Rev. Lett. 52 2115 (1984).

10) Pisarski, R.D. and Wilsczek, F., Phys. Rev. D29 338 (1984).

11) Kogut, J., Stone, M., Wyld, H. W., Gibbs, W. R., Shigemitsu, J., Shenker, S. H. and Sinclair, D. K., Phys. Rev. Lett. 50 393 (1983).

12) Celik, T., Engels, J. and Satz, H., Phys. Lett. 133B 427 (1984).

13) Heller, U. and Karsch, F., Nucl. Phys. B251 [FS13] 254 (1985); CERN preprint, CERN-TH 4078/84, Nucl. Phys. B to appear.

ON CRITICAL BEHAVIOUR IN STATISTICAL QCD

Helmut Satz

Fakultät für Physik
Universität Bielefeld
Germany

After discussing some conceptual aspects of the deconfinement transition, we show that the critical behaviour of the SU(2) system with dynamical quarks agrees with that of the pure gauge theory.

The thermodynamics of pure SU(N) gauge fields predicts the occurence of a deconfining phase transition[1]. At some temperature T_c , colour screening dissolves the confining bond and turns gluonium matter into a chromoplasma[2]. One of the most interesting problems in current statistical QCD is what effect the introduction of dynamical quarks has on this deconfinement.

On a phenomenological level, colour screening considerations appear to remain valid; the presence of many other colour charges should dissolve bound quark states just as well as gluonium states. On the other hand, the global symmetry under the center Z_N of the SU(N) gauge group is broken by the fermion term in the Lagrangian, and hence deconfinement can no longer be strictly characterized in terms of spontaneous Z_N symmetry breaking. The expectation value $\bar{L}$ of the thermal Wilson loop does not any more constitute an order parameter for the deconfinement transition, since it now does not vanish in the confinement region. However, it is unclear how much $\bar{L}$ differs from zero there - i.e., how strongly the Z_N symmetry is actually broken. We must therefore ask by what mechanism $\bar{L}$ is determined in the confinement zone.

The deconfinement transition is the chromodynamic analog of the metal-insulator transition in systems with electromagnetic forces. In insulating solids, the electric conductivity σ is not strictly zero, but only exponentially small[3],

$$\sigma \sim \exp\{-E_i/T\} \quad , \tag{1}$$

where E_i denotes the ionisation energy. Above the transition point to a metal, σ is non-zero because Debye screening has globally dissolved the Coulomb binding between ions and electrons; but even below this point, ionisation can locally provide some free electrons, and thus make $\sigma > 0$. The corresponding phenomenon in QCD with dynamical quarks is the production of $q\bar{q}$ pairs. In pure gauge theory below the deconfinement temperature, an infinite energy is needed to break up a static quark-antiquark pair; this corresponds to an infinite ionisation energy in eq. (1). With dynamic quarks, such a break-up becomes possible when the separation yields a binding energy equal to that needed to produce a $q\bar{q}$ pair - i.e., a hadron. The newly formed quarks neutralize the static quarks and thus allow their separation. We therefore expect that L will now no longer vanish in the confinement zone, but that

$$L \sim \exp\{-m_H/2T\} \quad , \tag{2}$$

where m_H is the mass of the dominant $q\bar{q}$ bound state. Letting $m_H \to \infty$ in eq. (2), we recover the pure gauge result.

Both the conventional Mott transition in solids and the deconfinement transition in hadronic matter thus lead from a regime, in which the binding can locally be broken (by ionisation or $q\bar{q}$ formation, respectively) to one, where it is globally removed by a collective screening of the binding force. Although this provides us with an intuitive picture of how an $L \neq 0$ arises in the confinement region, it does not allow us to estimate it quantitatively. Thus also the question of the sharpness of the transition, or of its order, if it still is a genuine phase transition, remains open.

To study this question in more detail, we consider the deconfinement order parameter of the pure gauge system and check what happens when dynamical quarks are introduced. We shall here treat only the SU(2) case; for this, there now exist fairly extensive lattice results both without[4,5] and with dynamical light quarks[6,7]. Moreover, the transition for the pure gauge system here appears to be continuous; hence the introduction of a symmetry breaking term does not have to overcome a "latent heat" discontinuity in order to modify the transition.

In the strong coupling limit, the SU(2) gauge system without quarks can be shown to have the same structure as the Z_2 spin system of the same spatial dimension[8]. This result has been extended to the conjecture that SU(N) gauge and Z_N spin systems should quite generally belong to the same universality class[9]. Specifically, the conjecture implies the same critical exponents for the deconfinement transition of the SU(2) gauge system and the order-disorder transition of the Ising model. A recent study[4] has confirmed this in the case of the deconfinement order parameter, and we want to investigate in particular how such behaviour is effected by dynamical quarks.

The deconfinement order parameter is given by the lattice average of the Polyakov loop[10,11]

$$L_{\vec{x}}(U) \equiv \frac{1}{2} \mathrm{Tr} \prod_{\tau=0}^{N_\tau-1} U_{\vec{x};\tau,\tau+1} \quad , \tag{3}$$

where $U_{\vec{x};\tau,\tau+1}$ is the SU(2) matrix associated to the link connecting the temporal layers τ and $\tau+1$ at the spatial point $\vec{x}$. We consider a lattice of overall size $N_\sigma^3 \times N_\tau$, with N_σ (N_τ) sites per space (temperature) direction. First we average $L_{\vec{x}}(U)$ for a given (equilibrium) configuration of U's over the entire lattice, obtaining $\bar{L}(U)$. For the order parameter $\bar{L}$, $\bar{L}(U)$ is then averaged over successive configurations.

For the pure gauge system, the average

$$\bar{L} = \int \prod_{\text{links}} dU\, e^{-S_0(U)}\, L(U) \Big/ \int \prod_{\text{links}} dU\, e^{-S_0(U)} \tag{4}$$

is performed with the Wilson action

$$S_0(U) = \beta \left\{ \sum_{P_\sigma} \left(1 - \frac{1}{2} \mathrm{Re\,Tr}\, UUUU\right) + \sum_{P_\tau} \left(1 - \frac{1}{2} \mathrm{Re\,Tr}\, UUUU\right) \right\} \quad , \tag{5}$$

where the summation runs over all space-space and space-temperature plaquettes, respectively; we have defined $\beta \equiv 4/g^2$ and assumed an isotropic lattice spacing a. It is intuitively clear that $\bar{L}$ will exhibit different behaviour for different β. If β is small, the weight $\exp\{-S_0(U)\}$ will be large for arbitrary U's ; this makes $\bar{L} \sim 0$. For large β, however, the U's must align and hence $\bar{L}$ will approach unity.

For β approaching from above the transition point β_c between the two phases, we write

$$\bar{L} = c(\beta - \beta_c)^b \quad ; \quad \beta > \beta_c \quad ; \quad b,c = \text{const.} \tag{6}$$

Since $\bar{L}$ corresponds to the spontaneous magnetisation of the Ising model, the mentioned spin-gauge universality[9] predicts

$$b \simeq 0.33 \tag{7}$$

for the critical exponent of deconfinement. Hence $\bar{L}^3$ should be linear in $(\beta - \beta_c)$. In fig. 1 we show recent lattice results, obtained on $10^3 \times 4$ [4] and $12^3 \times 4$ [5] lattices. They are seen to agree well with this prediction.

If the inverse coupling β is large enough, it is connected to the lattice spacing a by the renormalisation group relation

$$a\,\Lambda_L = \exp\left\{-\left(\frac{3\pi^2}{11 - N_f}\right)\beta + \frac{3(272 - 49N_f)}{16(11 - N_f)^2}\ln\left[\left(\frac{6\pi^2}{11 - N_f}\right)\beta\right]\right\} \tag{8}$$

where N_f denotes the number of quark species. Using eq. (8), we can study the critical behaviour in terms of the temperature $T = (N_\tau\, a(\beta))^{-1}$,

$$\bar{L} = c'(T - T_c)^b \quad , \tag{9}$$

with b the same as in eq. (6). Let us note, however, that the prediction (6/7) does not require us to be in the scaling regime; in fact, the relation between spin and gauge systems was first established in the strong coupling limit[8], where eq. (8) is certainly not applicable. The more general universality test is therefore provided by eq. (6). In fig. 2 and 3, we see that it is satisfied by calculations with $N_\tau = 3$ [5,12,14] and 2 [5,12,13] as well; the transition region for $N_\tau = 2$, relation (8) is known to be not applicable[12]. In table 1, the results for the pure gauge system are summarized. While the critical couplings for $N_\tau = 3$ and 4 lead within errors to the same critical temperature, that for $N_\tau = 2$ does not, since $\beta_c = 1.85$ lies below the region of validity of eq. (8).

Let us now see what happens when quarks are introduced. After integrating out the quark fields, we then obtain once more a gauge field partition function of the type in eq. (4); but the action (5) is now replaced by

$$S(U) = S_0(U) - N_f \log(\det Q) \quad , \tag{10}$$

where N_f denotes the number of quark species and $Q(U)$ the lattice operator describing the quark interaction. Calculations of the average of the Polyakov loop, using this action, have been performed for "staggered" fermions[15] in the pseudo-fermion scheme[16] with $N_f = 2$ [6] and in the microcanonical scheme[17] with $N_f = 4$ [7]. The resulting values of $\bar{L}$ are shown in figs. 4 and 5; they are obtained on $8^3 \times 4$ lattices for quark mass parameters $ma = 0.08$ and 0.1. We note that in both cases $\bar{L}^3$ becomes linear in β, thus giving us the universal exponent (7). In ref. 6) it had already been noted that the introduction of two quark species results mainly in a shift of $\bar{L}$ as function of β; it is seen here that this pattern holds true on a much more quantitative level and for $N_f = 4$ as well. The range of couplings very near the critical point is for the cases with quarks not as well covered as for the pure gauge system, so that a more detailed study would be of considerable interest. Nevertheless, the available points are in full agreement with universal critical behaviour also in the presence of two and four species of light dynamical quarks. In table 2 we list the parameters resulting from the form (6/7). We see that even the amplitude of $\bar{L}$ in eq. (6) is essentially the same for $N_f = 0$, 2 and 4; the very slight decrease of c for $N_f \neq 0$ may be an effect of the slightly smaller lattice size in the calculation with quarks. In the last columns of table 2 we list the critical temperature T_c obtained by using the renormalisation group relation (8); first we show $T_c/\Lambda_L^{N_f}$, then the values T_c/Λ_L^0 obtained by converting the lattice scale with N_f staggered fermions to that with $N_f = 0$, in the weak coupling limit[18]. In the scaling region, the values in the last column should agree; the spread in fact seen there seems to indicate that the renormalisation group relation, the $\Lambda_L^{N_f}/\Lambda_L^0$ relation, or both, are not yet so well satisfied. Such deviations for the perturbative relation of lattice scales would not be so surprising, as they are observed already in the pure gauge case[14]. Moreover, the critical coupling for $N_f = 4$ would in the case of $N_f = 0$ clearly be outside the scaling region. Hence any conclusions concerning the transition region as function of the temperature[7] appear to be questionable until the scaling behaviour is clarified. This does not effect, however, the universality of the critical behaviour observed as function of $\beta - \beta_c$.

To obtain some idea about the range of compatible values for the deconfinement exponent b, we show in fig. 6 the numerical results for $N_f = 0$, 2 and 4 in the same plot, as function of $\beta - \beta_c$. The set of presently available values seems to exclude deviations from 0.33 by more than about 20%. A more precise analysis would require a denser coverage of the region near β_c for the case with quarks.

We conclude: all presently available calculations indicate that a universal critical behaviour for SU(2) QCD persists in the presence of light, dynamical quarks. This suggests that quarks lead only to a very weak breaking of the global confinement symmetry, and that - as expected in the insulator-conductor picture given above - the phenomenon of deconfinement remains much the same with and without quarks[19].

REFERENCES

1) A.M. Polyakov, Phys. Lett. 72B (1978) 477;
L. Susskind, Phys. Rev. D20 (1979) 2610;
C. Borgs and E. Seiler, Nucl. Phys. B215 (1983) 125
2) H. Satz, Nucl. Phys. A418 (1984) 447c
3) N.F. Mott, Rev. Mod. Phys. 40 (1968) 677 and references given there
4) R.V. Gavai and H. Satz, Phys. Lett. 145B (1984) 248
5) G. Curci and R. Tripiccione, Pisa Preprint IFUP-TH-84/26, October 1984
6) U. Heller and F. Karsch, CERN Preprint TH-4078/84, December 1984
7) J. Kogut, J. Polônyi, H.W. Wyld and D.K. Sinclair, Illinois Preprint ILL-(TH)-85-6, January 1985
8) J. Polônyi and K. Szlachânyi, Phys. Lett. 110B (1982) 395
9) B. Svetitsky and L.G. Yaffe, Nucl. Phys. B210 [FS6] (1982) 423
10) L. McLerran and B. Svetitsky, Phys. Lett. 98B (1981) 195
11) J. Kuti, J. Polônyi and K. Szlachânyi, Phys. Lett. 98B (1981) 199
12) J. Engels, F. Karsch, I. Montvay and H. Satz, Phys. Lett. 101B (1981) 89
13) E. Kehl, Diploma-Thesis, University of Bielefeld 1984 (unpublished)
14) R.V. Gavai, F. Karsch and H. Satz, Nucl. Phys. B220 [FS8] (1983) 223

15) J. Kogut and L. Susskind, Phys. Rev. D11 (1975) 395;
L. Susskind, Phys. Rev. D16 (1977) 3031
16) F. Fucito, E. Marinari, G. Parisi and C. Rebbi, Nucl. Phys. B180 [FS2] (1981) 369
17) J. Polônyi and H.W. Wyld, Phys. Rev. Lett. 51 (1983) 2257
18) H.S. Sharatchandra, H.J. Thun and P. Weisz, Nucl. Phys. B192 (1981) 205
19) H. Satz, Phys. Lett. in press (BI-TP 85/07)

Table 1 : Deconfinement transition parameters for the SU(2) gauge system without quarks

N_τ	N_σ	β_c	c	T_c/Λ_L^0	References
2	8 - 10	1.85 ± 0.02	0.84	27.6 ± 1.4	5, 12, 13
3	7 - 10	2.16 ± 0.02	0.72	39.7 ± 2.0	5, 12, 14
4	10 - 12	2.28 ± 0.02	0.55	40.2 ± 2.1	4, 5

Table 2 : Deconfinement transition parameters for SU(2) QCD with different numbers of quark species; from lattice calculations with $N_\tau = 4$.

N_f	β_c	c	$T_c/\Lambda_L^{N_f}$	T_c/Λ_L^0	Ref.
0	2.28	0.55	40.2	40.2	4, 5
2	2.0	0.52	63.8	35.1	6
4	1.8	0.51	229.4	49.7	7

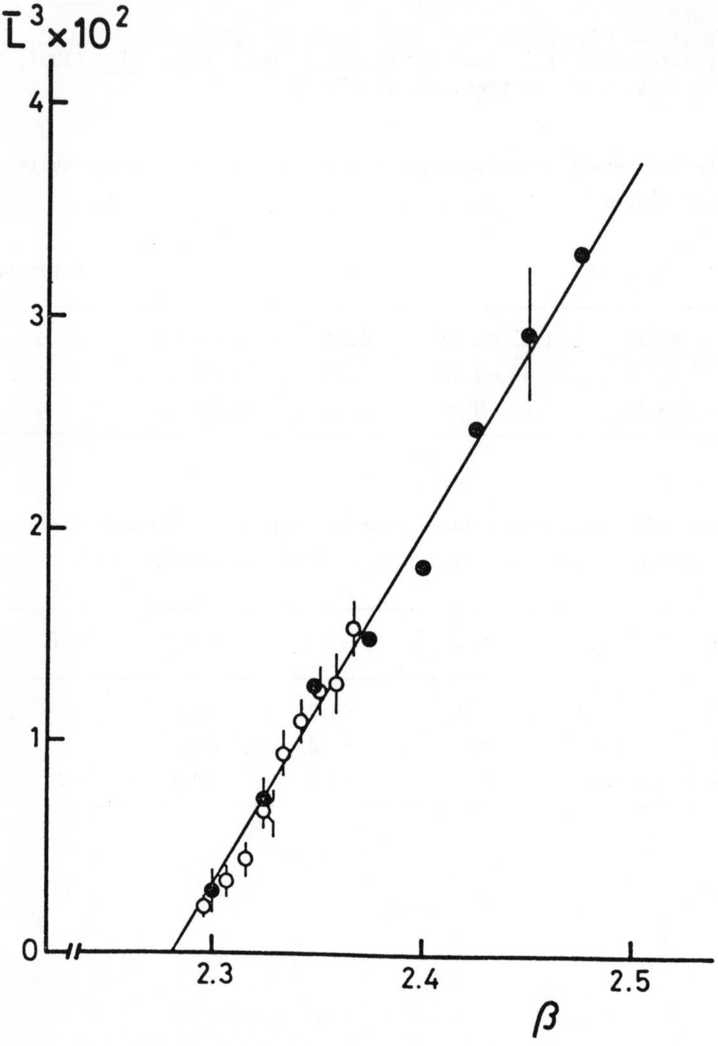

Figure 1 : Cube of the average Polyakov loop vs. $\beta = 4/g^2$ in pure SU(2) gauge theory, calculated on lattices with $N_\tau = 4$; points are from ref. 4 (○) and ref. 5 (●).

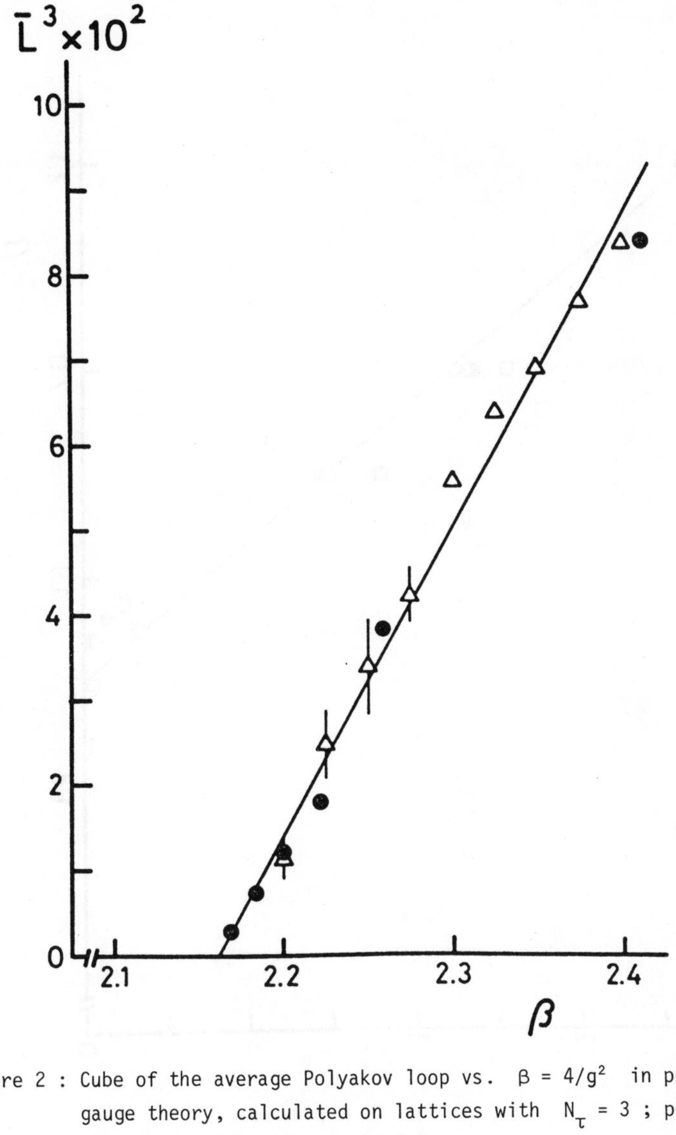

Figure 2 : Cube of the average Polyakov loop vs. $\beta = 4/g^2$ in pure SU(2) gauge theory, calculated on lattices with $N_\tau = 3$; points are from ref. 5 (Δ) and ref. 14 ($\bullet$).

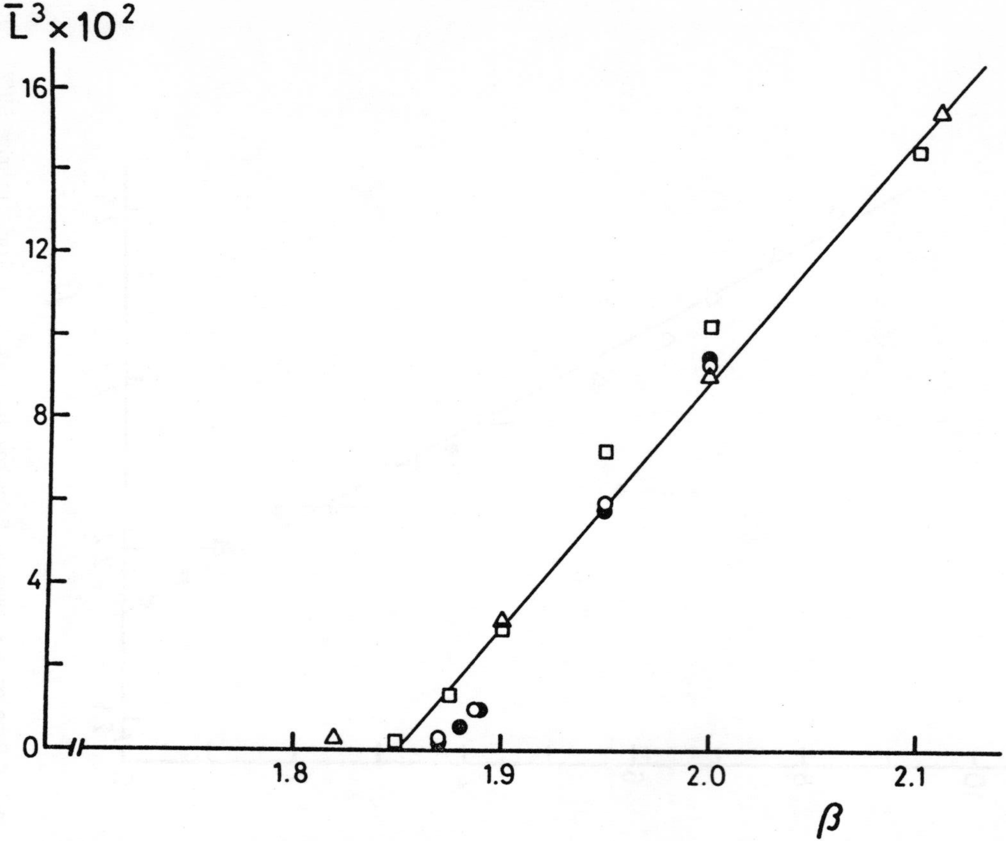

Figure 3 : Cube of the average Polyakov loop vs. $\beta = 4/g^2$ in pure SU(2) gauge theory, calculated on lattices with $N_\tau = 2$; points are from ref. 5 (□), ref. 12 (Δ) and ref. 13 (○ and ●).

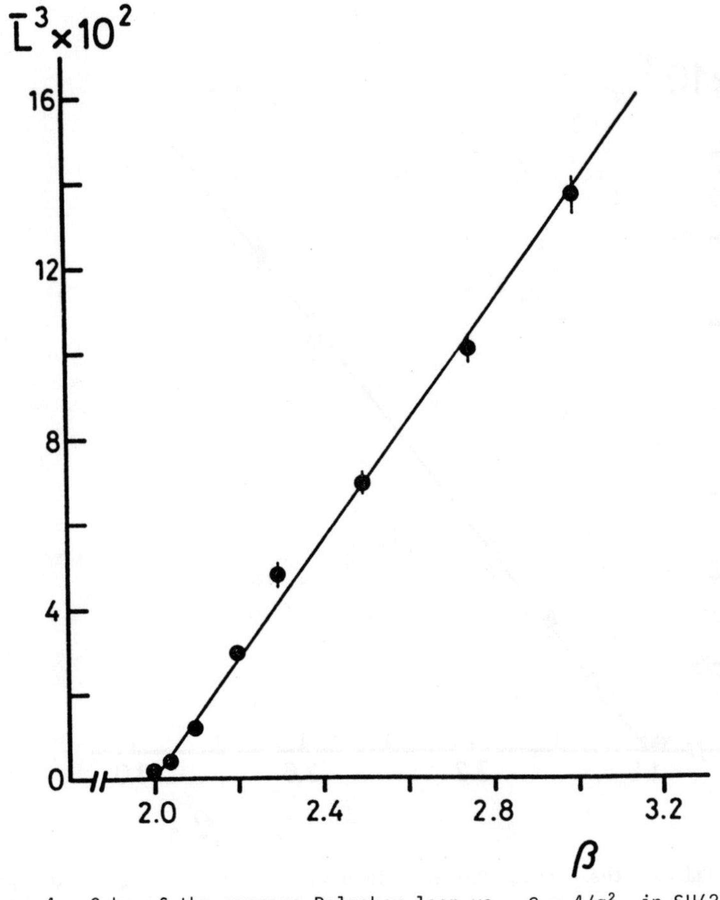

Figure 4 : Cube of the average Polyakov loop vs. $\beta = 4/g^2$ in SU(2) QCD with two species of dynamical quarks (ma = 0.1), calculated on an $8^3 \times 4$ lattice (from ref. 6).

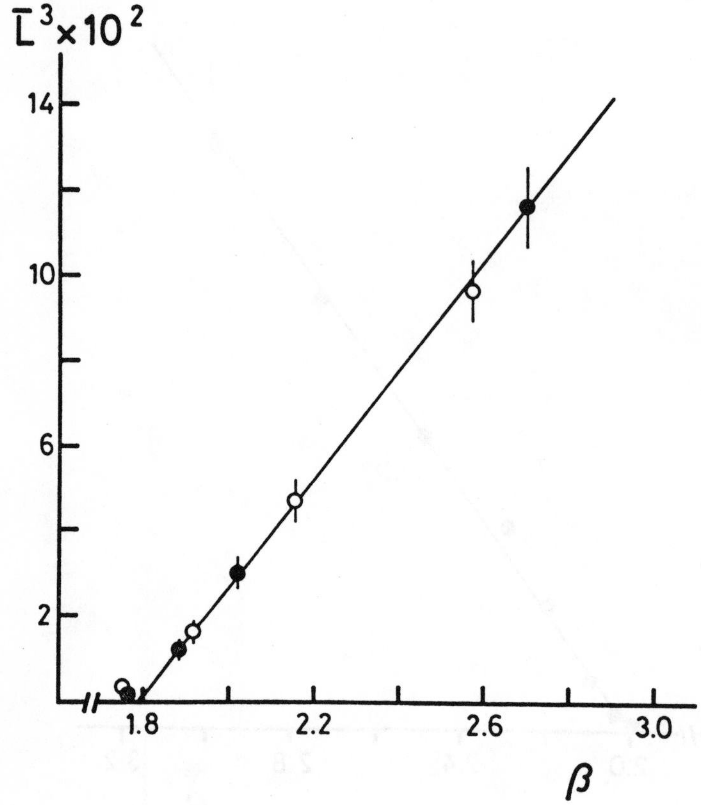

Figure 5 : Cube of the average Polyakov loop vs. $\beta = 4/g^2$ in SU(2) QCD with four species of dynamical quarks (● ma = 0.08, ○ ma = 0.01), calculated on an $8^3 \times 4$ lattice (from ref. 7).

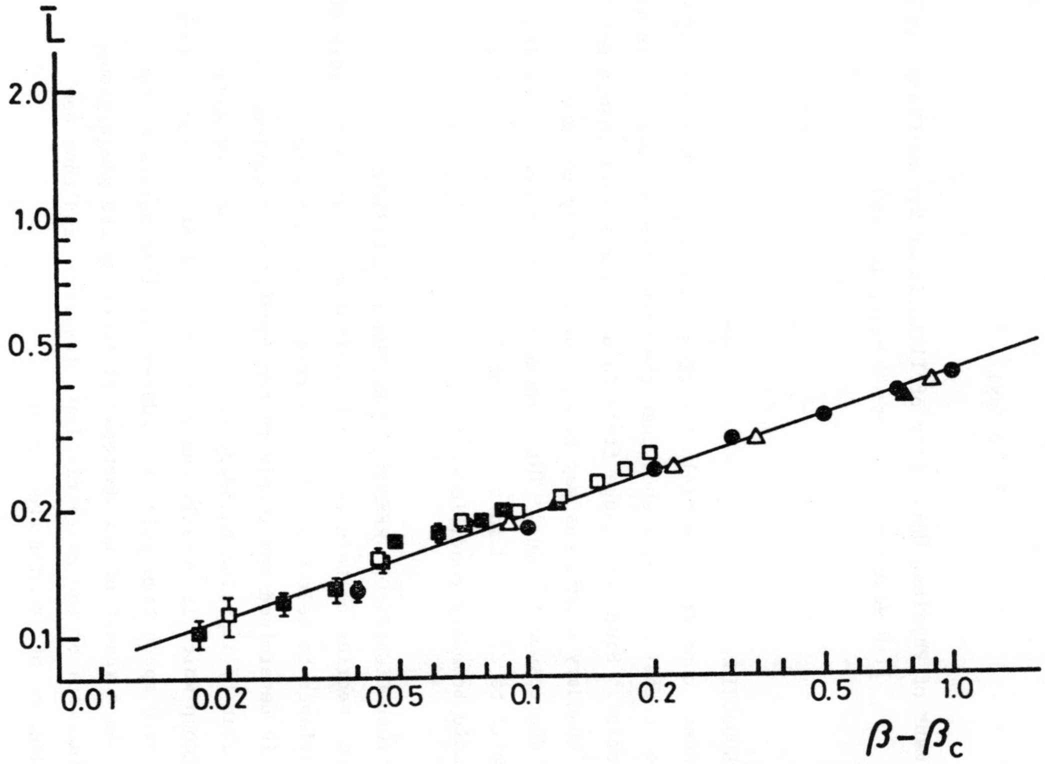

Figure 6 : Average Polyakov loop vs. $\beta - \beta_c$, with β_c given in table 2, for zero (■, □), two (●) and four (△, ▲) species of dynamical quarks. The straight line corresponds to the critical exponent $b = 1/3$.

SYMMETRY BREAKING THERMAL FLUCTUATIONS IN GAUGE THEORIES†

J. Polonyi*

Department of Physics, University of Illinois at Urbana-Champaign
1110 West Green Street, Urbana, IL 61801

I. INTRODUCTION

Boundary conditions in field theories usually play the role of an infrared cutoff. We will show that the situation is more subtle in gauge theories. Since the gauge field is a string rather than a point variable, boundary conditions may create qualitatively new gauge invariant degrees of freedom. This kinematical construction and its dynamics will be discussed in the case of SU(2) and SU(3) theories with periodic boundary conditions.[1)]

II. PHYSICAL DEGREES OF FREEDOM AND BOUNDARY CONDITIONS

Gauge theories contain redundant variables and it is the role of gauge invariance to select physical degrees of freedom whose time evolution is determined completely by the equations of motion. We will show that by imposing boundary conditions the complete gauge symmetry group which is usually the direct product of the local gauge group for each space time point is reduced: we lose points on the boundary. As a result of the decrease of the complete gauge group more combinations of the gauge field are invariant and thus new physical degrees of freedom are formed.

---

*On leave from Central Research Institute of Physics, Budapest, Hungary.
†Research was supported in part by the grant NSF PHY82-01948.

Consider Yang-Mills theory in the temporal gauge. The Hamiltonian is of the form

$$\hat{H} = \int d^3x \frac{1}{2} \{trE^{\vec{2}} + TrB^{\vec{2}}\}, \qquad \hat{B}^i = \frac{1}{2} \varepsilon^{ijk} \hat{F}_{jk},$$

$$\hat{F}_{ij} = D_i\hat{A}_j - D_j\hat{A}_i + [\hat{A}_i,\hat{A}_j], \; \hat{A}_i = \hat{A}^a_i \frac{\lambda^a}{2i} \qquad (1)$$

and the nonvanishing canonical commutators are $[A^a_i(\vec{x}), E^b_j(\vec{y})] = \delta(\vec{x}-\vec{y})\, \delta_{ij}\, \delta^{ab}$. We shall use a field diagonal representation $\vec{A}(\vec{x})|\vec{A}(y)\rangle = \vec{A}(\vec{x})|\vec{A}(\vec{y})\rangle$. Any physical (= gauge invariant) states can be obtained from a family of basis vectors $|\vec{A}^r\rangle$ by averaging over gauge transformations

$$|A^r\rangle_{phys.} = \prod_{\vec{x}} \int dg(\vec{x})\hat{g}(\vec{x})|\vec{A}^r\rangle \qquad (2)$$

where $\hat{g}|\vec{A}\rangle = |g(\vec{\nabla}+\vec{A})g^+\rangle$. The labeling $|A^r\rangle_{phys.}$ becomes unique from imposing suitable gauge fixing conditions

$$F_r(\vec{A}^r(x)) = 0 \qquad (3)$$

on the representative field configurations.

Matrix elements of the time evolution operator are written in the form

$$\begin{aligned} &\langle A^r_{fin.}|_{phys.} \exp\{-iTH\}|A^r_{init.}\rangle_{phys.} \\ &= \int D[g]D[h]\langle A^r_{fin.}|\hat{g}\exp\{-iT\hat{H}\}\hat{h}|A^r_{init.}\rangle \\ &= \int D[g]D[g^+h]\langle A^r_{fin.}|\hat{g}^+\exp\{-iT\hat{H}\}\hat{g}\hat{g}^+h|A^r_{init.}\rangle \\ &= \int D[g]\int D[h']\langle A^r_{fin.}|\exp\{-iT\hat{H}\}\hat{h}'|A^r_{init.}\rangle \end{aligned} \qquad (4)$$

where the invariance $[\hat{H},\hat{g}] = 0$ was used. It is well known[2)] that the path integral representation for (3) gives

$$\int D[A_\mu(\vec{x},t)]\exp\{i/4g^2\int dtd^3x\ tr(F_{\mu\nu})^2\} \tag{5}$$

where the domain of integration consists of field configurations which satisfy the boundary conditions $\vec{A}(\vec{x},o) = A^r_{init.}(\vec{x})$, $A_o(\vec{x},o) = 0$, $\vec{A}(\vec{x},T) = A^r_{fin.}(\vec{x})$, $A_o(\vec{x},T) = 0$. The projection onto the physical subspace is achieved by integration over A_o which is thought of as a representation of the projection operator

$$\hat{\pi}(t) = \int D[h]\hat{h} = \int D[A_o(\vec{x},t)]\exp\{-i\int d^3x\ \vec{E}(\vec{x})\vec{D}A_o(\vec{x})\}. \tag{6}$$

The relative gauge difference between the initial and final time in (5) is given by the Polyakov loop

$$\Omega(\vec{x}) = g^+(\vec{x})h(\vec{x}) = h'(\vec{x}) = P\exp\{\int dtA_o(\vec{x},t)\}. \tag{7}$$

The only gauge invariance at $t = 0$ and $t = T$ is the change of the gauge fixing condition (3) which appears as a periodic gauge transformation in the path integral. The Polyakov loop transforms as a matter field in the adjoint representation $\Omega(\vec{x}) \to g^+(\vec{x})\Omega(\vec{x})g(\vec{x})$ since

$$\int D[h']\langle A^r_{fin.}|\exp\{-it\hat{H}\}\hat{h}'|A^r_{init.}\rangle$$
$$= \int D[h']\langle A^r_{fin.}|\hat{g}^+\exp\{-it\hat{H}\}\hat{g}\hat{h}'\hat{g}^+\hat{g}|A^r_{init.}\rangle. \tag{8}$$

In particular the eigenvalues of the Polyakov loop are gauge invariant and represent the invariant degrees of freedom formed by boundary conditions.

The Polyakov loop can be considered as an order parameter for global (space independent) symmetries of the path integral. The symmetry group is $G \times C$, the direct product of the gauge group and its center. The latter factor corresponds to space-independent time-dependent gauge transformations which are periodic in time up to a center element (We tacitly assume that only fields in the adjoint representation occur in the path integral. Other fields such as fermionic ones have already been integrated out. The appearance of a complicated gauge invariant effective action does not alter this kinematical argument.). We say that the invariance under G is broken by the "vacuum configuration" of (5) if the Polyakov loop in the vacuum Ω_{vac} is not completely degenerate, i.e. has at least two different eigenvalues. Invariance under C is broken if $tr\Omega_{vac} \neq 0$.

We have to be careful in saying that the invariance with respect to global gauge transformations is broken. In fact Elitzur's theorem guarantees the absence of dynamical breakdown of local gauge invariance. Consequently we have to use a global order parameter to discuss breakdown of global gauge invariance. The relation between such nonlocal operators and local ones which describe measurements in quantum physics is not clear. The lesson of the study of spontaneously broken gauge theories in a given gauge is that as a result of this symmetry breaking the spectrum of the theory has massive gauge vector particles.

We shall see that as long as the path integral (5) can be approximated with a weak coupling expansion there is no such signature of gauge symmetry breaking caused by the Polyakov line. What is left as an operative definition for symmetry breaking is the presence of topologically stable (chromo)magnetic monopoles. In fact it has been noted that the eigenvalues of the Polyakov loop approach to a direction independent spectrum and the covariant derivatives of the Polyakov loop vanish, $\vec{D}\Omega(\vec{x}) \to 0$ for $\vec{x} \to \infty$, on finite action configurations.[2] As in the case of a Higgs system the first property leads to topologically stable quasiparticles and the second one relates the winding number of the Polyakov line to the magnetic charge of the Yang Mills field.

The argument of this section showed that the local gauge symmetry of the path integral (5) is reduced by imposing gauge invariant boundary conditions. This reduction of local gauge symmetry brings new physical degrees of freedom: static (chromo)electric charge densities. In fact nontrivial Polyakov loop and nonvanishing A_0 in the vacuum configuration can be interpreted as nonzero electric charge in Gauss's law.

III. THERMAL FLUCTUATIONS

Our discussion has been limited to kinematical considerations so far. Now we turn to the question of under which circumstance will the dynamics select a symmetry breaking realization.

The answer depends on the physical states prescribed at the boundary and the time elapsed T. For large T the boundary condition should be irrelevant (the true vacuum dominates the path integral) and there should be no symmetry breaking. In order to simplify the discussion we average over physical states at the boundary and consider $Z = \mathrm{tr}\hat{\pi} \exp\{-\beta\hat{H}\}$. A Wick rotation was carried out to give physical meaning to this average: it is the partition function at temperature $1/\beta$. The domain of integration consists of periodic configurations with period length β.

The effective potential for A_0 has been calculated in finite temperature SU(2) Yang Mills theory in two loop order.[3] Using the parametrization $A_0 = v\sigma^z/2\beta i$, the effective potential has the form

$$V_{eff}(v) = \beta^{-4}\{\frac{\pi^2}{3} v^2 - \frac{g^2(T)}{6} \pi v - 0.015622\, g^2(T) v \log v^2\},$$

$$g^2(T) = \frac{24\pi^2}{11 \log T^2/\Lambda^2} \qquad (9)$$

for small positive v. Consequently the gauge symmetry is broken to $SU(2) \to U(1)$ by thermal fluctuations. This perturbative result is reliable for sufficiently high temperature (asymptotic freedom) and small three volume (controlled infrared contributions).

Although the Polyakov loop plays a role analogous to an adjoint Higgs field in this construction, a novel feature is that no Higgs mechanism occurs. To see this consider the theory in physical "unitary" gauge where A_o is time independent and diagonal:

$$S = -\frac{1}{2}\int dtd^3x\ \mathrm{tr}\{F_{ij}^2 + 2(\partial_o A_i - D_i A_o)^2\} + H(A_o)$$

$$= \int dtd^3x\ \mathrm{tr}\{\frac{1}{2}F_{\mu\nu}^2 + [\Phi,\vec{A}]^2 + 2F_{oi}[\Phi,A_i]\} + H(A_o) \qquad (10)$$

where $H(A_o(\vec{x}))$ represents the contribution of the local Fadeev-Popov determinant resulting from the diagonalization and an external source coupled to A_o in order to generate $\langle A_o\rangle \neq 0$ in a consistent way.[4)] The variable A_o was shifted $A_o = A_o^1 + \Phi$ in the second equation to ensure $\langle A_o^1\rangle = 0$. Assuming that the system is perturbative we can extract qualitative results about the spectrum from the free dispersion relations $\omega + Q\Phi = +|\vec{p}|$ where Q is the electric charge of the gauge fields with respect to the unbroken U(1) factors. This describes massless particles in external potential. Naturally infrared modes may alter this picture in the thermodynamical limit leading to screening.

IV. MONTE CARLO SIMULATION OF SU(3) GAUGE THEORY

Perturbation expansion indicates that gauge invariance is broken dynamically at sufficiently high temperature. In order to see how this symmetry becomes restored at low temperature (and to study the case where the space extent can be much larger than the time one) Monte Carlo simulation was carried out for SU(3) lattice gauge theory.

The distribution of the phase in the eigenvalues of the Polyakov loops was measured during the calculation. Removing the Haar measure contribution $\rho(\theta_1,\theta_2,\theta_3) = \sin^2(\theta_1-\theta_2/2)\sin(\theta_1-\theta_3/2)\sin(\theta_2-\theta_3/2)$ from the distribution, we obtain $\exp\{-\beta V_{eff}(\Omega(\vec{n}))\}$ for a single site as a function of the phases of the eigenvalues.

The distribution of the eigenvalues has invariance under multiplication by the cubic roots of unity if the three volume is finite. We were not concerned with the breakdown of the Z_3 invariance, so in order to simplify the distributions and to gain statistics we made a Z_3 rotation: the Polyakov line of each configuration was multiplied by a constant space independent center element which maximized Re $\sum_n \mathrm{tr}\Omega(n)$.

This calculation gives the distribution of the local order parameter. It cannot be used to detect the breakdown of global gauge invariance directly. But our definition for this transition is the appearance of topologically stable quasiparticles. Since the eigenvalues of the Polyakov loops are close to their asymptotic values except in a finite region in space, the appearance of a peak at a noncenter group element in the the single site distribution signals the topological stability of Wu-Yang magnetic monopoles.

In non-gauge theories a nontrivial peak in the local distributions may be washed out by summation (more precisely convolution) over space points. This is not necessarily the case for the eigenvalues of the Polyakov loops because the effective action for $\Omega(n)$ is gauge invariant and the directional degrees of freedom of $\Omega(n)$ which are mainly responsible for the disappearance of the peak can be gauged away.

The distribution of the phase of the eigenvalues of the Polyakov loops is shown in Fig.1 for two temperatures, slightly below and above of the deconfinement point. The sharp peaks are the result of specially poor statistics in the region of the SU(3) group where the Haar measure is small: these are unlikely events but their contribution which is 1/Haar measure is large. There is a clear qualitative change in the distribution when deconfinement occurs: the flat effective potential develops minima at nonsymmetric (= not cubic root of unity) values. Since no other critical temperature than the deconfinement one was found in finite temperature studies of lattice gauge theories, we interpret this result to imply that the restoration of the symmetry with respect to G and C happens at the same point.

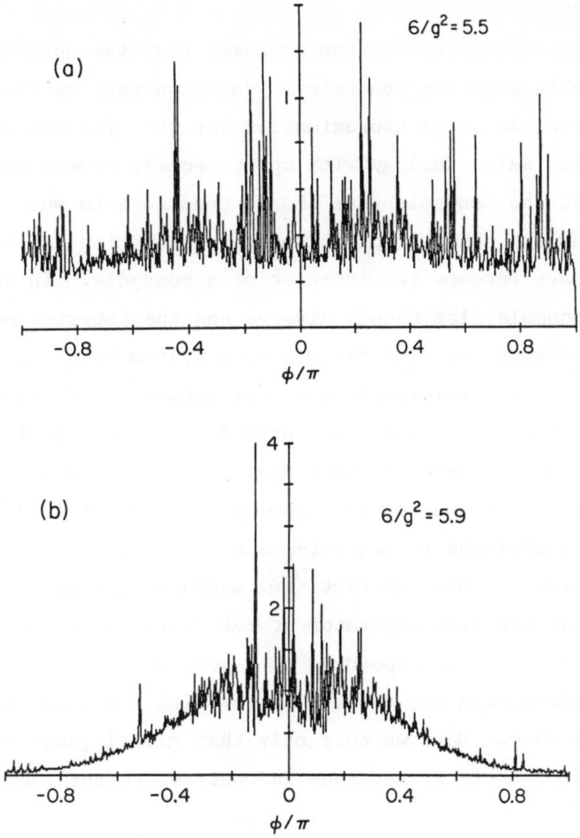

Fig. 1 Distribution of the phases of the eigenvalues of Ω after removing the Haar measure contribution. The calculation was done on $8^3 \times 4$ lattice. The values of the coupling constant are (a) $6/g^2 = 5.5$, (b) $6/g^2 = 5.9$ (the critical value is $6/g^2_{dec} \approx 5.7$).

There are two possible breaking patterns for G = SU(3): SU(3) → SU(2) × U(1) or SU(3) → U(1) × U(1). The former is excluded because its vacuum breaks charge (complex) conjugation symmetry as well. The latter case is supported by other results namely that none of the eigenvalues of the Polyakov loop is degenerate in the high temperature phase.[1] Thus deconfinement consists of losing invariance under non-Abelian directions of the gauge group.

V. OUTLOOK

Our Monte Carlo simulation suggests that the deconfined phase supports stable magnetic monopoles. Unfortunately we know no self consistent semiclassical approximation for the dynamics of this monopole gas. Naive analogy with spontaneously broken gauge theories suggests that the monopole mass should tend to zero when we approach the deconfinement point. This symmetry restoration can be imagined as the whole space becomes the interior of a monopole. In fact at this point the monopole size should diverge and the interior of the monopole is symmetric. The two monopole system has attractive interaction in the semiclassical approximation.[5)] It is natural to expect that this attractive force drives the system to form a monopole condensate when the monopole mass vanishes. This monopole condensation alone gives account of confinement as an electric Meissner effect.[6)] In addition it generates a nonzero chiral condensate in the quenched approximation. In fact $\langle\bar{\psi}\psi\rangle$ measures the level density of eigenvalues of the Dirac operator at zero eigenvalue and there is a localised zero mode corresponding to a monopole.

The calculation for gauge theories with dynamical fermions has not yet been done. Here we note only that global gauge symmetry is not broken explicitly by fermions in contrast to the case of the center.

Perturbation expansion predicts no symmetry breaking for periodic boundary condition in Minkowski space-time.[3)] So it seems that the spacelike nature of the compact direction is important in the symmetry breaking dynamics. This leads to the question whether one can have an analogous construction for boundary conditions in real space, e.g. in the case of the bag model.

Another interesting aspect of this symmetry breaking that it may be operating in higher dimensional unified theories as well. In this case it provides a new economical mechanism to reduce the symmetry of the models.

REFERENCES

1) Polonyi, J. and Wyld, H.W., Gauge Symmetry and Compactification, Urbana Preprint, ILL-TH-85-23.

2) Gross, D., Pisarski, R., Yaffe, L., "QCD and Instantons at Finite Temperature", Rev. Mod. Phys. 53, 43 (1981).

3) Anishetty, R., "Chemical potential for SU(N)-Infrared Problem", J. of Phys. G10, 423 (1984); "Colour Singlet Ensemble", G10, 439 (1984).

4) I thank Michael Peskin for a conversation on this point.

5) Coleman, S., The Magnetic Monopole Fifty Years Later, Lecture Notes of the 1981 International School of Subnuclear Physics, "Ettore Majorana".

6) Mandelstam, S., "Vortices and Quark Confinement in Non-Abelian Gauge Theories", Phys. Lett. 53B, 476 (1975);
't Hooft, G., "The Topological Mechanism for Permanent Quark Confinement in a Non-Abelian Gauge Theory", Phys. Scr. 25, 113 (1982).

The Approach to the Continuum Limit of Lattice QCD from Gluon Thermodynamics[1]

S. MEYER[2]
Universität Kaiserslautern, Fachbereich Physik
D–6750 Kaiserslautern, West Germany

May 15, 1985

ABSTRACT. The critical temperature of the deconfining phase transition and the free energy of a static quark–antiquark pair at finite temperature are investigated by means of Monte Carlo methods. The renormalization group β–function for lattice quantum chromodynamics with Wilson's action is calculated from gluon thermodynamics. We find a large dip in the β–function near $6/g^2 = 6$ which indicates that the onset of the asymptotic scaling regime is at much weaker coupling than used to be thought.

INTRODUCTION

In this talk I shall report some recent work in finite temperature lattice quantum chromodynamics [**1**] done in collaboration with A. D. Kennedy, J. Kuti and B. J. Pendleton [**2,3,4**].

It is of considerable interest to understand how the continuum limit of lattice QCD is approached. It is only in the asymptotic scaling regime of the theory that there is a unique prescription, given by the renormalization group equation with the universal two-loop β–function. At stronger coupling, where most present Monte Carlo simulations have been performed, large contributions

[1] Invited talk presented at the international conference on *"Advances in Lattice Gauge Theory,"* Supercomputer Computations Research Institute, Florida State University, Tallahassee, Florida (April 10–13, 1985).
[2] Work supported by Deutsche Forschungsgemeinschaft and NATO (grant No. RG.85/0136).

from irrelevant lattice operators are quite conceivable. This lack of asymptotic scaling obviously has serious consequences: If the onset of perturbative scaling is delayed by ≈ 0.6 in $6/g^2$, then it becomes necessary to refine the lattice spacing by a factor of two in order to undertake reliable non-perturbative hadron calculations with Wilson's lattice action.

Recently three methods of probing the scaling regime in the gluon sector of lattice QCD have been used [**5**]: block spin MCRG methods, studies of the behaviour of Wilson loop ratios under renormalization group transformations, and the determination of physical quantities like the transition temperature T_c or the string tension σ.

We have used two methods of studying gluon thermodynamics to measure the β-function. The first method uses our accurate measurement of the temperature of the deconfining phase transition [**2,3**] to probe scaling in the range $5.1 < 6/g^2 < 6.1$. Our second method is a renormalization group analysis of finite-temperature Green's functions, which we use to probe scaling up to $6/g^2 = 7.2$ [**4,7**].

THE β-FUNCTION FROM THE DECONFINEMENT TEMPERATURE

It is generally recognized that a phase transition occurs between the low temperature confining phase and the deconfined hot gluon plasma phase in finite temperature quantum chromodynamics [**1**]. At the transition temperature T_c the confining force between a static $Q\bar{Q}$-pair suddenly disappears. Locating T_c is particularly easy because quenched QCD undergoes a first order phase transition where rounding effects are small. At finite temperature quenched lattice QCD with the Wilson action has an additional global $Z(3)$ symmetry [**1**], which is not shared by the gauge invariant order parameter, the Polyakov loop

$$P(\mathbf{x}) = \mathrm{Tr}\left(\prod_{t=1}^{n_t} U_t(\mathbf{x},t)\right). \tag{1}$$

This order parameter satisfies

$$\langle P(\mathbf{x})\rangle = e^{-F_Q/T}, \tag{2}$$

where F_Q is the free energy of an isolated static quark. Therefore confinement occurs when $\langle P\rangle = 0$ (below T_c), and quark liberation occurs with $\langle P\rangle \neq 0$ (above T_c) where the $Z(3)$ symmetry is spontaneously broken.

The deconfining transition temperature T_c is a measurable physical quantity as the continuum limit is approached. In this limit, therefore, T_c is renormalization group invariant; that is $\frac{d}{da}T_c(g(a)) = 0$ (a is the lattice cutoff). This leads to the form $T_c = \tau(g)/a$ where the dimensionless function $\tau(g)$ is determined from the differential equation

$$\frac{d\tau}{dg} = -\frac{\tau(g)}{\beta(g)}. \tag{3}$$

In Eq. 3 $\beta(g)$ is the renormalization group β–function on the lattice,

$$\beta(g) = -a\frac{\partial g}{\partial a}. \tag{4}$$

In the vicinity of the fixed point $g = 0$, the lattice β-function can be calculated perturbatively in the two-loop approximation,

$$-\beta(g) = b_0 g^3 + b_1 g^5 + \cdots \tag{5}$$

with $b_0 = 11/16\pi^2$ and $b_1 = 102/(16\pi^2)^2$. The higher order loops are scheme dependent and as yet unknown for lattice regularization.

From Eq. 3 T_c depends on the lattice coupling g as

$$T_c = \frac{\text{const}}{a} e^{-\int_0^g \frac{dg'}{\beta(g')}}. \tag{6}$$

The constant in Eq. 6 has to be determined from non-perturbative calculations of T_c in the scaling regime.

We use a Monte Carlo measurement of the function $\tau(g)$ to determine the β-function. The results are to be compared with the two-loop form of Eq. 5 to verify asymptotic scaling.

Recall that finite temperature QCD is realized on the lattice with periodic boundary conditions in the Euclidean time direction with a periodicity of n_t

links. The system is then at a finite temperature $T = (n_t a)^{-1}$, and for the thermodynamic limit the spatial volume $V = (n_s a)^3$ should be taken to infinity. In our Monte Carlo calculation we measure the average value of the Polyakov loop $\langle P \rangle$ for several lattice sizes and for different temperatures $T = (n_t a(g))^{-1}$. As we increase the number of temporal links it becomes more difficult to distinguish the value of the order parameter P from the noise in the system. This is illustrated in Fig. 1 for a system with spatial volume V of 17^3 and $n_t = 10$. In the left part the Polyakov loop values cluster around the origin, which is indicative of the confined phase. In the middle part the distribution found has moved into one of the $Z(3)$ sectors of the spontaneously broken phase. This is confirmed by the distribution shown in the right part for a slightly higher temperature. Our results **[2,3]** for $\tau(g)$ are summarized in Table. 1, where we include the measurement for $\tau^{-1} = 3$ reported elsewhere **[6]**.

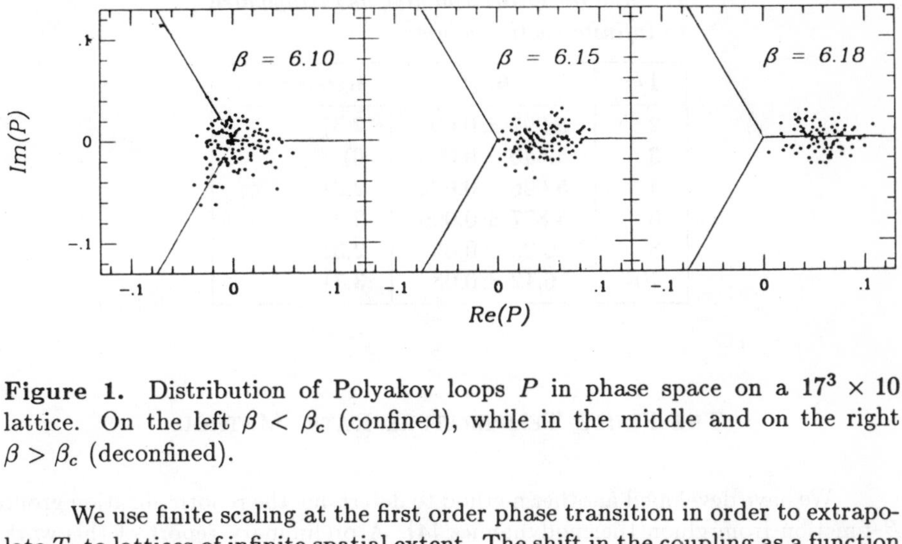

Figure 1. Distribution of Polyakov loops P in phase space on a $17^3 \times 10$ lattice. On the left $\beta < \beta_c$ (confined), while in the middle and on the right $\beta > \beta_c$ (deconfined).

We use finite scaling at the first order phase transition in order to extrapolate T_c to lattices of infinite spatial extent. The shift in the coupling as a function of the spatial volume V is expected to scale as $1/V$. This is confirmed by recent calculations **[7]** where it is found that $\beta_c = 5.692(2)$ for $n_t = 4$, $V = 13^3$, and $\beta_c = 5.872(3)$ for $n_t = 6$, $V = 17^3$. These values are in complete agreement with the earlier finite size extrapolations **[2,3]**.

In order to estimate the β–function, we determine the shift Δg in the coupling constant required to keep the system at T_c as we change n_t by a factor λ. This scaling requires that the lattice spacing $a \to a/\lambda$, as the physical temperature $T = n_t a$ must stay fixed, and thus $\tau \to \tau/\lambda$ too. Integrating Eq. 3 over this scale range we obtain

$$\int_{g}^{g+\Delta g} \frac{dg'}{\beta(g')} = -\int_{\tau}^{\lambda^{-1}\cdot\tau} \frac{d\tau'}{\tau'} = \ln \lambda, \tag{7}$$

which we compare with the perturbative prediction. As the continuum is approached ($a \to 0$) $g(a)$ should vary according to the continuum β–function, which is independent of T because there are no temperature-dependent divergences in the theory.

TABLE I. *The measured relation between τ^{-1} and the lattice coupling as extrapolated to infinite spatial volume.*

$1/\tau$	$6/g^2$	Reference
2	5.097 ± 0.001	2,3)
3	5.540 ± 0.005	6)
4	5.696 ± 0.004	2,3)
6	5.877 ± 0.006	2,3)
8	6.00 ± 0.02	2,3)
10	6.12 ± 0.03	3,7)

THE FINITE TEMPERATURE RATIO METHOD

We have developed another method to determine the renormalization group β–function from gluon thermodynamics [**4**]. A preliminary report of this work was presented earlier [**8**].

The free energy $F_{Q\bar{Q}}(\mathbf{x}, a, g, T)$ of a static $Q\bar{Q}$-pair at temperature T is the expectation value of the Polyakov loop correlation function,

$$\langle P_Q(0) P_{\bar{Q}}(\mathbf{x}) \rangle = e^{-\frac{1}{T} F_{Q\bar{Q}}(\mathbf{x},a,g,T)}. \tag{8}$$

The mean Polyakov loop $\langle P_Q(0) \rangle$ measures the free energy of the static quark located at the origin [**9**].

In the confined phase the Polyakov loop correlation function decays as

$$\langle P_Q(0)P_{\bar{Q}}(r)\rangle \sim e^{-\sigma(T)r}, \tag{9}$$

as $r \to \infty$, with $\sigma(T)$ being the string tension at finite temperature. In the broken symmetry phase it is expected that Debye screening takes place, with a screening length κ^{-1}:

$$\langle P_Q(0)P_{\bar{Q}}(r)\rangle \sim |\langle P\rangle|^2 exp\left(-\beta c\,\frac{e^{-\kappa r}}{r}\right). \tag{10}$$

For the renormalization group analysis it is simplest to subtract from $F_{Q\bar{Q}}$ the free energy of the $Q\bar{Q}$-pair at infinite separation, because it is only this subtracted free energy which is a physical observable. In the deconfined phase, therefore, we define the subtracted free energy $\Phi_{Q\bar{Q}}$ to be

$$\Phi_{Q\bar{Q}}(\mathbf{x}, a, g, T) = F_{Q\bar{Q}}(\mathbf{x}, a, g, T) - F_{Q\bar{Q}}(\infty, a, g, T). \tag{11}$$

Since $\Phi_{Q\bar{Q}}$ is a physical quantity, and $\Phi_{Q\bar{Q}}$, $\mathbf{x}$, and T are measured in physical units, $\Phi_{Q\bar{Q}}$ is renormalization group invariant,

$$a\frac{d}{da}\Phi_{Q\bar{Q}}(\mathbf{x}, a, g, T) = 0, \tag{12}$$

for fixed $\mathbf{x}$ and T. Integrating Eq. 12 for a finite scale change λ, and using dimensional analysis to recast it into "active" form, we get the matching condition:

$$\Phi_{Q\bar{Q}}(\mathbf{x}, a, g(a), T) = \lambda\Phi_{Q\bar{Q}}(\lambda\mathbf{x}, a, g\left(\frac{a}{\lambda}\right), \frac{T}{\lambda}). \tag{13}$$

In our Monte Carlo analysis we use Eq. 13 to calculate the renormalization group β–function from gluon thermodynamics. In Fig. 2 the Polyakov loop correlation function is shown as a function of the $Q\bar{Q}$ separation for different values of β. The measurements shown were obtained on lattices of size $15^3 \times 8$. The matching was done on pairs of lattices with 4 and 8 temporal links respectively, using small separations of the $Q\bar{Q}$–pair.

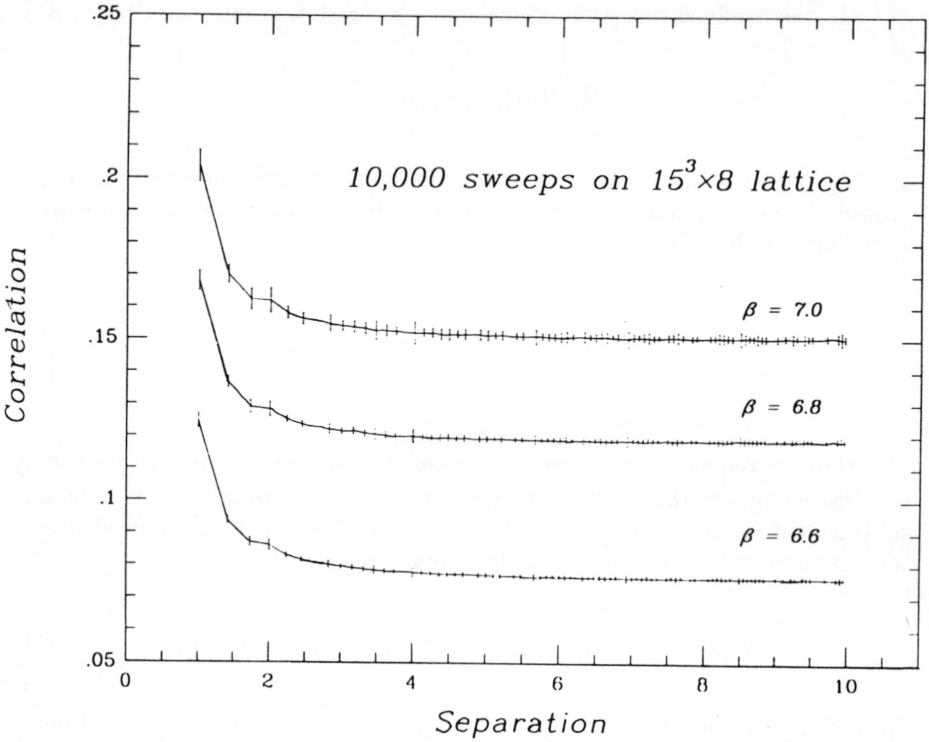

Figure 2. Polyakov loop correlation function as a function of the separation between the Q and $\bar{Q}$ for several values of β.

The results of this matching are summarized in Table. 2. We find a lower value for the shift $\Delta\frac{6}{g^2}$ at $6/g^2 = 6.8$ compared with our previous results on an $11^3 \times 8$ lattice [**4**], indicating that there were significant fluctuations in the system. For further details we refer the reader to Ref. 4.

Our computations were performed on the CYBER–205 at the University of Karlsruhe. Our highly vectorized program updates a link in 22 μs on a two-pipe 1 MWord machine. Details of our program are described elsewhere [**10**].

Discussion

The onset of asymptotic scaling is at much weaker coupling than was sug-

TABLE II. *The measured shift* $\Delta \frac{6}{g^2}$ *with two values of c measured on* $15^3 \times 8$ *and* $17^3 \times 8$ *(for* $6/g^2 = 7.2$*) relative to* $13^3 \times 4$ *lattices.*

$6/g^2$	$\Delta 6/G * G$
6.6	0.54 ± 0.03
6.8	0.50 ± 0.04
7.0	0.60 ± 0.03
7.2	0.58 ± 0.04

gested by early optimistic estimates. We find from gluon thermodynamics a large dip in the renormalization group β–function between $5.5 < 6/g^2 < 6.3$, which must be caused by non-perturbative contributions. Our results are in agreement with some recent determinations of the β–function from renormalization group studies at zero temperature [**11**].A remarkable consistent picture of the beta function is emerging from these calculations. Fig. 3 summarizes the available results in the coupling constant range $5.1 < 6/g^2 < 7.2$ It is important to explore the influence of dynamical quarks on this feature of the beta function. We can not yet decide whether there is universality of the nonperturbative beta function on the basis of the accuracy of data obtained so far.

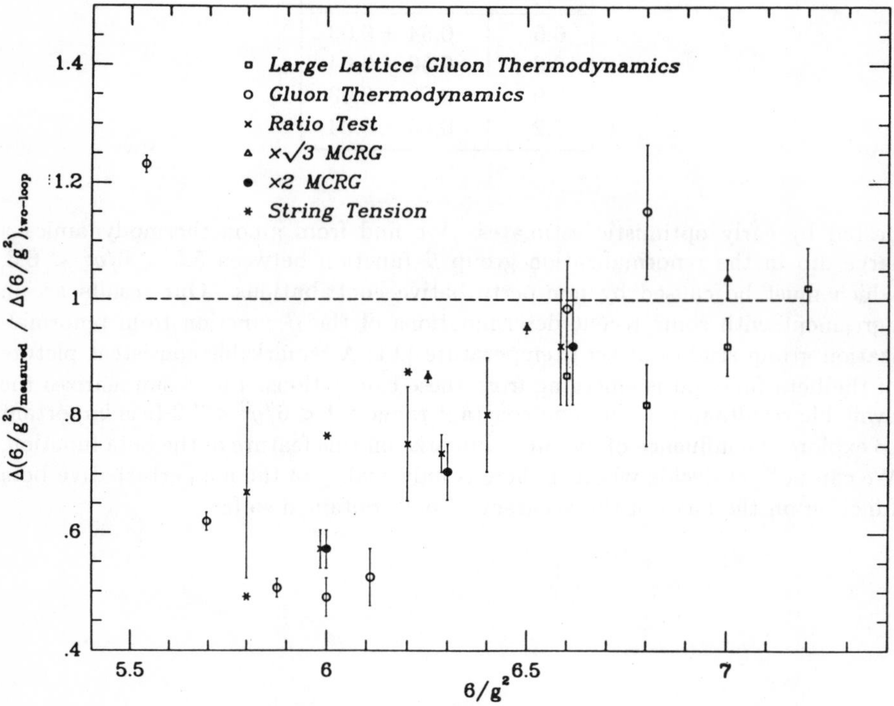

Figure 3. Summary of available results for the β–function.

ACKNOWLEDGEMENTS

I thank my collaborators Tony Kennedy, Julius Kuti, and Brian Pendleton with whom the work described above was carried out I also thank Dennis Duke for his kind invitation to this meeting, and Bob Sugar for inviting me to ITP in Santa Barbara. We greatly appreciate the assistance provided by the director and staff of the Rechenzentrum der Universität Karlsruhe.

This research was supported by the Deutsche Forschungsgemeinschaft, with additional funding from NATO and Control Data Corporation. This research was supported in part by the National Science Foundation under Grant No. PHY82–17853, supplemented by funds from the National Aeronautics and Space Administration, and by DOE under Grant No. DEAT0381ER40029. I also thank ITP and UCSD for their hospitality.

REFERENCES

1. A. M. Polyakov, Phys. Lett. **72B** (1978), p. 477.

   L. Susskind, Phys. Rev. **D20** (1979), p. 2610.

2. B. J. Pendleton, in "Proceedings of the Argonne National Laboratory Workshop on Lattice Gauge Theories", Argonne (April 5–7, 1984), p. 110.

3. A. D. Kennedy, J. Kuti, S. Meyer, and B. J. Pendleton, Phys. Rev. Lett. 54 (1985), p. 87.

4. A. D. Kennedy, J. Kuti, S. Meyer, and B. J. Pendleton, Phys. Lett. In press.

5. See: D. Barkai; R. Gupta; P. Hasenfratz; J. Kuti; A. Patel. Contributions at this conference.

6. B. Svetitsky and F. Fucito, Phys. Lett. 131B (1983), p. 165.

7. A. D. Kennedy, J. Kuti, S. Meyer, and B. J. Pendleton. In preparation.

8. S. Meyer. Contribution to XXIIth International Conference on High Energy Physics, Leipzig (July 19–25, 1984).

9. J. Kuti, J. Polónyi, and K. Szlachányi, Phys. Lett. 98B (1981), p. 199.

   L. D. McLerran and B. Svetitsky, Phys. Lett. 98B (1981), p. 195.

10. A. D. Kennedy, J. Kuti, S. Meyer, and B. J. Pendleton, J. Comp. Phys. In press.

11. A. Hasenfratz, P. Hasenfratz, U. Heller, and F. Karsch, Phys. Lett. 143B (1984), p. 193.

R. Gupta, G. Guralnik, A. Patel, T. Warnock, and C. Zemach, Phys. Rev. Lett. 53 (1984), p. 1721.

K. C. Bowler *et al.*, CERN preprint TH. 3952.

K. C. Bowler *et al.*, Edinburgh preprint 85/335.

SECOND GENERATION GLUEBALL MASS CALCULATIONS IN THE STRONG COUPLING REGION

Karsten Decker*

II. Institut für Theoretische Physik
der Universität Hamburg
2000 Hamburg 50, Luruper Chaussee 149
GERMANY

ABSTRACT

An algorithm which on principle allows to obtain an arbitrary number of terms in the systematical strong coupling expansion of the mass gap of the three-dimensional Z_2 lattice gauge theory is introduced. The importance of this algorithm for extended analytical calculations of the low-lying part of the glueball mass spectrum of four-dimensional pure Yang-Mills theories is pointed out. Finally, utilizing duality, the critical temperature and the critical exponent ν' of the Ising model are estimated.

1. INTRODUCTION

It is the aim of this talk to call your attention again to the analytical calculation of the glueball mass spectrum in the framework of Euclidean lattice gauge theory without matter fields. To my opinion, this is motivated quite naturally by a critical assessment of the current status of glueball mass calculations.

*supported by Bundesministerium für Forschung und Technologie, Bonn

In the Euclidean frame work, strong coupling expansions have already been obtained for the mass gap (corresponding to the lowest state in the $J^P = 0^+$ sector) as well as for some low-lying excited states for various gauge groups and actions [1)–4)]. Whereas the estimates for the mass gap are consistent with the results from Monte Carlo calculations (for a review see ref. 5)) the estimates for the 2^+ glueball are not in such a good shape:

For gauge group SU(2), Monte Carlo simulations predict

$$\frac{m_{2^+}}{m_{0^+}} \sim 1.8 \tag{1}$$

while the estimate from strong coupling expansions is

$$\frac{m_{2^+}}{m_{0^+}} \sim 1.25 \tag{2}$$

A ratio close to one is supported by finite volume calculations [6)]. However, the latest Monte Carlo investigations seem to indicate a trend to smaller values, too [7)]. The situation for gauge group SU(3) is quite similar.

This is clearly not very satisfying. But how can one do better? Progress in Monte Carlo calculations towards more reliable results is believed to depend essentially on the development of new methods.

Here I would like to convince you that it might pay out to put some more effort on the *analytical* determination of the mass spectrum of the glueballs. Analytical methods, in particular strong coupling expansions, possibly in *combination* with Monte Carlo simulations might provide us with a more accurate determination of the mass spectrum than pure Monte Carlo methods can do.

A fundamental problem of strong coupling calculations is that the physical region presumably has no overlap with the region of convergence of the strong coupling expansions. Although one may try to construct lattice actions such that the strong coupling region comes closer to the physical region, strong coupling expansions usually have to be supplemented by methods of analytic continuation in order to obtain predictions in the physical region. So far, however, the existing strong coupling series expansions are much too *short* to *reliably* apply extrapolation techniques such as Padé approximants, differential approximants, ..., possibly in combinations with series reexpansion techniques.

Unfortunately, the general beliefe is that longer series are *not* accessible by means of a *hand calculation*: the effort necessary to compute *one more* coefficient of a given series is roughly equivalent to the amount of work which was necessary to obtain *all* the known coefficients. Consequently, the natural solution is to *teach the computer* how to do the strong coupling expansion.

I would like to introduce to you a recently developed *computer algorithm* which on principle makes it possible to obtain *arbitrary* number of terms in the systematical strong coupling series expansion. This then will allow improved analytical estimates of glueball masses. In addition, our method can be used to study some aspects of the critical behavior of the 3d Ising model. The results of such an investigation will be reported in the second part of my talk.

2. THE ALGORITHM

2.1 The Algorithm For The Mass Gap Of The Three-Dimensional Z_2 Lattice Gauge Theory

This section serves as a description of the principles of the algorithm which has been developed yet.

We start as usual from a strong coupling *cluster expansion* of the connected correlation function Γ [1]:

$$< 0|\mathcal{O}_1(t)\mathcal{O}_2(0)|0 >_c \underset{t\to\infty}{\longrightarrow} const.\exp(-mt) \tag{3}$$

$\mathcal{O}_1$, $\mathcal{O}_2$ are local lattice operators which create states out of the vacuum $|0>$ which have *non-zero* matrix elements with a state of mass m and *zero momentum*. m denotes the mass gap.

Omitting all the technical details which can be found elsewhere [8] we restrict ourselves to a qualitative description in the following. This is sufficient for an intuitive understanding of our method.

The cluster expansion (3) may be represented by *graphs* on the lattice:

leading order.

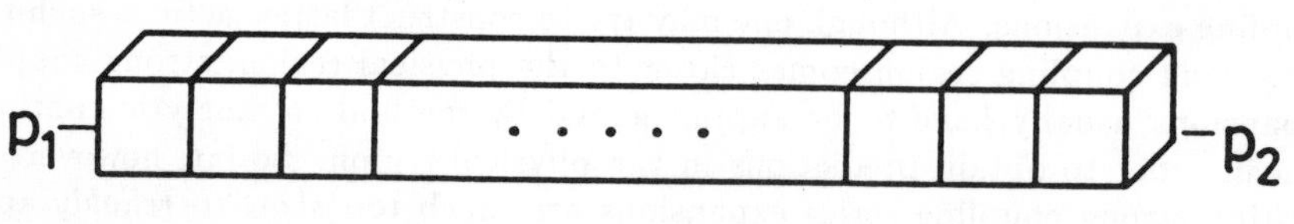

Fig. 1 The leading order contribution in the cluster expansion of Γ which is the 1-polymer cluster X_0. Time proceeds from left to right and p_1 and p_2 are supposed to be separated t lattice spacings in time direction. Tr $U_{p_{1,2}} \in \mathcal{O}_{1,2}$; each plaquette of X_0 carries the fundamental representation of the gauge group G.

geometrical corrections.

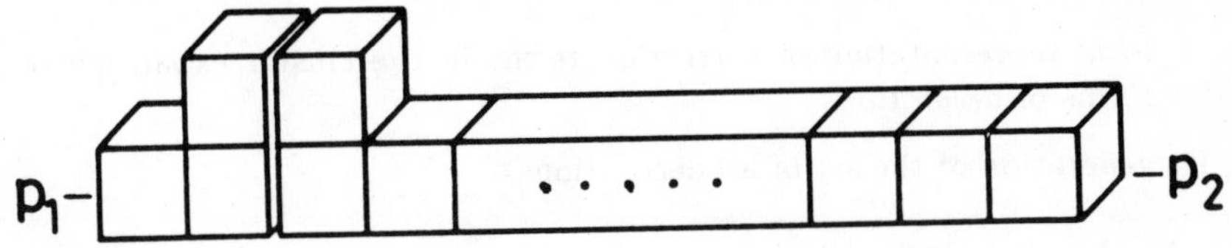

Fig. 2 An example for a higher order cluster (with 2 polymers) contributing to the cluster expansion of Γ.

All clusters which contribute to the strong coupling expansion of m can be obtained by *local modifications* of X_0. The plaquettes of the basic tube parts connecting the corrections carry the fundamental representation of G. The irreducible representations assigned to the plaquettes of the corrections are not a priori fixed but they have to fulfill certain sets of relations depending on the geometrical and topological properties of the corrections and on the properties of the gauge group G. For more details we refer to ref. 8).

pure X_0 corrections.

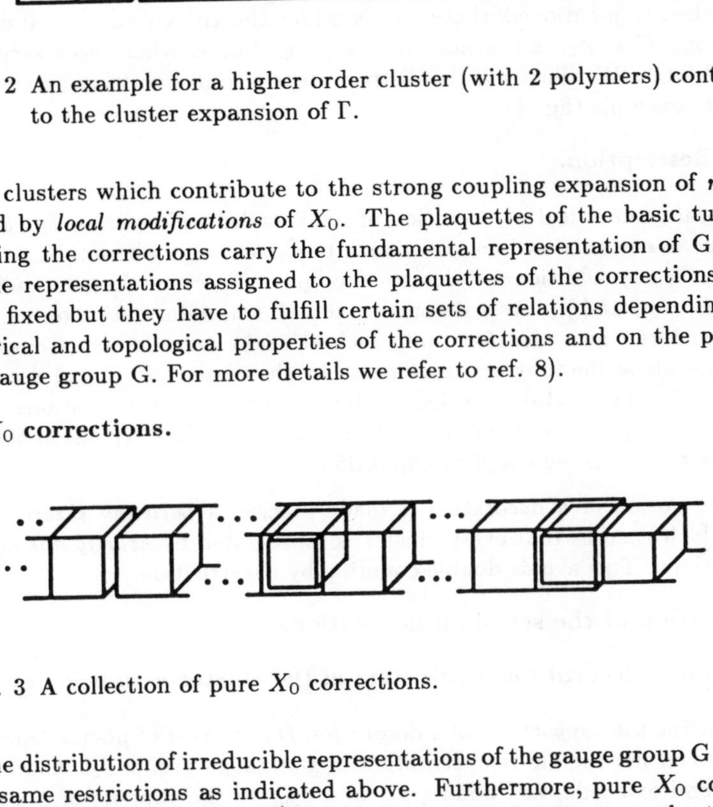

Fig. 3 A collection of pure X_0 corrections.

The distribution of irreducible representations of the gauge group G is subject to the same restrictions as indicated above. Furthermore, pure X_0 corrections can be studied in *tube models* [4]. As their contribution to m can be taken into account in closed form, the main problem are the geometrical corrections to which we restrict the discussion in the following.

The basic feature of our algorithm is that it can generate and process *all* the necessary *graphs* representing those clusters in the strong coupling cluster expansion of Γ which contribute to the strong coupling expansion of m.

We decided to develop an algorithm for the *graphical expansion* of Γ because it is widely believed that such an algorithm is much more efficient than one which evaluates the strong coupling cluster expansion in a purely algebraic fashion on a finite lattice.

The major steps of the algorithm are

I. local representation of correction terms in the cluster expansion of Γ in terms of *decorations*

II. generation of the set of all decorations

As already mentioned above, we consider the three-dimensional case with gauge group $G = Z_2$. Extensions to the general case, where necessary, will be indicated later. We illustrate the distinct steps of our algorithm by considering a concrete example (fig. 4).

I. local description.

A local geometrical modification of X_0 is called a *decoration D*; each decoration has an entrance and an exit plaquette p_{in} and p_{out} where tube parts enter which may carry additional decorations or pure X_0 corrections but ultimately connect D to p_1 and p_2. As a decoration contains no information on its position on X_0, clusters which transform into each other by *translations* of one or more decorations along the tube are represented by the same collection of decorations. Consequently, the local description of clusters in terms of decorations amounts to a *class decomposition* of the set of clusters in the cluster expansion of Γ which contribute to the strong coupling expansion of m.

If the notion of a decoration is made precise, a *partition theorem* can be proven which ensures that every cluster is represented by *exactly one* collection of decorations. This avoids double-counting by construction.

II. generation of the set of all decorations.

II.1 mapping decorations to clusters of the cluster expansion of log Z.

We define the *support* $|D|$ of a decoration D as the set of plaquettes carrying the nontrivial irreducible representation of the gauge group Z_2. Closing D at entrance and exit side, i.e. adding p_{in} and p_{out} to the support of D results in the support $|\hat{C}_D|$ of a cluster $\hat{C}_D$ which contributes to the cluster expansion of $\log Z$. Conversely, starting, from some $\hat{C}_D$, the set of plaquettes of the support of $\hat{C}_D$ which may serve as p_{in} and p_{out} is uniquely fixed by the partition theorem. Consequently, the generation of all decorations is mapped to the cluster expansion of $\log Z$.

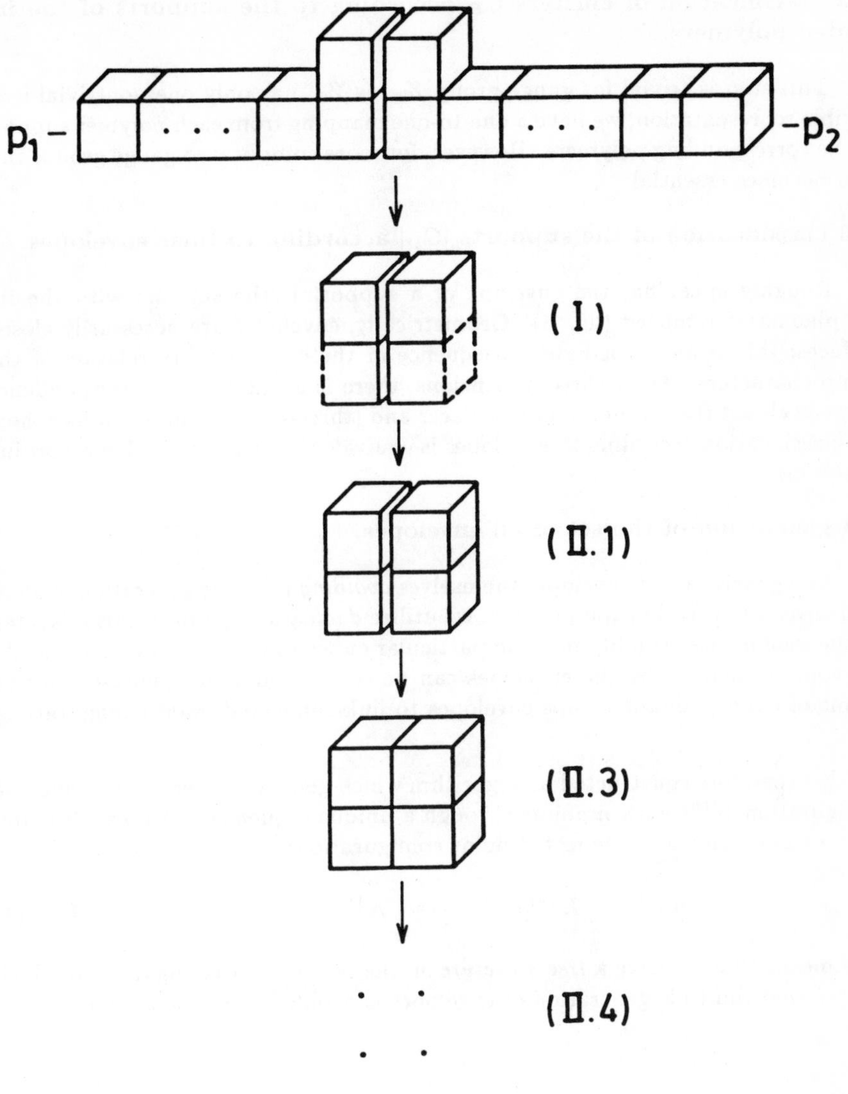

Fig. 4 The distinct steps of the algorithm illustrated with the help of the 2-polymer cluster of fig. 2. The numbers in parentheses refer to the steps of our algorithm as dicussed in the text.

II.2 classification of clusters $\hat{C}_D$ according to the supports of the included polymers.

This step is trivial for gauge group Z_2: as Z_2 has only one nontrivial irreducible representation, we have a one-to-one mapping from each polymer support to the corresponding polymers. However, for more complicated gauge groups this step becomes essential.

II.3 classification of the supports $|\hat{C}_D|$ according to their envelopes.

Roughly speaking, the envelope of a support is the support with the inner plaquettes removed (fig. 4). Geometrically, envelopes are necessarily closed surfaces; this is an immediate consequence of the orthogonality relation of the group characters. As in three dimensions, there is a one-to-one correspondence between closed (two-dimensional) surfaces and (three-dimensional) volumes, here the classification according to envelopes is equivalent to a classification according to volumes.

II.4 generation of the set of all envelopes.

As a generation of envelopes themselves *avoiding* multiple generation of identical envelopes is rather complicated, we utilize *duality* and perform this final step on the *dual lattice*. Duality maps in particular cubes to points and vice versa. As the volumes defined by the envelopes can be considered as plaquette-connected chains of cubes,[1] duality maps envelopes to link-connected point configurations K.

We therefore constructed an algorithm which generates every link-connected configuration $K^{(n)}$ with n points through a unique sequence of 1-point, 2-point, ..., $(n - 1)$ - point link-connected point configurations,

$$K^{(1)} \longmapsto K^{(2)} \longmapsto \cdots \longmapsto K^{(n-1)} \longmapsto K^{(n)} \tag{4}$$

This means that we have a *tree structure* on the set of point configurations which implies that multiple generation of envelopes is avoided by construction.

[1] to be precise, the chains are *plaquette-connected* only up to a certain order of the strong coupling expansion. However, this causes no limitations from a practical point of view because this order is already beyond the present and near future reach due to lack of computer power. Moreover, the algorithm may be easily extended to the general case.

2.2 Extension To The Case Of Four-Dimensional Nonabelian Lattice Gauge Theories

First of all we emphasize that the local description of correction terms, step I, is correct to *all orders* in the strong coupling expansion. Furthermore, for the definitions we do not make any reference to the geometrical properties of the clusters which eventually may not be fulfilled in *more* than three dimensions. Consequently, the local description *remains valid without any changes* to mass calculations in four-dimensional pure lattice gauge theories; only minor modifications are necessary to extend the validity to *arbitrary gauge groups.*

The generalization of step II to the case of arbitrary gauge groups also causes no fundamental difficulties. Generalizing this step to four dimensions is more complicated, because in four dimensions the one-to-one correspondence between envelopes and volumes which we made use of in II.3 is clearly lost. If this problem can be solved to *any order* of calculation by an additional ordering imposed on the set of volumes has not been proven yet.

I close this part of my talk with several comments:

1) Once the currently available algorithm has been generalized to cover also mass calculations in four dimensions, it easily can be extended to be also applicable to mass calculations of the *low-lying excited states* by simply implementing a book-keeping scheme for the relative (space-like) orientations of p_1 and p_2.

2) As it may already be clear from the above sketch of the whole algorithm, a subset can be used to derive high order series expansions for the free energy $\log Z$.

3) Efficient algorithms for the generation of point configurations may have some *direct* physical application in all lattice theories which deal with (additional) fields attached to the *sites* of the lattice, for example fermionic and scalar bosonic theories.

4) On the computer, the graphs are represented by various *lists* and *incidence tables.* From a purely technical point of view our algorithm consists mainly of list and table search and processing. Due to the rather complicated *logical structure* and due to the *data structure* the algorithm exhibits optimal performance on an array of scalar processors or on a machine which allows for multi-tasking.

My conclusion is that despite of the complications mentioned above a generalized algorithm will become available in the near future. This again will admit considerably extended strong coupling series expansions for the mass spectrum of the low-lying glueball states of four-dimensional pure Yang-Mills theories.

3. Application to the Three-Dimensional Ising Model

By duality, the three-dimensional Z_2 lattice gauge theory is related to the usual Ising model with nearest-neighbor interaction,

$$-H^{\star}/kT = \beta^{\star} \sum_{<ij>} s_i s_j, \qquad s_k = \pm 1 \tag{5}$$

The sum extends over all nearest-neighbor lattice points. In particular, the strong coupling or high temperature phase of the Z_2 lattice gauge theory is mapped to the low temperature phase of the Ising model and the mass gap of the gauge theory corresponds to the *inverse true correlation length* κ.

Using our algorithm, we computed the *next three* coefficients of the $O(u^5)$ low temperature series expansion of κ which had already been published in 1975 [9]. Here u is the usual low temperature variable

$$u := \exp(-4\beta^{\star}) \tag{6}$$

Relying on this extended series, we determine both the *critical exponent* ν' of the true correlation length and the *critical temperature* of the Ising model. While an estimate for ν' based on the series analysis of the series expansion of the second moment correlation length has already been given in ref. 9), estimates for ν' and the critical temperature determined from the analysis of the low temperature series expansion of κ have not yet been reported in the literature.

In lattice units (lattice spacing a) we obtain

$$\begin{aligned}\kappa a &= -2\log u + F(u)\\ &= -2\log u + u - \frac{15}{2}u^2 + \frac{7}{3}u^3 - \frac{87}{4}u^4 + \frac{251}{5}u^5\\ &\quad - \frac{331}{2}u^6 + \frac{8793}{7}u^7 - \frac{55927}{8}u^8 + O(u^9)\end{aligned} \tag{7}$$

We remind you that u is related to the *strong coupling* variable u_{gauge} of the Z_2 lattice gauge theory which is probably more familiar to you by

$$u = u_{gauge}^2 = [\beta + O(\beta^3)]^2 \tag{8}$$

$\beta = 2/g_0^2$ is the coupling parameter in the lattice action and g_0 is the bare coupling constant of the gauge theory.

To obtain the eight coefficients of $F(u)$ presented in (7), more than $1.5 \cdot 10^6$ configurations have been generated and investigated within approximately 45000 s of CPU-time on an IBM 3081-D. The corresponding numbers for the first five coefficients of $F(u)$ which have first been determined by Tarko and Fisher, ref. 9) are about 1800 configurations and roughly 20 s of CPU-time.

We begin the series analysis by a qualitative observation. As the expansion coefficients of $F(u)$ alternate in sign, the dominant singularity is located on the real *negative* u-axis. Stated differently, the ***true critical behavior*** is ***masked*** by an ***unphysical singularity*** which will complicate the quantitative analysis considerably.

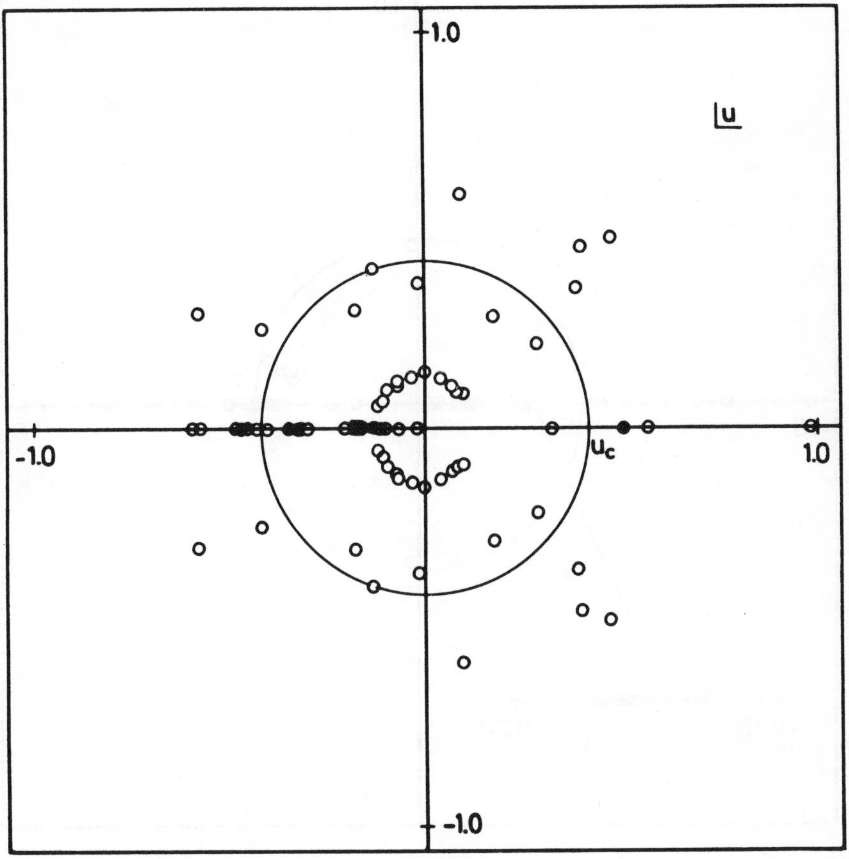

Fig. 5 The poles of all the Padé approximants to $F(u)$. Open circles represent poles which have been found in one approximant only. Poles indicated by a full circle, a full triangle or a full box have been obtained in two, three or four and more Padé approximants respectively.

The quantitative estimates rely on ratio methods [10], Shanks extrapolation [11] and Padé approximants. Where necessary, $F(u)$ is assumed to have pure power law singularities. Taking into account the results from all methods, we have indication for two *unphysical* singularities $u_1 \approx -0.13$ and $u_2 \approx -0.31$. The singularity structure as determined by the Padé analysis of $F(u)$ is displayed in fig. 5. Moreover, as can be concluded from fig. 5, the analysis of $F(u)$ admits no estimate of the location of the physical singularity u_c.

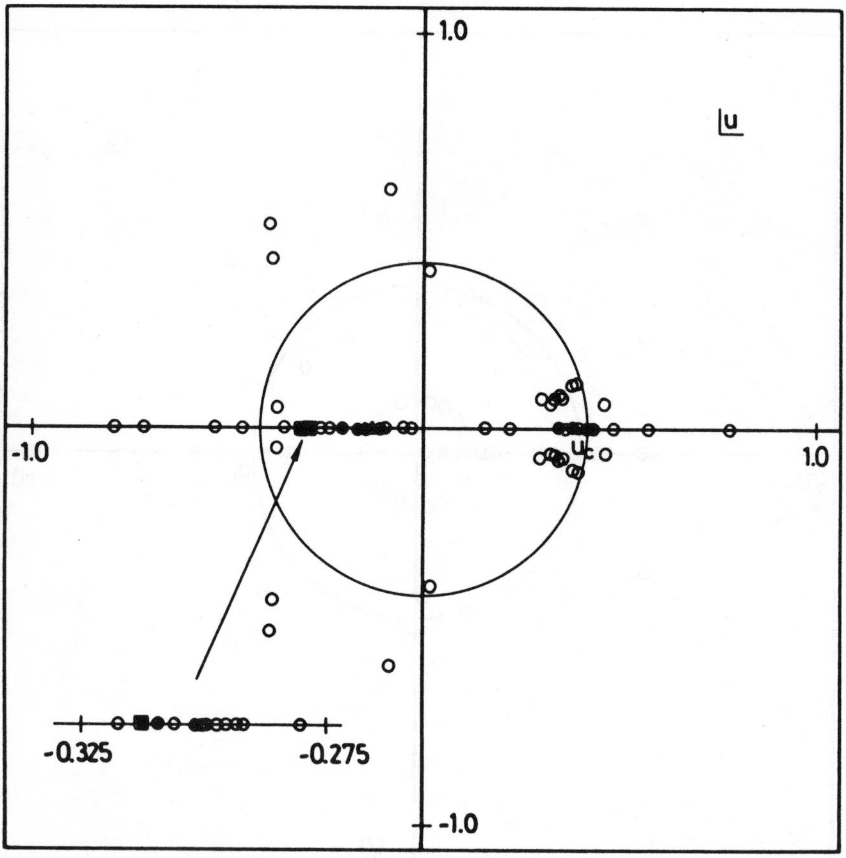

Fig. 6 The poles as determined by the Padé analysis of Λ_2'. Symbols as defined for fig. 5.

To obtain independent estimates for u_1 and u_2 and to get predictions for the location of u_c as well as estimates for the *critical exponent* ν' at u_c and at the unphysical singularities, we investigated the quantities [9),12)]

$$\Lambda_2' := \frac{1}{2[\cosh(\kappa a) - 1]} \equiv \frac{\exp(-\kappa a)}{[1 - \exp(-\kappa a)]^2} = (\xi/a)^2[1 + O((\xi/a)^{-2})] \qquad (9)$$

$$\Lambda_1' := \frac{1}{1 - \exp(-\kappa a)} = (\xi/a)[1 + O((\xi/a)^{-1})] \qquad (10)$$

where ξ is the true correlation length. The corresponding singularity structure is displayed in fig. 6 and fig. 7.

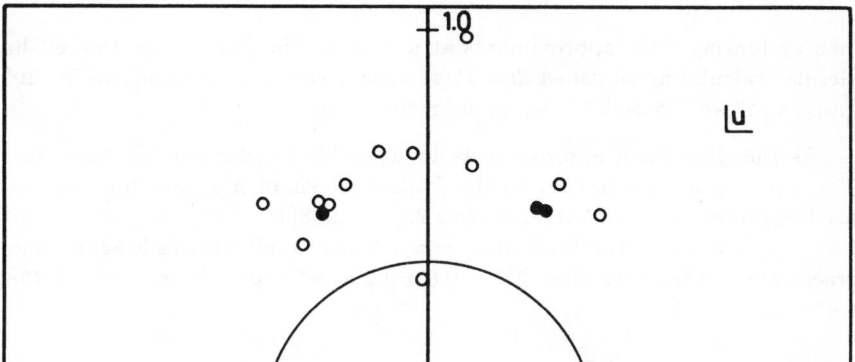

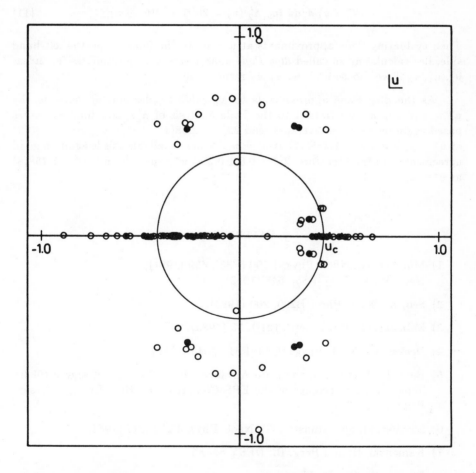

Fig. 7 The poles as determined by the Padé analysis of Λ_1'. Symbols as defined for fig. 5.

In addition to the methods mentioned above, we mapped away the dominant singularity by means of a *conformal transformation* and analyzed the transformed series. Concerning the location of the singularities u_1, u_2 and u_c, the final result based on *weighted averages* is [13] $u_1 = -0.1368(74)$, $u_2 = -0.296(11)$ and $u_c = 0.4079(58)$ in good agreement with the literature, $u_2 = -0.2860(2)$ [14] and $u_c = -0.412045(5)$ [15]. The new observation is the unphysical singularity u_1.

For the determination of ν' at $u^\star = u_1,\ u_2,\ u_c$ we assume that ξ has a pure power law singuality at $u^\star$ and that the amplitude function is analytic at $u = u^\star$. this implies for example for Λ_2':

$$[(u^\star - u)\ d/du\ log\ \Lambda_2'](u) = 2\nu'[1 + O(u^\star - u)] \tag{11}$$

Then evaluating Padé approximants at $u = u^\star$ to the function on the left-hand side, i.e. calculating so-called *dlog Padé approximants*, give estimates for $2\nu'$ at u_c, u_1, u_2 if we choose $u^\star = u_c, u_1, u_2$ respectively.

As the dlog Padé approximants to Λ_1' exhibit poles and/or zeros nearby $u^\star = u_1,\ u_2,\ u_c$, we restrict to the Padé analysis of Λ_2'. Our final estimates based again on weighted averages read $2\nu' = 0.008(1)$ at $u^\star = u_1$, $2\nu' = 0.80(2)$ at $u^\star = u_2$ and $2\nu' = 1.255(17)$ at $u^\star = u_c$. Where available,this is again in good agreement with the literature, $2\nu' = 0.8 \pm 0.2$ at $u^\star = u_2$ [16] and $2\nu' = 1.258(8)$ at $u^\star = u_c$ [17].

References

1) Münster, G., Nucl. Phys. B190 [FS3], 439 (1981);
Nucl. Phys. B205 [FS5], 648 (1982) (E).

2) Seo, K., Nucl. Phys. B209, 200 (1982).

3) Münster, G., Phys. Lett. 121B, 53 (1983).

4) Decker, K., Nucl. Phys. B240 [FS12], 543 (1984).

5) Berg, B., Lectures at the Nato Advanced Study Institute, Cargése 1983;
Teper, M., Proceedings of the EPS Conference on High Energy Physics, Brighton 1983.

6) Lüscher, M. and Münster, G., Nucl. Phys. B232, 445 (1984).

7) Kamenzki, H. and Berg, B., DESY 85-009.

8) Decker, K., Ph. D. Thesis.

9) Tarko, H. B. and Fisher, M. E., Phys. Rev. B11, 1217 (1975).

10) Gaunt, D. S. and Guttmann, A. J., in 'Phase Transitions and Critical Phenomena', Domb, C. and Green, M. S. (eds.), Vol.3 (New York, Academic Press, 1974).

11) Shanks, D., J. Math. and Phys. 34, 1 (1955);
Wynn, P., MTAC 10, 91 (1956).

12) Fisher, M. E. and Burford, R. J., Phys. Rev. 156, 583 (1967).

13) Decker, K., Nucl. Phys. B257 [FS14], 419 (1985).

14) Thompson, C. J., Guttmann, A. J. and Ninham, B. W., J. Phys. C2, 1889 (1969).

15) Adler, J., J. Phys. A16, 3585 (1983).

16) Itzykson, C., Pearson, R. B. and Zuber, J. B., Nucl. Phys. B220 [FS8], 415 (1983).

17) Pawley, G. S., Swendsen, R. H., Wallace, D. J. and Wilson, K. G., Phys. Rev. B29, 4030 (1984).

MCRG RESULTS FOR SU(2) AND SU(3) LATTICE GAUGE THEORIES

APOORVA PATEL[+]

Physics Department, University of California, La Jolla, CA 92093.

RAJAN GUPTA[*]

Physics Department, Northeastern University, Boston, MA 02115.

ABSTRACT

We present detailed results for the non-perturbative β-function along the Wilson axis for the SU(2) and SU(3) pure gauge theories, using the Monte Carlo Renormalization Group (MCRG) method and the $\sqrt{3}$ block transformation. We compare these results with the "β-functions" extracted from physical observables like the string tension, the deconfinement transition temperature and the 0^{++} glueball mass. The analysis shows that there is no asymptotic scaling for the coupling $K_F \equiv 4g^{-2} < 2.5$ for SU(2) and $K_F \equiv 6g^{-2} < 6.1$ for SU(3). We also give the results for the MCRG improved lattice action for both the theories.

The most reliable method for a detailed calculation of the non-perturbative β-function is the Monte Carlo Renormalization Group (MCRG)[1]. In this procedure, the discrete β-function is calculated using Wilson's 2-lattice method[1] and an appropriate Renormalization Group Transformation (RGT). We choose to work with the $\sqrt{3}$ scale factor RGT[2,3]. The quantity measured on the lattice is ΔK_F, such that the integral of the β-function from K_F to $K_F - \Delta K_F$ corresponds to a scale change $b = \sqrt{3}$. Since this is calculated by matching Wilson loops on the blocked lattices, we expect the string tension to change by a factor of 3 between K_F and $K_F - \Delta K_F$. The β-function is expected to be universal only at sufficiently weak coupling, therefore the important question is : How sensitive is the β-function to the physical quantity held fixed in the renormalization procedure at couplings where Monte Carlo calculations are done? Universality of the β-function implies that all dimensionful quantities measured in lattice units scale according to the scale factor of the RGT and consequently all mass ratios stay constant and equal to their continuum values. Therefore a test of this universality and a determination of the position of onset of scaling is a prerequisite for any reliable estimate of the continuum properties of the theory. We test for this universality indirectly by calculating ΔK_F from the recent string tension[4,5], the deconfining transition temperature[6,7] and the 0^{++} glueball mass data[8,9], as well as comparing our results with those of the $b = 2$ MCRG

+ Work supported in part by the DOE Contract DE-AT03-81-ER40029.
* Work supported in part by the NSF under the Grant Nos. PHY-83-05734 and PHY-82-15249.

calculations[10,11].

In our previous works[2,3], we had obtained results for the integrated ΔK_F, for $K_F \in [2.5,3.5]$ in case of SU(2) and for $K_F \in [6.25,7.0]$ in case of SU(3). We briefly mention the important features :

(a) There are significant deviations of the β-function from asymptotic scaling for $K_F < 2.5$ in case of SU(2)[2,10] and $K_F < 6.1$ in case of SU(3)[3,11]. These deviations are attributed to the phase structure in the fundamental-adjoint action plane[12]. At weaker couplings, the observed ΔK_F are consistent with the 2-loop perturbative results.

(b) The renormalized action on the blocked lattices showed that both the 6-link operators and the higher representations of the plaquette are needed to improve the scaling behavior of the theory.

We learnt from these calculations that in MCRG one has a good control over systematic and statistical errors, however some features of the data still remained unresolved. So detailed calculations were undertaken with the following goals :

(1) To provide an accurate guide for the change of scale under a change of coupling in the range $K_F \equiv 4g^{-2} \in [2.5,3.5]$ for SU(2) and $K_F \equiv 6g^{-2} \in [6.0,7.0]$ for SU(3). Such results are essential in the determination of the point of onset of scaling, i. e., the smallest value of K_F beyond which all physical observables scale in unison. The control over finite size effects in the determination of physical observables like the string tension, the hadron masses and the O^{++} glueball mass is poor, so an independent accurate MCRG determination of the β-function is a reliable test of scaling.

(2) In the case of SU(2), the ΔK_F were larger than the 2-loop result for $K_F > 3.0$. For SU(3), there wasn't enough data available at weaker couplings. It was also not clear if after the dip the β-function approached its asymptotic value smoothly or whether there was any additional structure in the β-function. It was therefore essential to repeat the calculations and understand these features, and rule out as the probable causes : (a) The finite size effects, and (b) Poor convergence for the $b = \sqrt{3}$ RGT at weaker couplings, requiring many blockings to estimate the ΔK_F.

(3) To find a better estimate the renormalized action generated after one $b = \sqrt{3}$ RGT[13] in case of SU(3). The aim is to provide an approximate single parameter trajectory that eliminates a major fraction of the unwanted effects of the dominant irrelevant operators. This first step estimate for the improved action will subsequently be refined using the methods of Ref.[14].

The usefulness of these calculations becomes evident when the "β-functions" calculated from observables like the string tension[4,5], the deconfining transition temperature[6,7] and the 0^{++} glueball mass[8,9] are compared with the MCRG results. Though the error bars on the data are large, we find significant dispersions between the various "β-functions", and hence consider any conjectures about scaling for $K_F < 2.5$ in case of SU(2) and for $K_F < 6.1$ in case of SU(3) premature. We also present a theoretical argument based on the fixed point structure of the theory for a non-perturbative breakdown of scaling (i.e. non-universal β-function) in the region of the dip. It is our conclusion that calculations should be done at K_F values larger than the bounds given above, for which there is approximate ($<10\%$ error) asymptotic scaling. Better still, an improved action (with respect to both early scaling and reduced effects of the lattice artifacts) should be used for both the gauge and fermion degrees of freedom.

The MCRG method[1] is described in detail in Ref.[2]. Features specific to SU(3) are discussed in Ref.[3]. The method to find the renormalized action is discussed in Ref.[13]. We direct readers unfamiliar with these methods or those interested in the details to these references.

1. CHECKING THE SCALING BEHAVIOR

To extract the continuum properties of the theory, one needs to know how a particle mass, measured in units of the lattice spacing a, changes with the coupling g_{bare} , i.e., how a physical quantity scales. In the asymptotic region (near $g_{bare} \sim 0$), this scaling is given by the 2-loop perturbative β-function :

$$\frac{\partial(g^{-2})}{\partial(ln\ a)} = -\frac{11N}{24\pi^2} - \frac{17N^2}{192\pi^4} g^2 + \cdots \tag{1}$$

The quantity we calculate using MCRG is,

$$\Delta K_F = -\frac{\Delta(2Ng^{-2})}{\Delta(ln\ a)} \cdot \ln\sqrt{3}\ , \tag{2}$$

i.e., the discrete β-function at K_F evaluated for a scale change of $\sqrt{3}$. The same quantity can be extracted from any long distance physical observable m as follows : Let ma_1 (ma_2) be the measured values at coupling K_F^1 (K_F^2). Then renormalizing the theory holding m fixed implies that the length scale has changed by the ratio a_1/a_2 for a change K_F^2–K_F^1 in the coupling. For comparison, the ΔK_F can be scaled to correspond to a scale change b ,

$$\Delta K_F = (K_F^2 - K_F^1) \frac{\ln(b)}{\ln(ma_1/ma_2)} \ . \tag{3}$$

In Wilson's 2-lattice MCRG method[1], the ΔK_F is calculated by matching the long distance behavior (block Wilson loops) of the two theories defined by K_F and $K_F - \Delta K_F$. The comparison is made in such a way that the theory starting at K_F is renormalized once more than the theory starting at $K_F - \Delta K_F$. Then at matching, the two starting theories differ by the scale factor b of the RGT. To control the finite size effects, the block expectation values are compared on the same physical size lattices. The fundamental assumption of this method is that under blocking all lattice versions of the same physical theory converge to the Renormalized Trajectory (RT). This convergence is only asymptotic and improves as the starting point is picked closer to the fixed point. Therefore, in practice, systematic errors are present due to incomplete convergence of actions at the level of comparison. The check is to require matching at a number of blocking levels simultaneously. This requires a large starting lattice or a starting action close to the RT. We find that with a 9^4 lattice and the $b = \sqrt{3}$ RGT, the last two levels show simultaneous matching with less than 10% error. Also our SU(2) results obtained using a starting 9^4 lattice agree with those obtained using a starting 18^4 lattice. A final remaining check we would like to make is to use a starting $(9\sqrt{3})^4$ lattice and test convergence at one more blocking step.

SU(2) :

The operators (with the corresponding couplings defined in parentheses), used in the matching of the block expectation values and in the determination of the improved action, were the simple plaquette U_p in the $2(K_F)$, $3(K_3)$ and $4(K_4)$ representations, the rectangular 1×2 loop U_{6p} in the $2(K_{6p})$ and 3 representations, and the L-shaped and twisted 6-link operators designated as $U_{6l}(K_{6l})$ and $U_{6t}(K_{6t})$ respectively. The SU(2) action in the $[K_F, K_3, K_4, K_{6p}, K_{6l}, K_{6t}]$ space is defined to be,

$$S = K_F \sum TrU_p + K_{6p} \sum TrU_{6p} + K_{6l} \sum TrU_{6l} + K_{6t} \sum TrU_{6t} + K_3 \sum[\frac{4}{3}(TrU_p)^2 - \frac{1}{3}] \} + K_4 \sum[2(TrU_p)^3 - TrU_p] \,. \quad (4)$$

Here the higher representations have been constructed from U_p, all the traces are normalized to unity and the sums are over all sites and positive orientations of the loops.

Our previous results for the ΔK_F were obtained by matching block loops from starting 9^4 and starting $(3\sqrt{3})^4$ lattices[2], and are reproduced in Table 1. The present calculation was done on the 64 node hypercube computer at Caltech[15]. Larger lattices were used to study finite size effects. The update and blocking were done separately due to memory constrains, and we achieved a healthy 93% communication efficiency. The two starting lattices used in the matching process were 18^4 and $(6\sqrt{3})^4$, which were updated using a 7-hit Metropolis algorithm. We used the 120-element icosahedral subgroup approximation for SU(2) in the update, storing each element as an 8 bit quantity. The group elements were converted to SU(2) matrices prior to blocking. This trick speeds up the computation by a factor of 3 and the results are the same as for the full group[2]. To increase the statistics on the block lattices, we summed over all 9 possible choices of the block sites at each level of blocking. The number of sweeps discarded to ensure thermalization were 800 (18^4) and 500 ($(6\sqrt{3})^4$) respectively. The total number of configurations (separated by 10 update sweeps) used in the analysis were 200 (300) for the 18^4 ($(6\sqrt{3})^4$) lattices.

The results for ΔK_F are shown in Table 1 and Fig. 1 and are consistent with the previous values[2]. The main advantage of a larger lattice was that all 7 operators could be used to find the ΔK_F instead of just the 1×1 plaquette in the 2 and 3 representations as in Ref.[2]. We calculated the ΔK_F for each of the 7 operators and the quoted value is the maximum likelihood estimate, i. e., the mean weighed by the square of the reciprocal of the respective statistical errors.

We observed the following features in the analysis :

(1) The values of ΔK_F obtained from matching on the block 2^4 lattices are smaller than that from the block $(2\sqrt{3})^4$ lattices. This behavior was also present in Ref.[2,3] and we believe that for the $b=\sqrt{3}$ RGT, the estimates from successive levels of blocking oscillate about the limiting value. Therefore we assume that the true values lie in between the two estimates from the $(2\sqrt{3})^4$ and 2^4 block lattices.

(2) The dip in the β-function is still present for $K_F < 2.5$. The value of ΔK_F at $K_F = 2.5$ shows a decrease with the number of blocking steps, changing from $0.189(4) \rightarrow 0.181(4) \rightarrow 0.160(7)$. It is not known if the last two values are in accord with feature (1) discussed above; therefore the estimate of the limiting

Table [1] : The values of ΔK_F at different levels of matching for different values of the couplings. The matching $(6\sqrt{3})^4$ K_F were determined by linear interpolation and the errors in parentheses are based on a 1σ fit. Also shown are the values of ΔK_F corresponding to 2-loop asymptotic scaling, and the previous results obtained using 9^4 vs. $(3\sqrt{3})^4$ lattices.

SU(2)						
18^4	ΔK_F from matching on			2-loop	9^4 results	
K_F	6^4	$(2\sqrt{3})^4$	2^4	ΔK_F	$(\sqrt{3})^4$	1^4
2.5	.189(4)	.181(4)	.160(07)	.218	.182(4)	.172(07)
2.625	.193(2)	.191(4)	.178(07)	.218		
2.75	.207(3)	.215(7)	.196(11)	.217	.225(8)	.223(10)
2.875	.192(4)	.222(6)	.189(16)	.216		
3.0	.199(3)	.240(7)	.227(17)	.216	.226(4)	.200(15)
3.25	.166(5)	.223(5)	.213(10)	.215	.235(6)	.210(20)
3.5	.153(6)	.233(6)	.223(19)	.214	.245(6)	.239(15)

value has an undetermined systematic error.

(3) For $K_F \geq 3.0$, the estimates of the ΔK_F from block 6^4 lattices are significantly smaller than the limiting values and the deviation increases with K_F. This is because the starting coupling (Wilson axis) lies farther away from the RT and a few RGTs are required to reach the RT. The estimates were also sensitive to the loop size; using larger loops in the matching yielded smaller deviations from the limiting values. This is expected since larger loops depend less on the irrelevant short distance lattice artifacts.

(4) The ΔK_F at $K_F = 3.0$ and 3.5 are larger than the 2-loop values, but within errors consistent with it. Also the change in the ΔK_F values from their previous ones[2] suggests that the discrepancy is primarily a statistical effect. We therefore conclude that the $b=\sqrt{3}$ RGT preserves the scaling properties of the theory and provides a reliable estimate of the β-function from small (9^4) lattices. We cannot rule out the presence of any fine structure, in addition to the dip at $K_F < 2.5$, from this work. Note that such fine structure is expected if there is additional phase structure, for example, in the 3 representation of the 6-link planar operator or in the 5 representation of the plaquette, with an end point close to the Wilson axis.

In comparing the $b=\sqrt{3}$ MCRG results with the "β-functions" obtained from observables like the string tension[4], the deconfining transition temperature[6] and the 0^{++} glueball mass[8], one encounters the problem of adjusting scales. The operational solution we have adopted is to only choose pairs of data points with a scale change a_1/a_2 within 20% of $\sqrt{3}$, rescale the ΔK_F to $b=\sqrt{3}$ and then plot the scaled

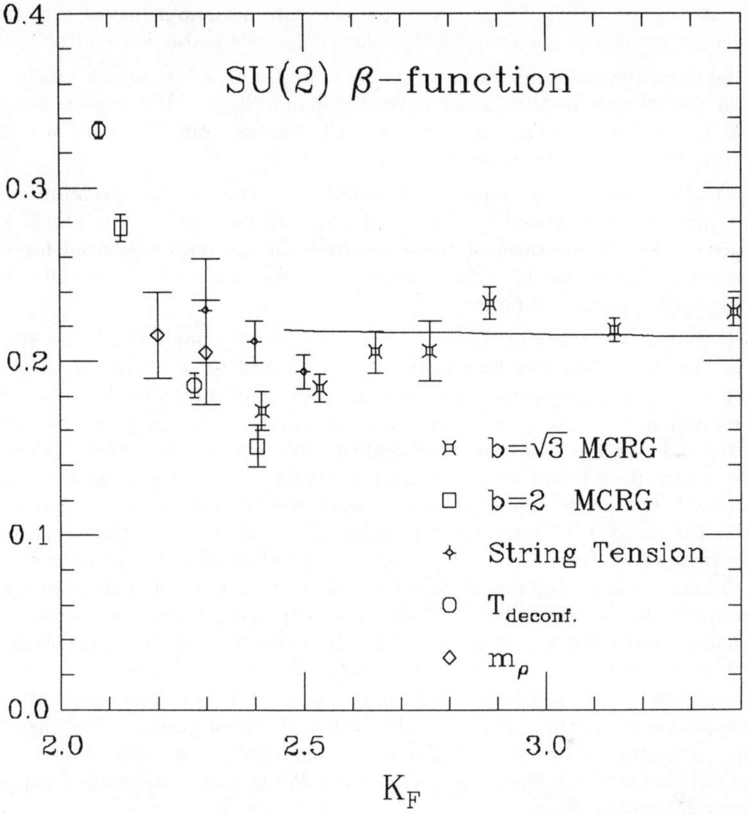

Fig. [1] : Comparison of the ΔK_F extracted from the MCRG calculations and various physical observables, with the 2-loop perturbative result (solid line) for SU(2). In all cases the ΔK_F are scaled for a scale change $b = \sqrt{3}$ and plotted at the midpoint of the two couplings matched.

ΔK_F at the midpoint of the original interval. We find $b = \sqrt{3}$ large enough so that the errors in the individual measurements are small compared to the differences. However, to detect fine structure in the β-function a still smaller value of b might be necessary.

The MCRG work of Mackenzie[10] covers the region $K_F = 1.7$–2.5. The data point at 2.7 suffers from a lack of convergence of block expectation values since matching is done after 1 RGT on the 4^4 lattice. We feel that the ΔK_F has been underestimated even at 2.5, but it should be reliable at other points. We shall therefore use his results as a qualitative guide in that range of K_F.

As shown in Fig. 1, the measurements of the deconfining transition temperature[6] yield reliable estimates of ΔK_F which are in good agreement with the MCRG results. On the other hand, though the string tension results of Ref.[4] are consistent with each-other, they yield a distinctly different β-function. The dip appears to be occurring at a larger value of K_F, and more accurate measurements of the string tension at weaker couplings are needed before any definitive statement can be made.

In the currently available data for $m_{0^{++}}$[8], we found a large discrepancy, both in the lattice values and in the ΔK_F extracted from them. We expect a significant change in the value of $m_{0^{++}}$ with better calculations, and therefore we have not included the corresponding results in Fig. 1.

Fig. 1 also shows the β-function extracted from the hadron spectrum calculated in the valence approximation[16]. We used m_ρ extrapolated to the chiral limit for this purpose. The comparison of these results with the ones measured for the pure gauge theory is questionable. Surprisingly, the deviation of the results from the asymptotic scaling curve is smaller.

Even though we consider the comparison to be only qualitative, the presence of the dip in the β-function can be seen in all the plotted data. The non-perturbative β-function crosses the 2-loop asymptotic scaling result in the region $K_F \sim 2.3$. This interval should *not* be regarded as a scaling window due to the presence of the subsequent dip. There are two sources of scaling violation in this region : (1) Perturbative (power law in g) due to the non-universality of the 3-loop and higher order terms, and (2) The phase structure in the fundamental-adjoint action plane which is responsible for the dip. (Our previous analysis[2] had shown that the deviation of the measured β-function from the expected one is much smaller in the negative adjoint plane.) There are no estimates of how large these violations should be except in the strong coupling limit, where the "β-functions" extracted from the glueball and the string tension results differ by a factor of 2. It is therefore to be expected that for g greater than some value, the non-universality of the β-function can be measured. Since the present data has large error bars we cannot definitely conclude whether there is approximate scaling or not in the region of investigation. We therefore feel that it is necessary to use $K_F > 2.5$ to extract continuum mass ratios or use an improved action for which the scaling violations due to lattice artifacts disappear at a smaller correlation length.

SU(3) :

The operators used in the matching of the block expectation values and in the determination of the improved action, were the simple plaquette in the 3, 6, 8, 10, 15 and 15′ representations, and the planar, L-shaped and twisted 6-link operators. The SU(3) action in the $[K_F, K_6, K_8, K_{6p}, K_{6l}, K_{6t}]$ space is defined to be, with the same notation as in the SU(2) case,

$$S = \mathrm{Re}\,\{\, K_F \sum TrU_p + K_{6p} \sum TrU_{6p} + K_{6l} \sum TrU_{6l} + K_{6t} \sum TrU_{6t} + K_6 \sum [\frac{3}{2}(TrU_p)^2 - \frac{1}{2} TrU_p] \,\} + K_8 \sum [\frac{9}{8} \mid TrU_p \mid^2 - \frac{1}{8}] \,. \tag{5}$$

The update of the pure gauge theory was done using a 20-hit Metropolis algorithm. The first 500 sweeps were discarded to ensure thermalization. Thereafter, the block lattices were constructed every 15th (10th) sweep on the 9^4 ($(3\sqrt{3})^4$) lattice. The number of configurations varied from 600 to 1300 (2400 to 4800). In order to improve the statistics, we again exploited the freedom in the choice of the block site and summed over all possible constructions of the block lattices at each level.

Table [2] : The values of ΔK_F at different levels of matching for different values of the couplings. The matching $(3\sqrt{3})^4$ K_F were determined by linear interpolation and the errors in parentheses are based on a 1σ fit. Also shown are the values of ΔK_F corresponding to asymptotic scaling.

SU(3)				
9^4	ΔK_F for matching on			2-loop
K_F	3^4	$(\sqrt{3})^4$	1^4	ΔK_F
6.0	.337(5)	.323(5)	.308(06)	.489
6.125	.387(5)	.376(5)	.351(06)	.488
6.25	.421(4)	.424(5)	.401(09)	.488
6.35	.431(4)	.452(5)	.445(09)	.487
6.45	.432(4)	.464(6)	.423(12)	.487
6.5	.435(4)	.464(6)	.449(15)	.487
6.75	.430(4)	.485(5)	.443(09)	.485
7.0	.422(7)	.503(11)	.488(20)	.484

The results for the ΔK_F, again determined as the maximum likelihood estimate, are presented in Table 2 and Fig. 2. The change in ΔK_F between the last two levels of matching is a measure of the systematic error present in the estimate and is much larger than the statistical error. The significant feature again is the dip in the β-function at $K_F \sim 6.0$. This dip is caused by the repellent phase structure (with a critical end point and an associated fixed point) in the extended $[K_F, K_8]$ plane. The flow from $g_{bare}=0$ to $g_{bare}=\infty$ slows down on the weak coupling side of the cross-over region, and accelerates once it is past the phase structure. The dip is larger here than the SU(2) case in agreement with the fact that the critical end point is closer. It is believed that as one approaches this end point, the string tension remains finite while the correlation length diverges. Such a scaling behavior of the associated extraneous fixed point is radically different from that of the continuum theory (fixed point at $g_{bare}=0$). Since the large dip is evidence that in the region $5.6 < K_F < 6.1$ the unphysical fixed point has a significant effect, scaling cannot be assumed in this region. We would also like to point out the possibility of additional critical points in the multi-dimensional coupling constant space. If they are close to the Wilson axis then the β-function will have more structure.

In Fig. 2, we compare our results with the "β-functions" obtained from the string tension[5], the deconfining transition temperature[7] and the 0^{++} glueball mass[9] data. These "β-functions" were obtained following the same prescription as for SU(2). The $b=2$ MCRG data[11] falls slightly below our results. This difference could be due to the rescaling of the ΔK_F. The three sets of data for the string tension[5] have about 20% spread in their lattice values. We therefore calculated ΔK_F separately for each set and these results were found to be more consistent with each-

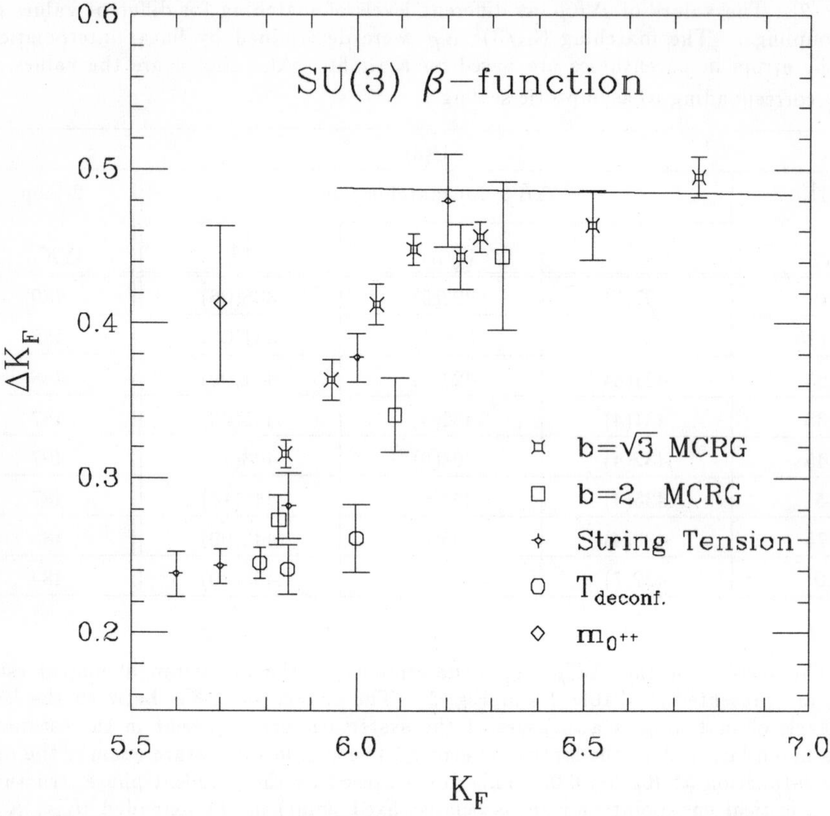

Fig. [2] : Same as Fig. 1 but for SU(3).

other. We regard the spread as a realistic measure of the systematic errors still present in the calculations. As shown in Fig. 2, the $b = \sqrt{3}\ \Delta K_F$ do agree within errors with the string tension results. The ΔK_F from the deconfining transition temperature data show a dip extending to weaker couplings and are in better agreement with the $b = 2$ MCRG results. The ΔK_F from the 0^{++} glueball mass data is in disagreement with all the rest and does not show the same pronounced dip. We hope that the situation will be clarified as better data becomes available at weaker couplings for all these long distance observables.

Given the errors in the individual measurements and the errors introduced in rescaling, the only safe conclusion we can draw is that asymptotic scaling (within 10% error) sets in after $K_F = 6.1$. At this coupling the correlation length $(m_{0^{++}}a)^{-1}$ is only ~ 2. We therefore advocate that all attempts to extract continuum mass ratios be made at weaker couplings.

We again emphasize the two main conclusions of our calculations :

(a) Along the Wilson axis, asymptotic scaling does not exist for $K_F < 2.5$ in case of SU(2) and for $K_F < 6.1$ in case of SU(3).

(b) MCRG is crucial in order to test scaling. Even a 9^4 lattice combined with a $b = \sqrt{3}$ RGT gives a reliable estimate of the β-function.

2. ESTIMATE OF THE IMPROVED ACTION

The multi-coupling renormalized action $\{K^{A^1}\}$ was calculated by comparing the expectation values Ω_i on the once blocked $(3\sqrt{3})^4$ lattice, with those on a $(3\sqrt{3})^4$ lattice updated with couplings $\{K^B\}$. To first order,

$$<\Omega_i>_{A^1} \sim <\Omega_i>_B + <:\Omega_i \Omega_j:> (K_j^{A^1} - K_j^B) , \tag{6}$$

where $<:\Omega_i \Omega_j:>$ is the connected correlation function. We evaluated the correlation function both on the lattice A^1 and on the lattice B, and compared the results to ensure the validity of the linearization. The mean $<:\Omega_i \Omega_j:>$ were used in the final analysis to obtain results correct to 2nd order.

For SU(2), we carried out this analysis on the block $(6\sqrt{3})^4$ lattices. We found the statistics too small to measure any significant deviations from the previous measurements[2] :

$$\frac{K_3}{K_F} = -0.19 \ , \ \frac{K_4}{K_F} = 0.04 \ , \ \frac{K_{6p}}{K_F} = -0.06 \ . \tag{7}$$

The SU(3) results are shown in Table 3. Since there is considerably more statistics and also the number of $\{K^B\}$ points used per K_F is larger than the earlier estimates[3], these results replace them. We find that the results depend strongly on the number of operators kept in the analysis. Our restriction to the space $[K_F, K_6, K_8, K_{6p}]$ is enforced by the statistics and the failure in agreement between the two correlation functions when more operators are included. Our first step estimate for a single parameter improved action is :

$$\frac{K_6}{K_F} = \frac{K_8}{K_F} = -0.12 \ , \ \frac{K_{6p}}{K_F} = -0.04 \ . \tag{8}$$

The feature to take note in Eqs.(7) and (8) is that the contributions of the 6-link loops and the higher representations of the plaquette are not small. By comparing the SU(2) and the SU(3) results, we find that the various operators have the same signs and behave in a qualitatively similar fashion. The fact that TrU_p for SU(3) is complex introduces two quadratic operators, K_6 and K_8. The lowest order weak coupling perturbative analysis and the large-N resummation technique do not distinguish between these two operators. Therefore it is a good approximation to regard them as equivalent and treat their combined degrees of freedom as indistinguishable modes contributing equally (equipartition theorem). Since we have normalized these operators to unity, this argument predicts $K_6 = K_8$. The results in Table 3 indeed show that the mean of K_6 and K_8 is very stable and the difference fluctuates around zero. We also note that the heat kernel action produced by the Migdal-Kadanoff approximate RGT[17] agrees with our conjecture ($K_6/K_8 = 0.98$), although their value of K_6/K_F (-0.284) is quite different from ours†.

† In obtaining the best approximation to the heat kernel action, the contribution of various representations is strongly dependent on the order of truncation of the character expansion. This and the fact that different renormalization transformations have different RTs can explain the difference.

Table [3] : Projections of the renormalized SU(3) action for different starting actions along the Wilson axis. The errors in parentheses were estimated from the fluctuations of the results for different $\{K^B\}$ points.

Projection of the improved SU(3) action in the $[K_F, K_{6p}, K_6, K_8]$ space				
Starting point	K_F	K_{6p}/K_F	$(K_6+K_8)/2K_F$	$(K_6-K_8)/2K_F$
6.125	7.08(1)	-.005(1)	-.129(1)	.009(2)
6.25	7.32(1)	-.013(2)	-.118(1)	-.036(3)
6.35	7.50(2)	-.015(1)	-.115(1)	.023(4)
6.45	7.63(1)	-.019(1)	-.108(1)	-.002(2)
6.50	7.84(1)	-.021(1)	-.111(1)	.002(2)
6.75	8.31(1)	-.027(1)	-.103(1)	-.027(2)
7.00	8.93(1)	-.031(1)	-.103(1)	.024(5)

From an eigenvalue-eigenvector analysis of the linearized transformation matrix, we find for both the SU(2) and SU(3) theories :

(1) The leading eigenvalue is consistent with 1.0, as expected of an asymptotically free theory.

(2) It is mostly K_{6p} that is necessary to get rid of the leading irrelevant operator.

Hence, we strongly recommend that future Monte Carlo calculations be done using the improved action estimated in Eqs.(7) and (8).

We thank our collaborators Steve Otto, Gerry Guralnik, Tony Warnock and Charles Zemach for a cohesive group effort. It is our pleasure to thank G. C. Fox for his constant support and encouragement.

REFERENCES

[1] Wilson K., in *Recent Developments in Gauge Theories (Cargese 1979)*, ed. G. 't Hooft et al, Plenum Press, New York (1980).

Ma S. K., Phys. Rev. Lett. 37 , 461 (1976).

Kadanoff L. P., Rev. Mod. Phys. 49 , 267 (1977).

Swendsen R. H., Phys. Rev. Lett. 42 , 859 (1979).

Swendsen R. H., in *Phase Transitions (Cargese 1980)*, ed. M. Levy et al, Plenum Press, New York (1982), and references therein.

Shenker S. H. and Tobochnik J., Phys. Rev. B22 , 4462 (1980).

Cordery R., Gupta R. and Novotny M., Phys. Lett. 128B , 425 (1983).

[2] Patel A., Cordery R., Gupta R. and Novotny M., Phys. Rev. Lett. 53 , 527 (1984).
Patel A. and Gupta R., Nucl. Phys. B251 [FS13], 789 (1985).

[3] Gupta R., Guralnik G., Patel A., Warnock T. and Zemach C., Phys. Rev. Lett. 53 , 1721 (1984).
Gupta R. and Patel A., *Proceedings of the 1984 D.P.F. Meeting at Santa Fe,* World Scientific (1985).

[4] Gutbrod F. and Montvay I., Phys. Lett. 136B , 411 (1984).
Berg B., Billoire A., Meyer S. and Panagiotakopolous C., Comm. Math. Phys. 97 , 31 (1985).

[5] Barkai D., Moriarty K. J. M. and Rebbi C., Phys. Rev. D30 , 1293 (1984).
Otto S. W. and Stack J. D., Phys. Rev. Lett. 52 , 2328 (1984).
de Forcrand P., Schierholz G., Schneider H. and Teper M., DESY preprint, DESY-84/116 (1984).

[6] Curci G. and Tripiccione R., Phys. Lett. 151B , 145 (1985).

[7] Kennedy A. D., Kuti J., Meyer S. and Pendleton B. J., Phys. Rev. Lett. 54 , 87 (1985).

[8] Schierholz G. and Teper M., Phys. Lett. 136B , 64 (1984).
Meyer-Ortmanns H. and Montvay I., Phys. Lett. 145B , 251 (1984).
Berg B., Billoire A., Meyer S. and Panagiotakopolous C., Comm. Math. Phys. 97 , 31 (1985).

[9] de Forcrand P., Schierholz G., Schneider H. and Teper M., Phys. Lett. 152B , 107 (1985).

[10] Mackenzie P. B., *Proceedings of the Argonne National Lab. Workshop "Gauge Theory on a Lattice: 1984"*, 171 (1984).

[11] Hasenfratz A., Hasenfratz P., Heller U. and Karsch F., Phys. Lett. 143B , 193 (1984).
Bowler K. C. et al, Edinburgh Preprint No. 85/335 (1985).

[12] Bhanot G. and Creutz M., Phys. Rev. D24 , 3212 (1981).
Bhanot G., Phys. Lett. 108B , 337 (1982).

[13] Gupta R. and Patel A., Phys. Rev. Lett. 53 , 531 (1984).
Swendsen R. H., *"Real Space Renormalization"*, in *Topics in Physics, Vol. 30,* Springer Verlag (1983).

[14] Gupta R. and Cordery R., Phys. Lett. 105A , 415 (1984).
Creutz M., *Proceedings of the Argonne National Lab. Workshop "Gauge Theory on a Lattice: 1984"*, 1 (1984).

[15] Fox G. and Otto S., Physics Today 37 , 53 (May 1984).
Brooks E. et al, Phys. Rev. Lett. 52 , 2324 (1984).

[16] Billoire A., Lacaze R., Marinari E. and Morel A., Nucl. Phys. B251 [FS13], 581 (1985).

[17] Bitar K. M., Duke D. W. and Jadid M., Phys. Rev. D31 , 1470 (1985).

IMPROVED MONTE CARLO RENORMALIZATION GROUP

Rajan Gupta

Physics Department
Northeastern University
Boston, MA 02115

ABSTRACT

I review the Cordery-Gupta Monte Carlo Renormalization Group method ($MCRG$) to calculate the Renormalized Hamiltonian in a many coupling space and to determine the Linearized Transformation Matrix (LTM). The important advantages of this method, *i.e.* there are no long time correlations even on the critical surface and secondly the block 2-point correlation functions $\langle S^1_\alpha S^1_\beta \rangle - \langle S^1_\alpha \rangle \langle S^1_\beta \rangle$ are calculable numbers, are stressed. The method allows a careful error analysis in the determination of the renormalized couplings and in the LTM. Using the exact solution of the $d = 2$ Ising model ($IM2$) we deduce a number of properties of the LTM and show why great care is needed in the truncation.

OUTLINE

In this talk I will describe and dissect an improved $MCRG$ method ($IMCRG$) [1]. The topics addressed are arranged in sections as follows and the unsolved problems are stated at the appropriate places.

(1) A brief review of the Monte Carlo Renormalization Group method [3,4,5,6,7].

(2) The derivation of the Improved $MCRG$ method. I show that in the linear region this method has no truncation errors in the determination of the renormalized couplings and the LTM.

(3) Using the exact solution of the $d = 2$ Ising model ($IM2$) a number of properties of the LTM are obtained. The left eigenvector of the LTM is shown to be the normal to the critical surface and for $IM2$ it is shown that its elements g_α grow with the range of the coupling as $\sim x^{\frac{3}{4}}$. These elements g_α are shown to be to a good approximation the response functions in the coupling constant

space *i.e.* the row elements of the LTM. The fact that these diverge has two important consequences; 1) the columns of the LTM have to fall off faster than a certain bound to make $MCRG$ sensible and 2) a small error in a long range coupling will cause a diverging compensating change in the renormalized nearest-neighbor coupling K_{nn}. Thus very careful attention has to be paid to the small long range couplings in order to keep the system on the critical surface in any iterative scheme.

(4) We compare the $IMCRG$ and the $MCRG$ methods and discuss the truncation errors in the LTM. Even though the construction of the LTM in $IMCRG$ is free of truncation errors, we find for $IM2$ that $MCRG$ gives 'better' estimates of the critical exponents. However, the reason for this has to be found before the results can be trusted. We agree with Wilson's observation that the statistical accuracy of Swendsen's method derives from as yet ununderstood cancellation of correlations in the ratio block-block to spin-block correlation matrices.

(5) The calculation of the first two renormalized couplings for the $d = 2$ Ising model is presented. The iterative $IMCRG$ method to calculate the fixed point Hamiltonian is not very stable so an extension of the method is described.

1) INTRODUCTION TO *MCRG*

Renormalization Group (RG) [2] is a general framework for studying systems near their critical point where all length scales are important. The scaling properties associated with second order phase transitions and the universal critical exponents have been calculated for many systems either analytically or by the Monte-Carlo RG ($MCRG$) [2,3,4,5] method. The idea is as follows: Consider a magnetic system consisting of spins $\{s\}$ on the sites of a $d-$ *dimensional* lattice L described by a Hamiltonian H with all possible couplings $\{K_\alpha\}$. All thermodynamic quantities can be found from a detailed knowledge of the partition function

$$Z = Tr\ e^{-H} = Tr\ e^{K_\alpha S_\alpha} \tag{1}$$

where S_α are the interactions.

In the standard $MCRG$ method [3,4,5,6], the spin configurations are generated with the Boltzmann factor e^{-H}. The renormalized theory –– interaction of block spins $\{s^1\}$ defined on the sites of a sublattice L^1 with lattice spacing b times that of L –– is defined by

$$e^{-H^1(s^1)} = Tr\ P(s^1, s)\ e^{-H(s)} \quad . \tag{2}$$

where the projection operator $P(s^1, s)$ (alias the renormalization group transformation (RGT)) should integrate out the short distance fluctuations but leave the long distance physics unchanged. It satisfies the constraint

$$Tr^1\, P(s^1, s) = 1 \tag{3}$$

independent of the state $\{s\}$. This guarantees that the two theories have the same partition function.

Renormalization group is the study of the transformation $H^1 = R(H)$ defined on the space of coupling constants, $\{K_\alpha\}$, of the model. At all fixed points H^* which have a divergent correlation length, the theory is scale invariant. This is the source of the scaling functions observed in thermodynamic systems. A certain neighborhood, the set of critical points in the coupling constants space, forms the domain of attraction of the fixed point. The long distance physics of all theories attracted by a given fixed point is the same and under a RGT a critical Hamiltonian flows to the fixed point with the rate of flow given by the irrelevant eigenvalues. The relevant eigenvalue(s) give the rate of flow away from the fixed point (along the unstable direction) and are related to the critical exponent(s) ν. In the standard $MCRG$ method [6] these are calculated from the eigenvalues of the linearized transformation matrix $T^n_{\alpha\beta}$ which is defined to be

$$T^n_{\alpha\beta} = \frac{\partial K^n_\alpha}{\partial K^{n-1}_\beta} = \frac{\partial K^n_\alpha}{\partial \langle S^n_\sigma \rangle} \frac{\partial \langle S^n_\sigma \rangle}{\partial K^{n-1}_\beta} \quad . \tag{4}$$

Each of the two terms on the right is a connected 2-point correlation function

$$\frac{\partial \langle S^n_\sigma \rangle}{\partial K^{n-1}_\beta} = \langle S^n_\sigma S^{n-1}_\beta \rangle \; - \; \langle S^n_\sigma \rangle \langle S^{n-1}_\beta \rangle \, . \tag{5}$$

and

$$\frac{\partial \langle S^n_\sigma \rangle}{\partial K^n_\beta} = \langle S^n_\sigma S^n_\beta \rangle \; - \; \langle S^n_\sigma \rangle \langle S^n_\beta \rangle \, . \tag{6}$$

Here $\langle S^n_\sigma \rangle$ are the expectation values on the nth renormalized lattice and K^n_σ are the corresponding couplings. The exponent ν is found from the leading eigenvalue λ of $T^n_{\alpha\beta}$ as

$$\nu = \frac{\ln b}{\ln \lambda} \tag{7}$$

where b is the scale factor of the RGT. The accuracy of the calculated exponents improves if they are evaluated close to the fixed point. This can be achieved by starting from a critical point and blocking the lattice a sufficient number of times. The convergence is therefore limited by the starting lattice size and can be improved if H^n is used in the update. Thus it is important to determine the renormalized couplings $\{K^n\}$. The Achilles heel of this method is that as yet no way is known to determine the errors in the exponents obtained from a truncated set of matrices. This is discussed in more detail in section 4.

2) IMPROVED *MCRG*

In the Improved *MCRG* method the configurations $\{s\}$ are generated with the weight

$$P(s^1, s)e^{-H(s)+H^g(s^1)} \tag{8}$$

where H^g is a guess for the H^1. Using both site and block couplings eliminates the long time correlations due to a divergent correlation length. If $H^g = H^1$, then the block spins are completely uncorrelated and

$$\langle S^1_\alpha \rangle = 0 \qquad \langle S^1_\alpha S^1_\beta \rangle \ = \ n_\alpha \delta_{\alpha\beta} \tag{9}$$

where for the Ising model (and most other models) the integer n_α is simply a count of the number of terms (multiplicity) of interaction type S_α. When $H^g \neq H^1$, then to first order

$$\langle S^1_\alpha \rangle \ = \ \langle S^1_\alpha S^1_\beta \rangle_{H^g=H^1} \ (K^1 - K^g)_\beta \tag{10}$$

and using Eq. (9), the renormalized couplings $\{K^1_\alpha\}$ are determined with no truncation errors as

$$K^1_\alpha \ = \ K^g_\alpha \ + \ \frac{\langle S^1_\alpha \rangle}{n_\alpha} \ . \tag{11}$$

This procedure can be iterated using H^{n-1} as the spin H in Eq. (8) to find H^n. If the irrelevant eigenvalues are small, then after two or three repetitions of the *RGT*, the sequence H^n converges to the fixed point Hamiltonian H^* which is assumed to be short ranged. The only limitation of this method is the approximation that one is in the linear regime, Eq. (10), and the use of a truncated H^{n-1} for the spin Hamiltonian in the update. In section 5, I describe a modification that overcomes the second limitation. This modification is necessary because errors in long range couplings due to finite statistics and a truncation in the spin H^{n-1} get magnified and the system rapidly flows away from the fixed point.

The calculation of the *LTM* proceeds exactly as in the standard *MCRG i.e.* Eqs. (4) to (6). However, in the limit $H^g = H^1$, the block-block correlation matrix is diagonal and given by Eq. (9). Thus it has no truncation errors, can be inverted with impunity and the final *LTM* elements are also free of all truncation errors. The only error is in finding the eigenvalues of a truncated matrix. I will discuss how to estimate errors due to this truncation in section 4.

To summarize, we find that simulating the system with several couplings and with both the block and site spins does complicate the program but there are three very important advantages of this method:

(1) Generating configurations according to Eq. (8) removes the long time correlations, so there is a very large gain in statistics, i.e. thermal equilibrium is reached quickly and the correlation between successive sweeps is limited to a few $(O(1))$ passes .

(2) The hardest part of such methods, an accurate calculation of $\langle S^1_\alpha S^1_\beta \rangle$ for many long range interactions, is known exactly. Also in the evaluation of the H^1 and LTM, a truncation in the coupling constant space does not affect the results because this matrix is diagonal.

(3) This method extends easily to Lattice Gauge Theories and other spin models [11].

3) EXACT RESULTS FROM THE 2*d* ISING MODEL AND CONJECTURES [9]

In a recent calculation, Shankar and Murthy [8] found an equation for the tangent to the critical surface at the nearest-neighbor Ising critical point K^c_{nn}. On this surface deviations, δK_α, from K^c_{nn} satisfy the equation

$$\delta K_\alpha \, g_\alpha \; = 0 \; . \tag{12}$$

where the *metric* g_α, defined in Eq. (13), was calculated from the exact expressions for the expectation values $\langle S_\alpha \rangle$. From Eq. (12) it is also clear that the g_α are the components of a vector normal to the critical surface. Now turning to the LTM, the right eigenvectors $\mid R\rangle$ corresponding to the irrelevant eigenvalues lie in the critical surface and span it. The left eigenvectors $\langle L \mid$ are the dual vectors and are orthogonal to $\mid R\rangle$. Thus for theories with 1 relevant eigenvalue, the corresponding left eigenvector is normal to the critical surface and its components are the g_α.

The behavior of the g_α for the 2-spin couplings can be estimated as follows: The g_α are defined as

$$g_\alpha \; = \; M \lim_{t \to 0} \frac{\frac{\partial \langle S_\alpha \rangle}{\partial t}}{\frac{\partial \langle S_{nn} \rangle}{\partial t}} \tag{13}$$

where M is the multiplicity (per site) of S_α. The 2-spin interactions S_α scale as

$$S_\alpha(x,t) \; \sim \; x^{2-d-\eta} \, G(xt^\nu) \tag{14}$$

therefore the g_α scale as $x^{3-d-\eta}$. For $IM2$ this predicts a powerlaw growth

$$g(x) \; \sim \; M \, x^{\frac{3}{4}} \; . \tag{15}$$

Similar arguments (which are harder to show) tell us that the g_α for all types of interactions grow with a powerlaw in an appropriately defined x. The second important property of the g_α is that they are positive definite. Finally note that these g_α can be determined from $MCRG$ for *any* model by evaluating the relevant left eigenvector.

The *LTM* (for the even coupling sector) can be written as

$$T = \sum \lambda_i \, |R_i\rangle\langle L_i| \tag{16}$$

where the sum is over all the eigenvalues. If the leading (thermal) eigenvalue λ_t dominates then

$$T \sim \lambda_t \, |R_t\rangle\langle L_t| \tag{17}$$

and each row of the *LTM* grows like the g_α. This growth gets modified by the irrelevant eigenvalues but we conjecture that the essential behavior is still a growth along the rows especially for the couplings that contribute significantly to the leading eigenvalue. This growth could have been anticipated from physical arguments: The first row elements are the response of the block nearest neighbour coupling K^1_{nn} to changes in the various spin couplings K^0_α. If H^* is short ranged, then tweaking a long range coupling will completely alter the short distance behavior and may even change the universality class. Therefore the short distance couplings must respond in a divergent manner to changes in the long range interactions to keep the physics unchanged. The second conjecture is that since the critical surface (and consequently the normal to it) is independent of the *RGT*, the *growth* along the rows cannot be modified by changing the *RGT*. If H^* is short ranged and the critical surface has small curveature then the above statements are still true in the sense of an average over the line of flow under *RGT*. These results are true for the odd (magnetic) sector also. For *IM*2 the single relevant magnetic eigenvalue is ~ 3.66. We find that even though the leading eigenvalue is more dominant, the elements along the first row are not a good approximation to the g_α given by the leading left eigenvector. The reason is that the 3 spin and longer range interactions have a large effect from the irrelevant eigenvalues so the corresponding elements differ from the g_α.

The columns of *LTM* give the response of long range couplings to changes in the short range couplings. Universality in fixed point theories tells us that the long distance critical behavior is independent of the details of the short distance couplings. Thus these responses should $\longrightarrow 0$. The important question here is how fast? Consider Eq. (17), then demanding that there exist a sensible norm *i.e.* the dot product $\langle L_t|R_t\rangle$ converge absolutely, gives that if elements L_α grow as x^y then the elements R_α have to fall at least as fast as x^{y+2} in 2 dimensions. So one has a bound. Next, after a little thought, it becomes clear that the fall off depends on the *RGT* and the distance from the fixed point. This makes analysis much harder and quantitative results are lacking.

4) TRUNCATION ERRORS IN *LTM*

The *LTM* has elements that grow along rows and fall along columns, therefore it can be arranged to look like

$$\begin{pmatrix} A & B \\ \varepsilon & D \end{pmatrix} \tag{18}$$

with A the minimal truncated $n \times n$ block matrix that should be calculated. The case $\varepsilon = 0$ is simple; there are no truncation errors in either method and diagonalizing A gives the n largest eigenvalues. Otherwise for $IMCRG$ the truncation error depends on the dot product of terms in ε and B. The requirement of absolute convergence in the estimate of the above bound only guarantees that this product is finite but it may be arbitrarily large *i.e.* $O(1)$. Therefore for each model, a careful study of the signs and magnitude of the elements in ε as a function of the RGT becomes necessary. This is being done at Cornell.

Post Script: I direct the readers to ref. [9] for the description of a very accurate perturbation theory method to calculate the corrections due to truncation in the $IMCRG$.

In Swendsen's $MCRG$ method, a finite number of the elements of the 2 matrices in Eq. (5) and Eq. (6) are calculated. The truncated matrix, Eq. (6), is inverted before A is calculated. This introduces errors that cannot be estimated and one would have expected that these extra truncations lead to poorer results. However, in practice one finds stability with respect to the truncation. Comparing the results for the first level of blocking in $IM2$ we find that his method does much better than $IMCRG$. The relevant eigenvalue from $IMCRG$ ($MCRG$) fluctuates by up to 10% (< 1%) depending on what couplings are kept in A. We do not understand this phenomenon yet. Secondly, as first noticed by Wilson, there is a cancellation of sweep to sweep correlations between the two matrices in the determination of LTM by $MCRG$, so statistically both methods do not suffer from critical slowing down. It is possible that the optimum procedure to determine the exponents may be to calculate the renormalized Hamiltonian using $IMCRG$ and then do a $MCRG$ calculation close to the fixed point using the same tuned RGT. However, it behooves us to first understand how to determine the error in the eigenvalues due to the truncation.

5) BLOCK HAMILTONIANS FOR THE 2*d* ISING MODEL

We used the heat-bath update in which the joint probability for changing the 4 spins that form a block spin was calculated. A tie was broken with equal probability. To improve the statistics (reduces the fluctuations in observables), the means $\langle S_\alpha \rangle$ were calculated using the weights generated in the update rather than just looking at the spins after each pass through the lattice. This is possible in the heatbath algorithm for a discrete state system because the probability of all possible spin configurations is calculated.

The results for the first and second renormalized couplings are given in Table

1. The type of interaction is shown on the left. Only even couplings that fit into a 3×3 square were programmed into H^g but a few longer range 2-spin couplings were also calculated. The couplings marked with $*$ and those not mentioned were set to zero in the update. Coupling with a value smaller than 1×10^{-4} are not reported. The lattice size and the statistics in thousands of sweeps is given in the fourth row. All errors quoted are statistical and based on a 1σ fit.

The starting H^0 was the known nearest-neighbor critical point $K^c_{nn} = 0.44068$. We find that the results for H^1 are independent (within statistical accuracy) of finite size effects for lattice sizes 16, 32, 64 and 128. For example compare columns 3 and 4 in Table 1 where the same H^1, determined in run 1 (column 1) and truncated after the 14 largest couplings, was used as the H^g. Secondly, the results for H^1 converged provided the couplings in H^g were correct to $O(10^{-3})$. This initial accuracy can be achieved with a few thousand sweeps on a 128×128 lattice. The errors in H^1 can be evaluated very reliably from statistics. Detailed binning analysis showed that each sweep is approximately independent and an accuracy of 10^{-5} is obtained in all couplings with $\sim 2 \cdot 10^6$ sweeps on a 64^2 lattice. This could be achieved with 3000 Vax 11/780 hours. We also find that the statistical errors in H^1 are quite independent of the strength of the coupling as shown in Table 1.

In run 1, H^1 given in column 1 was the spin Hamiltonian used to obtain the H^2 given in column 2. In run 2, the mean of H^1 in columns 3 and 4 was used with H^2 from column 2 (column 5) as H^g to obtain the results in column 5 (column 6). The principal observation is that while the two starting H^1 differ only to $O(10^{-5})$ plus different truncations, the resulting H^2 in columns 2 and 5/6 differ at $O(10^{-3})$. The errors shown in Table 1 are therefore statistical and we estimate the systematic shift due to the errors in the first step could be as large as $O(10^{-3})$. We conclude that the determination of H^2 is very sensitive to truncation errors in H^1. One possible reason is, as explained in section 3, small errors in the long range couplings get amplified in K^2_{nn} (in H^2) due to the metric g_α. In run 2 for H^2 we ignored about 5 known couplings of $O(10^{-4})$ with g_α as large as 10 (four of these are marked with $*$ in Table 1). We therefore do not have a realistic estimate of error in H^2 due to the truncation. This makes the method unstable for finding the fixed point Hamiltonian. It is therefore essential to isolate the ill-determined couplings with a large g_α and to find a better method to keep the system on the critical surface.

The solution is straightforward: In Eq. (8) use H^g as the guess for H^n. The update now involves the original spins and the nth *level* block spins in the Boltzmann weight

$$P(s^n, s^{n-1}) \;......\; P(s^1, s) e^{-H(s)+H^g(s^n)} \quad . \tag{19}$$

The three Eqs. (9-11) are unchanged except that the *level* superscipt is replaced by n, *i.e.* the nth *level* block-block correlation matrix is diagonal and given by Eq. (9). With this modification, the only limitation left is the size of the starting lattice. The results for $n = 2$ will be available soon and these will provide a true measure of the systematic errors.

Recently Swendsen has proposed a different method to calculate the renormalized Hamiltonian [10]. It is difficult to compare the results in detail since his renormalized couplings depend on the number of couplings kept, i.e. the matrix $\langle S^1 S^1 \rangle$ in Eq. (9) is not diagonal nor known exactly. This truncation is the most likely explanation for the difference between the two results.

To conclude, we believe that $IMCRG$ provides a complete framework to analyse the critical behavior of spin and gauge models. With the increased availability of supercomputer time we shall have very accurate and reliable results.

ACKNOWLEDGEMENTS

I would like to thank my collaborators R. Cordery, G. Murthy, A. Patel, R. Shankar, C. Umrigar and K. Wilson for the work presented here and for many long and fruitful discussions. I also thank Dennis Duke for inviting me to participate in an exciting conference. This work was Supported in part by the NSF under grant numbers PHY-83-05734 and PHY-82-15249

REFERENCES

[1] R. Gupta and R. Cordery, Phys. Lett. A105 (1984) 415.
[2] K. G. Wilson and J. Kogut, Phys. Rep. 12C, (1974) 76.
P. Pfeuty and G. Toulouse, *Introduction to the Renormalization Group and Critical Phenomenona*, (John Wiley & Sons, New York 1978).
[3] S. K. Ma, Phys. Rev. Lett. 37, (1976) 461.
[4] L. P. Kadanoff, Rev. Mod. Phys. 49, (1977) 267.
[5] R. H. Swendsen, Phys. Rev. Lett. 42, (1979) 859.
[6] R. H. Swendsen, in *Real Space Renormalization, Topics in Current Physics*, Vol 30, edited by Th. W. Burkhardt and J. M. J. van Leeuwen (Springer, Berlin, 1982) pg. 57.
[7] K. G. Wilson, in *Recent Developments in Gauge Theories*, Cargese (1979), eds. G. t' Hooft, *et al.* (Plenum, New York, 1980).
[8] G. Murthy and R. Shankar, Phys. Rev. Lett. 54, (1985) 1110.
[9] R. Shankar, R. Gupta and G. Murthy, Yale Preprint # YTP 85/13 (1985).
[10] R. H. Swendsen, Phys. Rev. Lett. 52, (1984) 1165.
[11] K. Binder, in Monte Carlo Methods in Statistical Physics, edited by K. Binder (Springer, Berlin,1979) Vol 7 , Pg. 1. and M. Creutz, L. Jacobs and C. Rebbi, Phys. Rep. 95, (1983) 201. These two references give a detailed introduction to Monte Carlo Methods in spin and gauge systems.

RENORMALIZED COUPLINGS STARTING FROM THE NEAREST NEIGHBOR CRITICAL POINT

K	RUN 1			RUN 2		
012A 345B 678C	H^1	H^2	H^1	H^1	H^2	H^2
	64(146)	64(100)	64(400)	128(87)	64(164)	64(90)
01	0.35357(5)	0.34638(6)	0.35359(3)	0.35359(3)	0.34308(4)	0.34297(7)
04	0.07502(5)	0.08889(6)	0.07499(3)	0.07504(3)	0.08668(4)	0.08667(7)
02	−0.00772(5)	−0.00954(6)	−0.00762(3)	−0.00770(3)	−0.00996(4)	−0.01003(7)
05	−0.00623(3)	−0.00699(4)	−0.00626(2)	−0.00624(2)	−0.00727(3)	−0.00725(5)
08	−0.00293(5)	−0.00285(6)	−0.00287(3)	−0.00289(3)	−0.00289(4)	−0.00288(7)
0A*	0.00056(3)	0.00052(4)	—	0.00053(2)	—	—
0B*	0.00020(2)	0.00030(3)	—	0.00017(2)	—	—
0134	−0.01525(5)	−0.01859(6)	−0.01517(3)	−0.01520(3)	−0.01876(4)	−0.01881(7)
0123	0.00158(2)	0.00188(3)	0.00157(1)	0.00158(1)	0.00193(2)	0.00186(3)
0135	0	−0.00009(3)	−0.00022(1)	−0.00020(1)	0.00000(2)	−0.00002(3)
0138	0	0.00011(3)	0.00004(1)	0.00004(1)	0.00010(2)	0.00009(3)
0148	0.00093(2)	0.00087(3)	0.00092(1)	0.00090(1)	0.00091(2)	0.00089(3)
0145	0.00098(2)	0.00157(3)	0.00103(1)	0.00105(1)	0.00139(2)	0.00142(3)
1345	0.00758(2)	0.00740(3)	0.00759(1)	0.00760(1)	0.00730(2)	0.00722(3)
1348	0.00296(2)	0.00283(3)	0.00298(1)	0.00297(1)	0.00281(2)	0.00278(3)
0128	0	−0.00014(3)	−0.00016(1)	−0.00017(1)	−0.00020(2)	−0.00021(3)
0127	0	−0.00030(3)	−0.00031(1)	−0.00033(1)	−0.00038(2)	−0.00037(3)
0167	0	−0.00022(3)	0.00008(1)	0.00005(1)	−0.00011(2)	−0.00016(3)
0178	0	−0.00029(3)	−0.00024(1)	−0.00023(1)	−0.00020(2)	−0.00027(3)
0158	0	−0.00024(3)	−0.00037(1)	−0.00036(1)	−0.00029(2)	−0.00032(3)
0156	0	−0.00019(3)	−0.00012(1)	−0.00013(1)	−0.00022(2)	−0.00022(3)
0247	−0.00115(2)	−0.00124(3)	−0.00116(1)	−0.00116(1)	−0.00114(2)	−0.00116(3)
0246	0.00047(2)	0.00065(3)	0.00047(1)	0.00044(1)	0.00062(2)	0.00061(3)
0268	0	0.00024(3)	0.00017(1)	0.00016(1)	0.00022(2)	0.00022(3)
1357	−0.00579(4)	−0.00479(5)	−0.00579(2)	−0.00572(2)	−0.00446(3)	−0.00443(5)
012* 345	0	0.00009(3)	0.00011(1)	0.00011(1)	0.00014(2)	0.00012(3)
012* 346	0	0.00012(2)	0.00012(1)	0.00013(1)	0.00011(1)	0.00013(2)

BLOCK-SPIN APPROACH TO THE FIXED-POINT STRUCTURE OF LATTICE Φ^4-THEORY

C. B. Lang

Institut f. Theor. Physik
Universität Graz
A-8010 Graz, Austria

ABSTRACT

A real space renormalization group investigation of Φ^4 theory is performed. The theory is simulated by MC techniques on a 16^4 lattice at several positive values of the bare Φ^4-coupling. Starting at the critical point block-spin transformations of two kinds (linear and non-linear) to size 8^4 and 4^4 allow to follow the flow of couplings on the critical surface. The values of the coupling constants corresponding to 24 types of even interaction terms are determined with a method recently suggested by Swendsen. The results demonstrate that the theories for different Φ^4-coupling have the same long distance behaviour: the trajectories approach each other rapidly and continue to move towards a common fixed point. Thus the continuum limit is that of a single, non-interacting gaussian theory.

1. MOTIVATION AND INTRODUCTION

The methods of real space renormalization group[1)] provide an intriguing way to identify the relevant couplings of a given lattice field theory. Our candidate in this investigation is the Euclidean Φ^4 theory in 4 space-time dimensions. It is regularized by formulation on a hypercubic lattice and its continuum limit has to be taken at a critical point. Approaching this point the relevant couplings have to be fine-tuned such that the physical properties (like masses) remain fixed while the lattice spacing decreases towards zero.

$$1/\xi \sim m_{phys} a(g) \xrightarrow[g \to g_{crit}]{} 0. \qquad (1.1)$$

Working directly at the critical point leads to a continuum theory with zero mass.

The physics is then determined by the long distance behaviour (with regard to the lattice unit a). Block spin RG transformations[2,3,4] provide a method to identify the universality class of the theory in question. Given a certain BST and starting at the critical point of the system the repeated application of the BST leads one to a fixed point. Its domain of attraction is the critical surface and its actual position in the parameter space of all possible interaction terms depends on the specific BST.

For asymptotically free theories like lattice QCD the fixed point position for a given BST could be determined by weak coupling expansions. In practice this involves laborious calculations of higher order in the coupling. Furthermore for that case MC-methods are applicable only away from the weak coupling point g=0 for values of $1/g^2$, say, of 0(1). Non-perturbative contributions are relevant and may influence the RG flow. Various gauge groups have been investigated by now within this context[5,6,7].

Renormalizable but not asymptotically free theories like d=4 Φ^4 theory have their critical point at a non-zero value of, say, the nearest neighbour coupling K, or, respectively, the mass term coefficient.

Different models correspond to different interaction terms to begin with. The irrelevance of some interaction term (like e.g. the Φ^4-coupling) may be demonstrated by following the RG-trajectories starting from different points on the critical surface due to different values of that

Figure 1: Renormalization flow lines on the critical surface starting at different values of some coupling parameter leading towards a common fixed point.

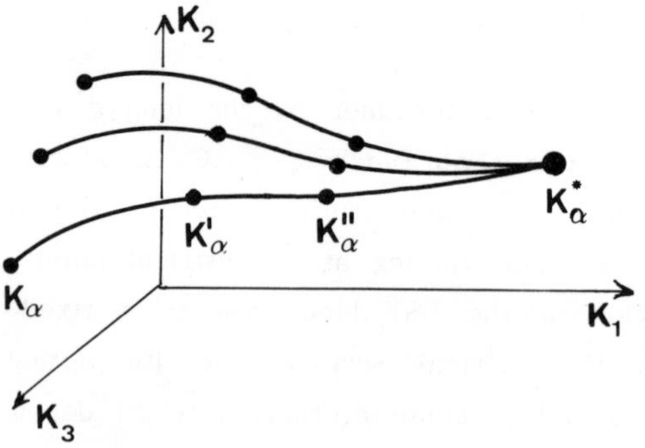

coupling. If all routes lead to a common f.p. that interaction is irrelevant for the long distance physics, the continuum limit (cf. Fig. 1).

For d>4 it has been proved rigorously that the regularized Φ^4 theory leads to a non-interacting free theory in the limit of vanishing regularization parameter (lattice spacing). In d=4 pertubation theory and (for small positive Φ^4-coupling rigorously controlled) renormalization group arguments give the same result[2,8].

Concerning the whole range of positive bare Φ^4-coupling there is some numerical evidence as well. The theory may be formulated in terms of non-intersecting random paths; MC-simulations in path-space along these lines[9] support the triviality conjecture for d=4. In another MC approach a physical quantity (the mass) was held fixed by readjusting the Φ^4-coupling while approaching the critical point. In consequence the 4-point coupling decreased in this limit[10].

There are RSRG results that the d=4 Ising model has mean field critical exponents[11]. Recent work[12] provides further evidence that this is the case also for Φ^4 theory.

Here I want to present results of a block-spin RG investigation to localize the couplings K_α at the fixed points of two different types of BST. One of them is linear in the block field variables, the other non-linear similar to the majority rule for Ising spins. A brief account of the results on the first type of BST was published elsewhere[13].

2. NOTATION AND METHODS

The general form of the action for the lattice Φ^4 theory may be written

$$S = \sum_\alpha K_\alpha S_\alpha , \quad S_\alpha = \sum_x S_{\alpha,x} , \tag{2.1}$$

where $S_{\alpha,x}$ denotes the contribution to S_α from site x, a polynomial in Φ_x and its neighbour fields. Fig. 2 shows the 6 types of quadratic and the 18 types of quartic interaction term that have been considered in this work. We assume periodic boundary conditions.

A minimal version of the action is

$$S = - K \sum_{x,\mu} \Phi_x \Phi_{x+\mu} + a \sum_x \Phi_x^2 + b \sum_x \Phi_x^4. \tag{2.2}$$

Due to the invariance of physical observables like the correlation length under an overall scale transformation of the fields one has to introduce a constraint in order to guarantee a unique relation between parameters and physics of the model. We choose the relation

$$a = 1 - 2b \tag{2.3}$$

corresponding to the action

$$S = - K \sum_{x,\mu} \Phi_x \Phi_{x+\mu} + \sum_x \Phi_x^2 + b \sum_x (\Phi_x^2 - 1)^2. \tag{2.4}$$

Figure 2: The interaction types considered, six of them are quadratic in the fields, the other 18 are quartic. The numbers at the sites (dots) denote the powers with which the corresponding fields contribute in the interaction term.

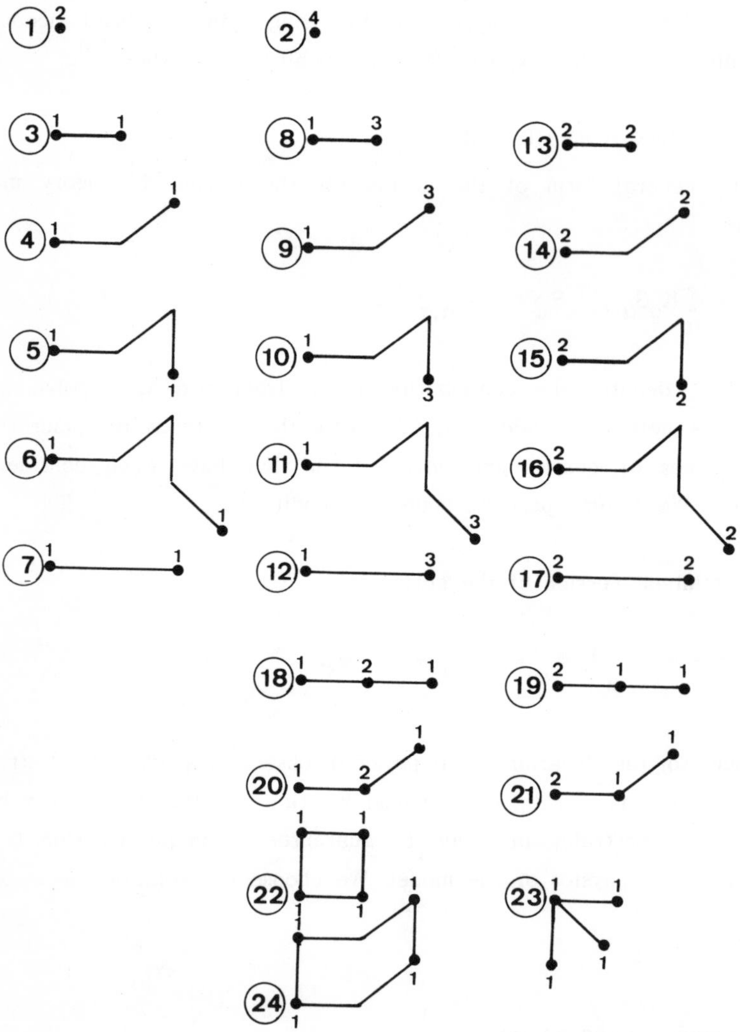

This relation (2.3) is simple to obey on the level of explicite knowledge of the coupling parameters. It is enforced a posteriori on the final results on the 8^4 and 4^4 lattices by a suitable scale transformation (cf. the discussion concerning parameter λ in the linear BST below).

Another possible constraint is to require

$$\langle\Phi^2\rangle = 1 \tag{2.5}$$

which may be satisfied again by a suitable scale transformation of the final results corresponding to a rescaling of the coupling constants. We choose this type of constraint for the non-linear BST (see below, cf. also Ref. 14 for further choices). Both constraints (2.4) and (2.5) remove one degree of freedom in the infinite dimensional parameter space of possible couplings, although in different ways. Any BST has to be designed such that one stays within the same class of theories, that obeying (2.4) or that obeying (2.5).

The ensemble of configurations on the 16^4 lattice was produced by standard MC methods discussed in more detail elsewhere[13,14]. The Φ^4 theory was simulated for three different values of the bare Φ^4-coupling. The corresponding critical values of the n.n. coupling K were taken from Ref.2, where they were obtained by Pade-analysis of high-temperature expansion. The numbers are ($b=K_2$, $K=-K_3$)

$K_2 = 0.2$	$K_2 = 0.4$	$K_2 = 2.0$
$K_3 = -0.2811$	$K_3 = -0.2844$	$K_3 = -0.2304$

and may be easily transformed to the somewhat different convention of Ref. 2 ($a+b=1$).

We studied the flow of coupling constants on the critical surface for two different block spin transformations. The first is the linear BST

$$\Phi'_{x'} = \lambda \sum_{x \varepsilon B_{x'}} \Phi_{x}, \tag{2.6}$$

where $B_{x'}$ denotes a block of 2^4 sites of the larger lattice located at site x' of the smaller lattice. Only the value $\lambda=1/8$ is compatible with a gaussian f.p. - however, we cannot prejudice our results. Therefore we have to adjust λ such as to stay within the same class of theories as discussed above. In practice λ enters in a definite way in all observables of the smaller lattice and one may decide on its actual value after the completion of the whole simulation program depending on the resulting values for the coupling constants or the obtained $<\Phi^2>$. Its compatibility with 1/8 provides another check as to whether the flow trajectories approach a gaussian f.p. or not.

The second BST considered leaves $<\Phi^2>$ invariant and has the form

$$\Phi'_{x'} = \text{sign} \sum_{x \varepsilon B_{x'}} \Phi_x \sqrt{\frac{1}{16} \sum_{x \varepsilon B_{x'}} \Phi_x^2} \quad . \tag{2.7}$$

For $\Phi^2=1$ (the Ising model) it corresponds to the majority rule.

The idea is to follow the flow of the coupling constants on the critical surface due to the applied BSTs. Starting on the critical surface one is bound to stay there and eventually will approach a fixed point. All theories (each of them corresponding to a possible starting point in parameter space) that approach a joint f.p. have the same long distance behaviour and thus the same continuum limit. The obvious problem is to find out the values of the coupling constants from the ensemble of configuration on the smaller size lattice obtained by the BST.

A method recently proposed by Swendsen[15)] allows to attack this problem. It is known[16)] that the expectation value of a local observable may be estimated by measuring its conditional expectation value in a fixed background configuration and then averaging over all these configurations. The "modified operator"

$$\overline{S}_{\alpha,x} \equiv \frac{\int d\phi_x \; S_{\alpha,x} \; (\phi_x, \ldots) \exp[-\sum_{\alpha} \tilde{K}_\alpha S_{\alpha,x}(\phi_x, \ldots)]}{\int d\phi_x \exp[-\sum_{\alpha} \tilde{K}_\alpha S_{\alpha,x}(\phi_x, \ldots)]} \tag{2.8}$$

is given by the integral over the site field variable ϕ_x with frozen neighbour fields. The expectation value of this modified operator equals that of the ordinary operator only if one has used the correct values $\tilde{K}$ for the ensemble of configurations considered.

For a small deviation of $\tilde{K}$ from the (in our case for lattice size 8^4 and 4^4 unknown) K we find

$$\langle S_\alpha \rangle = \langle \overline{S}_\alpha \rangle + \sum_\beta \frac{\partial \langle \overline{S}_\alpha \rangle}{\partial \tilde{K}_\beta} (K_\beta - \tilde{K}_\beta) + 0(K-\tilde{K})^2), \tag{2.9}$$

which leads to an iterative scheme to find a suitable value for K,

$$K_\beta \text{ (improved)} \simeq \tilde{K}_\beta(\text{old}) - \sum_\alpha [\frac{\partial \langle \overline{S}_\alpha \rangle}{\partial \tilde{K}_\beta}]^{-1} (\langle \overline{S}_\alpha \rangle - \langle S_\alpha \rangle) \tag{2.10}$$

The derivative term is given by to a formula similar to (2.8). Unlike other MC methods where one has to compare quantities obtained in independent MC runs here the values $\langle \overline{S} \rangle$ and $\langle S \rangle$ come from one run. Fluctuations are not independent and have a tendency to cancel in the difference. More details of the method and its implementation may be found elsewhere[15,13,14].

Several tests have been performed to check this method. One of them was to reproduce the values of a,b, and K that have been used for a simulation on an 8^4 lattice. Another was to investigate the sensitivity of the algorithm with regard to the neglect of certain interaction terms. The stability of Swendsen's method appeared to be very satisfactory[14].

3. RESULTS

The position of the f.p. on the critical surface itself is not universal, only the critical exponents are physical quantities and universal. Therefore it should not surprise that different BSTs lead to different flow structure. An important feature to be learned from this structure is, however, whether f.p. position depends on certain bare couplings or not.

The linear BST (2.6) moves the system towards smaller Φ^4-coupling. As an example cf. Tab. 1a where the coupling constants for starting value b=2.0 on 16^4 after one and two BST are given. The Φ^4-coupling decreases rapidly and the Φ^2 coupling approaches 1 simultaneously due to (2.3). Only the quadratic coupling constants are significantly different from zero, almost all quartic couplings are less than the typical error in their determination. The most important quadratic coupling seems to be the straight 2-link term number 7 (Fig.2).

Table 1a: Coupling constants on the critical surface as obtained on 8^4 and 4^4 lattices from BSTs of the 16^4 system for b=2.

α	16^4	8^4	4^4
1	-3.	.8220(149)	.9805(6)
2	2.	.0890(74)	.0098(3)
3	-.2304	-.2711(141)	-.3520(24)
4	0.	-.0041(22)	.0096(3)
5	0.	.0024(2)	.0035(2)
6	0.	.0037(7)	.0018(6)
7	0.	.0644(133)	.0455(12)
8-24	0.	$\|K_\alpha\| < 0.04$	$\|K_\alpha\| < 0.003$

In Tab.1b one finds the values obtained on 4^4 (after two BSTs from 16^4) for three different starting values of the bare Φ^4-coupling. One finds a remarkable agreement demonstrating that the trajectories are already very close to each other.

Table 1: Results for the coupling constants found after two BST (from 16^4 to 4^4 for three different starting values of the bare Φ^4-coupling b.

α	b=0.2	b=0.4	b=2.0
1	0.9918(15)	0.9955(82)	0.9805(6)
2	0.0041(8)	0.0023(41)	0.0098(3)
3	-0.3457(30)	-0.3487(29)	-0.3520(24)
4	0.0098(7)	0.0109(15)	0.0096(3)
5	0.0024(10)	0.0031(3)	0.0035(2)
6	0.0020(4)	-0.0001(19)	0.0018(6)
7	0.0445(17)	0.0388(71)	0.0455(12)
8	0.0000(1)	0.0002(2)	-0.0002(0)
9	0.0000(0)	0.0001(1)	-0.0000(0)
10	0.0000(0)	0.0001(1)	-0.0000(0)
11	0.0000(1)	0.0002(2)	-0.0000(0)
12	0.0005(2)	0.0011(10)	0.0001(1)
13	0.0000(0)	0.0001(1)	0.0000(0)
14	0.0000(0)	0.0000(1)	-0.0000(0)
15	0.0000(0)	0.0000(1)	-0.0000(0)
16	0.0000(0)	0.0001(1)	-0.0001(0)
17	0.0001(1)	0.0003(3)	-0.0001(0)
18	0.0013(6)	0.0032(29)	0.0006(3)
19	-0.0000(0)	-0.0000(0)	0.0000(0)
20	0.0000(0)	0.0001(1)	-0.0000(0)
21	0.0000(0)	0.0000(0)	0.0000(0)
22	0.0006(10)	0.0002(21)	0.0026(14)
23	-0.0001(3)	-0.0004(4)	-0.0002(3)
24	-0.0004(1)	-0.0007(3)	-0.0002(1)

We conclude that the trajectories tend towards a common f.p. independent of the bare Φ^4-coupling. For small enough b this is the gaussian f.p. and the observed behaviour indicates that it stays the only attractor (at least up to b=2).

Let us now turn to the non-linear BST (2.7). If the Ising model is in the same universality class we might expect that there is a BST with an Ising-type f.p. where $b\to\infty$. In this limit we find that Φ_x may assume only values 1 or -1. Many of the possible interaction terms become trivial (1,2,13-17) or reduce to other quadratic terms (8-12,18-21). Only 5 quadratic and 3 quartic interactions out of the 24 are still meaningful in this limit. In Ref. 11 it was shown that the d=4 Ising model ($b=\infty$) has critical exponents compatible with gaussian values.

Table 2: Coupling constants obtained after two BST (2.7) to size 4^4. The couplings not given in the table correspond to interactions that become in the Ising-limit identical to the ones determined. Due to this evolving linear dependence it was not possible to determine them separately.

α	b=0.0613	b=0.8788
1	-136.0	-213.9
2	67.9	105.6
3	-0.108	-0.112
4	-0.012	-0.011
5	-0.001	-0.001
6	0.001	0.002
7	0.007	0.006
22	-0.000	-0.000
23	-0.001	-0.001
24	-0.001	-0.001

It turns out that BST (2.7) shows that behaviour, i.e. the flow on the critical surface is towards an Ising model fixed point. For definiteness we now choose constraint (2.5) to start with. That requirement may be fulfilled by a rescaling of fields and coupling constants and leads to values of (K_1, K_2, K_3) = (a, b, -K) = (0.3321, 0.0613, -0.1556) and (-1.9886, 0.8788, -0.1527) for the earlier values b=0.2 and 2.0 respectively. As may be seen from Tab. 2 the coupling constants K_1 and K_2 approach the Ising-model values $-a/2b \to 1$.

This flow structure for K_1 and K_2 has been observed in Ref. 12 too. The renormalization flow becomes independent of the bare Φ^4-coupling after 2 BSTs.

We conclude from these results that the Ising model and the gaussian model in d=4 are in the same universality class. Φ^4 theories with different positive values of the bare Φ^4-coupling have the same long distance behaviour and thus the same continuum limit: that of the gaussian model.

Acknowledgement: This work was made possible by an invitation of Prof. Ludwig Streit to participate at "Project No. 2: Mathematics and Physcis" at ZIF, Universität Bielefeld. I want to thank for this opportunity and the access to the CYBER 205 of Universität Bochum. Support of the "Fonds zur Förderung der Wissenschaftl. Forschung in Österreich", project P5125 is also acknowledged.

4. REFERENCES

1 Th. W. Burkhardt and J. M. J. van Leeuwen, eds., "Real Space Renormalization", Topics in Current Physics 30 (Springer: Berlin 1982); Th. Niemeijer and J. M. J. van Leeuwen, in "Phase Transitions and Critical Phenomena", ed. by C. Domb and M. S. Green, Vol. 6 (Academic: New York, 1976).

2 K. G. Wilson and J. Kogut, Phys. Reports 12C, 76 (1974).

3 S. K. Ma, Phys. Rev. Lett. 37, 461 (1976).

4 R. H. Swendsen, in "Real Space Renormalization", Topics in Current Physics 30, ed. by Th. W. Burkhardt and J. M. J. van Leeuwen

(Springer: Berlin, 1982).

5 K. G. Wilson, in "Recent Developments of Gauge Theories", ed. by G.'t Hooft et al. (Plenum Press, 1980).

6 P. Hasenfratz, these Proceedings;
K. C. Bowler, A. Hasenfratz, P. Hasenfratz, U. Heller, F. Karsch, R. D. Kenway. I. Montvay, G. S. Pawley and D. J. Wallace, CERN preprint TH.3952 (1984).

7 R. Gupta and A. Patel, CALTECH preprint, CALT-68-1121 (1984); R. Gupta, G. Guralnik, A. Patel, T. Warnock and C. Zemach, CALTECH preprint, CALT-68-1143 (1984).

8 K. Gawedzki and A. Kupiainen, Phys. Rev. Lett. 54,92 (1985).

9 C. Aragao de Carvalho, S. Caracciolo and J. Fröhlich, Nucl. Phys. B215 [FS7], 209 (1983).

10 B. Freedman, P. Smolensky and D. Weingarten, Phys. Lett. 113B, 481 (1982).

11 H. W. J. Blöte and R. H. Swendsen, Phys. Rev. B22, 4481 (1980).

12 D. J. E. Callaway and R. Petronzio, Phys. Lett. 139B, 189 (1984), Nucl. Phys. B240 [FS12], 577 (1984).

13 C. B. Lang, Phys. Lett. 155B, 399 (1985).

14 C. B. Lang, in preparation.

15 R. H. Swendsen, Phys. Rev. Lett. 52, 1165 (1984).

16 H. B. Callen, Phys. Lett. 4B, 161 (1963).
R. L. Dobrushin, Theory Prob. Appl. 13, 387 (1969).
O. E. Lanford III and D. Ruelle, Commun. Math. Phys. 13, 194 (1969).
G. Parisi, R. Petronzio and F. Rapuano, Phys. Lett. 128B, 418 (1983).

PROPERTIES OF PHASE TRANSITIONS IN U(1) AND SU(2) LATTICE HIGGS MODELS

Jiří Jersák

Institute of Theoretical Physics, E
Technische Hochschule Aachen
D-51 Aachen
F.R.GERMANY

ABSTRACT

The U(1) and SU(2) lattice Higgs models with scalar fields in the fundamental representations have been studied in high statistics Monte Carlo simulation. We have completed the phase diagrams of both models by finding several new phase transitions at negative β, and determined the positions of the Higgs phase transitions with great accuracy. The weakening of the first-order Higgs phase transition with increasing λ or β has been demonstrated. At the points $\lambda = 0.5$, $\beta = 2.25$ and $\lambda = 3$, $\beta = 2.5$ in the U(1) and SU(2) models, respectively, we have found signs of critical behaviour.

1. INTRODUCTION

I would like to describe the recent results of Monte Carlo simulations of lattice Higgs models obtained by the Aachen-Graz group whose members are C.B.Lang and G.Vones (Graz) and H.G.Evertz, K.Jansen, T.Neuhaus and myself (Aachen). We have been studying both the U(1) and the SU(2) gauge fields coupled to scalar fields in fundamental representations of the gauge groups [1,2].

The usual explanation of nonzero masses of intermediate vector bosons in the U(1)×SU(2) gauge theory of electroweak interactions and in its generalizations is the Higgs mechanism for the generation of gauge boson masses. On the tree level this mechanism is based on the supposed spontaneous breakdown of a global symmetry in the ϕ^4_4 field theory. However, the accumulating evidence that this theory might be trivial (see e.g. the talk by C. B. Lang at this meeting or Ref.3) forces us to look at the Higgs mechanism more carefully, preferably by using nonperturbative methods. It should be shown that the coupling to a gauge field makes the ϕ^4_4 theory nontrivial and that the

Higgs mechanism really does exist. This might be true only in some regions in the space of the coupling parameters[4]. A resulting restriction on the Higgs boson mass would be extremely valuable also from the phenomenological point of view. Thus there is a good incentive to study the Higgs mechanism by the nonperturbative methods of lattice gauge theory.

A nonperturbative gauge invariant approach to the Higgs models, in particular on the lattice, reveals for the Higgs fields in the fundamental representation an analytic connection between the Higgs and the confinement regions of the phase diagram. This Susskind-Fradkin-Shenker phase identity[5,6] suggests that beyond the usual perturbative spectrum of the Higgs models also some excited states of gauge and Higgs bosons in analogy to the meson spectroscopy might exist. It would be very interesting to estimate the observability of these states.

Both the U(1) and SU(2) lattice Higgs models with the scalar field in the fundamental representation were studied by various analytic methods[5] and by numerous Monte Carlo simulations[7-10]. The phase diagrams are well understood at positive β, the inverse gauge field coupling squared. We have completed this analysis by detecting several phase transitions (PT's) also at negative β. We have also determined more precisely the position of the sheet of Higgs PT's in the 3-dimensional space of couplings.

Nevertheless, the most important information is still missing. We do not yet know whether and where the Higgs PT is of higher order, i.e. where we can attempt to construct the continuum limit. One obvious candidate is the line $\beta = \infty$ on the Higgs PT sheet, since the pure ϕ^4_4 theory on a lattice is believed to have a second-order PT. But for the purposes of numerical simulations it would be helpful to find a critical behaviour at moderate values of β, $\beta = 0(1)$.

It is easy to show that for sufficiently small values of the quartic coupling λ the Higgs PT at moderate values of β is in both models of first order. This has been expected on the basis of the Coleman-Weinberg phenomenon[11]. The transition weakens with increasing λ. But does it change its order as suggested for the U(1) model by Kleinert[12] and, if so, where? To clarify this by means of Monte Carlo simulations may turn out to be a very difficult task.

If the Higgs PT is of second order at some large but finite λ, then on the Higgs PTsheet there exists a line of the so-called tricritical points separating the first- and second-order PT areas. The experience both in statistical mechanics models (see e.g. Ref.13) and in pure U(1) lattice gauge theory[14] teaches us that one can easily misinterpret the order of a PT in the vicinity of tricritical points in calculations on finite, small lattices. False signals of coexistent states on a second-order PT as well as indications of critical behaviour on a first-order PT are common, causing a lot of confusion.

With our limited resources (Cyber 205 in Bochum having only 0.5 Megaword of memory) we have started a search for the regions on the Higgs PT sheet where the system gets at least as critical as the sizes of our lattices allow. For U(1) we find at $\lambda = 3$ and $\beta = 2.5$ on the lattices of sizes up to 8^4 an agreement with the finite size scaling theory, which indicates that some correlation length is of the order of the size of this lattice. In SU(2) we have localized at $\lambda = 0.5$ and $\beta = 2.25$ an extremely narrow interval ($\Delta\kappa \simeq .001$, κ being the gauge-Higgs field coupling) where the correlation length corresponding to the inverse Higgs boson mass achieves the value of 5 lattice spacings on an $8^3 \cdot 16$ lattice. This are encouraging results, but notice that we make no claim about what will happen at these points on larger lattices.

We also show some preliminary results of the calculations of the Higgs and gauge boson masses in the vicinity of the Higgs PT in the SU(2) model. Needless to say, at present we are presumably very far from the continuum limit and these mass values obtained on a lattice have no obvious relation to the masses of the physical particles.

2. THE MODELS

The action for the U(1) Higgs model reads

$$\begin{aligned} S = & -\beta \frac{1}{2} \sum_p (U_p + U_p^*) \\ & + \lambda \sum_x (\phi_x^* \phi_x - 1)^2 + \sum_x \phi_x^* \phi_x \\ & - \kappa \sum_{x,\mu} (\phi_x^* U_{x\mu} \phi_{x+\mu} + \text{c.c.}) . \end{aligned} \tag{1}$$

Here $U_{x\mu}$ are the U(1) link variables, U_p are their products around the plaquettes and ϕ_x is the complex scalar field living on the lattice sites. It includes the variable radial mode $|\phi_x|$. The action for the SU(2) model is analogous to (1), with $U_{x\mu}$ and ϕ_x being now quadratic and column matrices, respectively. We follow exactly the notation described in detail in Ref.9.

The action (1) is related to the form of the action used in the continuum theory

$$S_c = \int d^4x \, (|D_\mu \phi_c|^2 + m_c^2 |\phi_c|^2 + \lambda_c |\phi_c|^4) \tag{2}$$

by a set of reparametrizations

$$\phi_c = \phi \frac{\sqrt{\kappa}}{a} \quad , \quad \lambda_c = \frac{\lambda}{\kappa^2} \quad , \quad m_c^2 = \frac{1 - 2\lambda - 8\kappa}{\kappa a^2} \quad , \tag{3}$$

where a is the lattice constant.

Both models possess two symmetries in the 3-dimensional space of coupling constants β, λ, κ. The gauge-Higgs field coupling can be reflected, $\kappa \rightarrow -\kappa$. Thus it suffices to investigate the models for κ nonnegative. For $\kappa = 0$ also the sign of β can be reversed [9].

In both models we have approximated the continuous gauge groups U(1) and SU(2) by their discrete subgroups, Z(60) and $\tilde{Y}$, respectively. In the case of U(1) the approximation is perfect. But in the case of the icosahedral approximation the region of its validity has to be carefully determined, since the effects of this approximation can be detected close to the Higgs PT at values of β far below $\beta = 6$ (this is the position of the freezing PT in the pure $\tilde{Y}$ lattice gauge theory).

3. PHASES AND POSITIONS OF PHASE TRANSITION LINES

Let us start by displaying the phase structure of the U(1) model at fixed $\lambda = 0.01$ (Fig.1). The PT line (a) separates the confinement

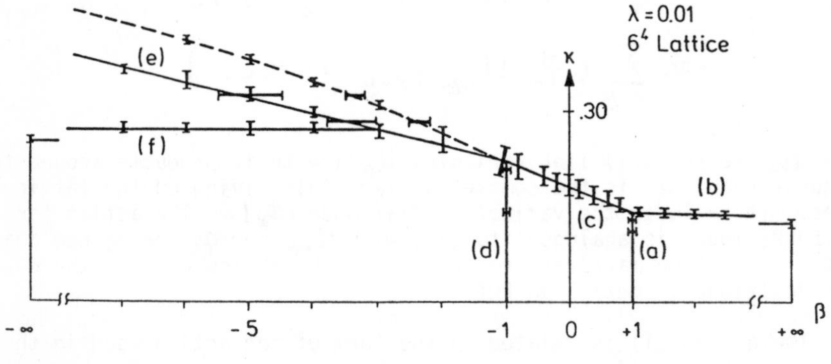

Fig.1. Phase structure of the U(1) model at $\lambda = 0.01$.

and the Coulomb phases. The line (b) is the Coulomb-Higgs PT. At this low value of λ the confinement and the Higgs regions are separated by the Higgs PT line (c) which extends to negative β. Following this line we have found several additional PT's: Line (d) is the twin of (a) (due to the $\beta \to -\beta$ symmetry at $\kappa = 0$) and separates the confinement phase from the plaquette frustration phase. In the frustration phase at large negative β the mean plaquette values $\langle U_p \rangle$ are negative and close to -1. This phase extends up to the PT line (e), where the mean plaquette changes sign. In addition, on the PT (f) the Higgs field expectation value $\langle \phi^* \phi \rangle$ starts to grow linearly with κ, similarly as above the Higgs PT lines (b) and (c). The dashed curve indicates the position of a cross-over behaviour of the mean plaquette when it approaches the maximal value $\langle U_p \rangle = 1$.

As can be seen in Fig.2, this picture changes with λ. The most interesting feature is the opening of the analyticity strip between the confinement and the Higgs regions at higher λ in accordance with the Susskind-Fradkin-Shenker phase identity. Even at low λ these regions are connected through a tunnel along the λ axis.

A typical picture of the Higgs PT at low λ is shown in Fig.3. A prominent hysteresis obtained in a fine thermal cycle indicates that the transition is of first order. The Higgs field expectation value rises as

$$\langle \phi^* \phi \rangle \simeq \frac{4}{\lambda} \kappa + \text{const.} \tag{4}$$

already quite early above the PT. The slope in (4) can be easily deduced from the action (1) for $\kappa \to \infty$.

The shapes of the PT's at negative β are illustrated in Fig.4. It shows a thermal cycle crossing both PT lines (e) and (f), as well as the cross-over line, at $\lambda = 0.1$. The frustration PT (e) is the most prominent event on the cycle.

The phase diagram of the SU(2) Higgs model for $\lambda = 0.01$ (Fig.5) turns out to be quite analogous to the phase diagram of the U(1) model, shown in Fig.1. But now the confinement phase extends to $\beta = \pm\infty$. On the dashed lines the icosahedral freezing transition or its cross-over-like remnant above the Higgs PT have been found.

The Coulomb-Higgs and/or the confinement-Higgs PT's in the U(1) and SU(2) models are, of course, physically the most interesting of all the PT lines. The two-dimensional sheet of these Higgs phase transitions is the lattice analogue of the Higgs PT in the continuum theory. A continuum limit should be constructed there. Therefore we have paid some attention to the accurate position of the Higgs PT sheet.

The position of the Higgs PT sheet at several values of λ in the U(1) model is shown in Fig.6. The accuracy can be estimated from the widths of the error bars on the PT lines (b) and (c) in Fig.1.

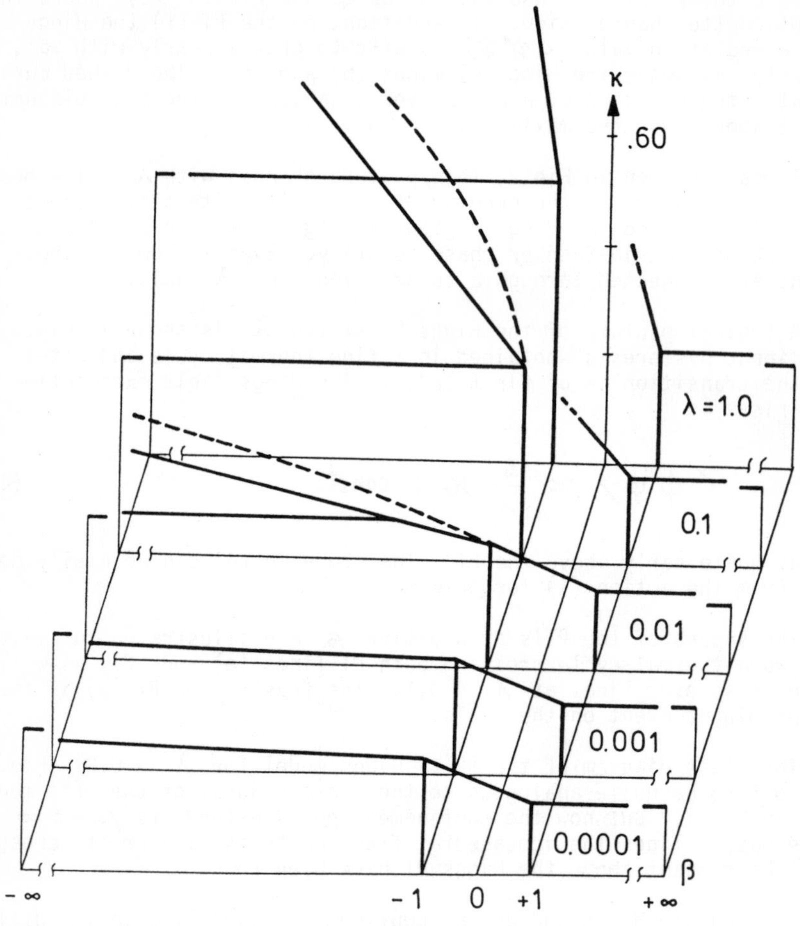

Fig. 2. Three-dimensional phase diagram of the U(1) model.

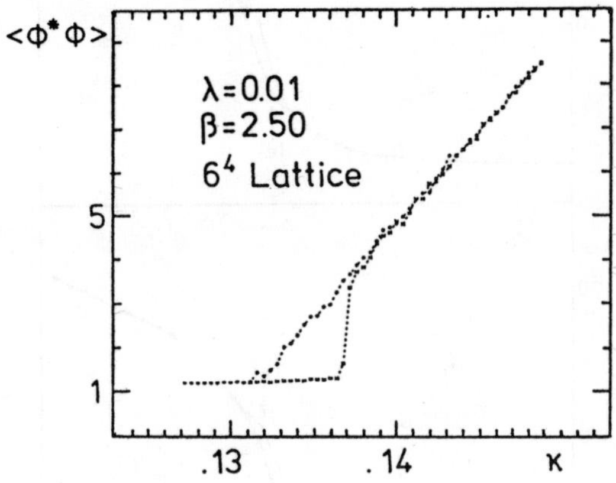

Fig. 3. Thermal cycle on the Higgs PT in the U(1) model.

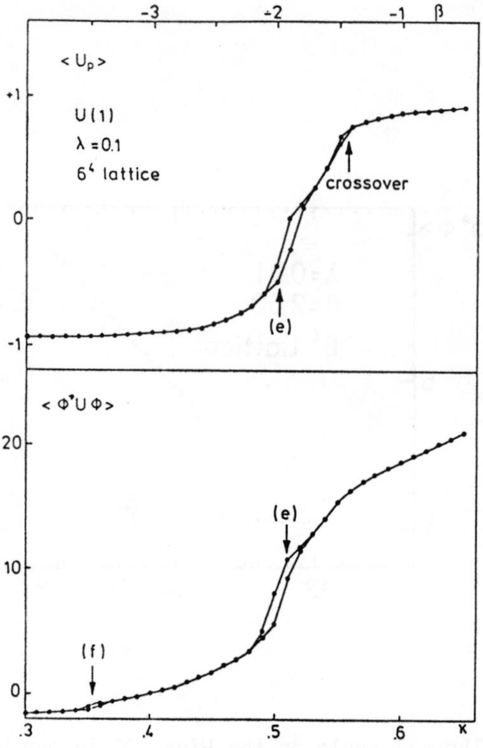

Fig. 4. Thermal cycle at negative β in the U(1) model.

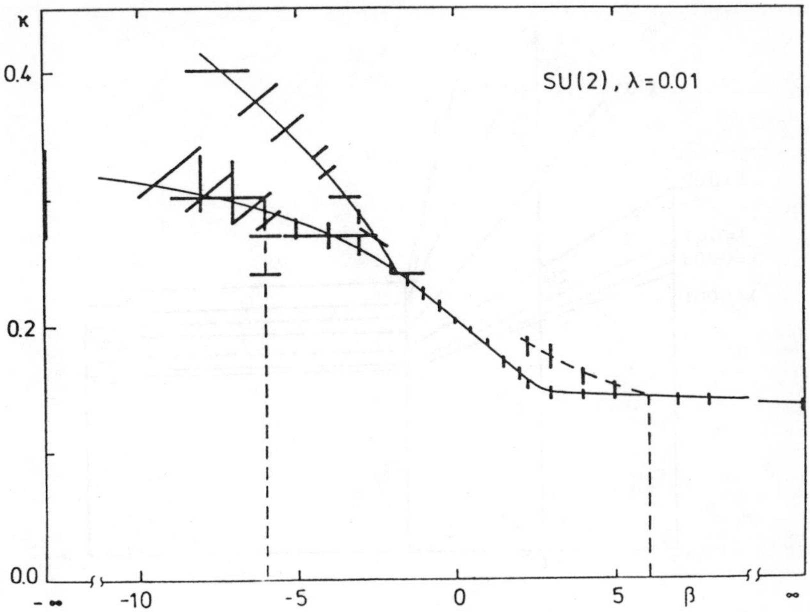

Fig. 5. Phase structure of the SU(2) model at λ = 0.01.

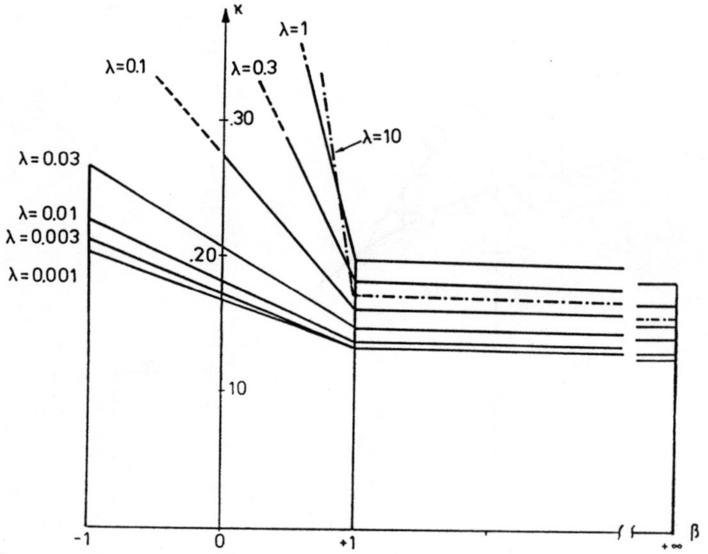

Fig. 6. Positions of the Higgs PT in the U(1) model for several fixed λ.

In Fig.7 we display two lines of the sheet for β = 2.5 and $\beta = \infty$.

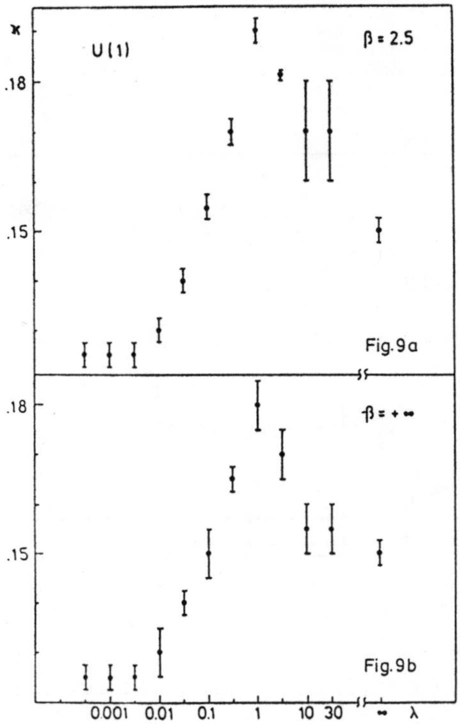

Fig. 7. Positions of the Higgs PT in the U(1) model for two fixed values of β .

Analogous results for the SU(2) model are given in the next two figures. In Fig.8 the four diamond points are results obtained by Montvay[10]. The dashed lines indicate the uncertainty at the positions of the end points of the lines at higher values of λ. Fig. 9 contains a PT line at fixed β. It is remarkable that this line rises monotonically with increasing λ , whereas in U(1) the analogous lines have a maximum height for $\lambda \simeq 1$ (Fig.7).

The dashed line in Fig.9 is the position of the cross-over caused by the freezing of the icosahedral subgroup of the SU(2) gauge group. At low values of β the position of this cross-over is quite incons-

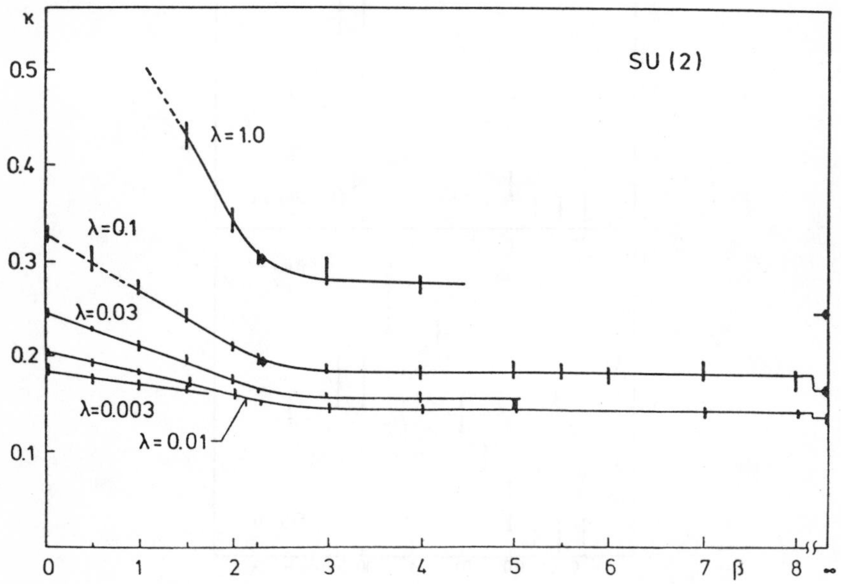

Fig.8. Positions of the Higgs PT in the SU(2) model for several fixed λ.

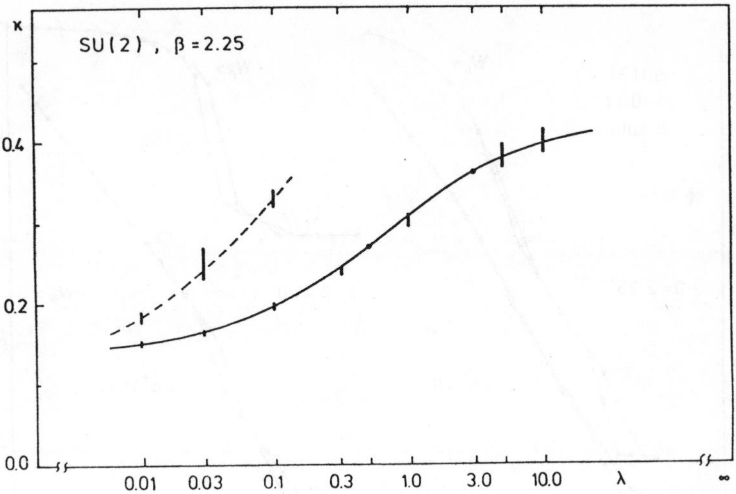

Fig. 9. Positions of the Higgs PT in the SU(2) model at β = 2.25.

picuous (see Fig.10) and can be most reliably detected by tracing the whole line of the freezing transitions as it weakens and changes into a cross-over with decreasing β. Such a line for $\lambda = 0.01$ is indicated in Fig. 5 (the dashed line at positive β). We conclude from Figs. 5 and 9 that the icosahedral approximation is reliable in the vicinity of the Higgs PT for large λ at all $\beta < 6$. At small λ some caution is necessary, however.

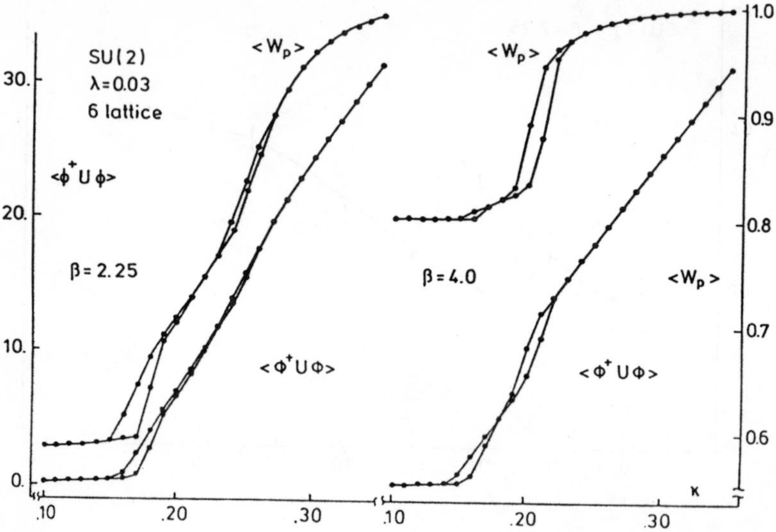

Fig.10. Effects of the icosahedral approximation of the SU(2) group.

4. WEAKENING OF THE HIGGS PHASE TRANSITION

Let us first discuss the β-dependence of the strength of the Higgs PT. We have compared the hystereses for the link product $\langle\phi^* U \phi\rangle$ produced by equivalent thermal cycles (i.e. those with the same number of sweeps at each point and with the same width of the steps in κ) for a given λ at various values of β. In the SU(2) model for λ = 0.03 (Fig.11) the weak hysteresis at β = 0 indicates the vicinity of the end point of the Higgs PT line. With increasing β the PT first builds up, but for large β it weakens again. At λ = 0.01 (Fig.12) the prominent hystereses at negative values of β decrease monotonically with increasing β. This evolution with β is consistent with the second order of the PT at $\beta = \infty$, where the Higgs model reduces to the pure lattice ϕ^4_4 theory.

At these values of λ for β=0(1) it is easy to show that the Higgs PT is of first order. For example, even on the smallest of the hystereses in Fig.12, at the point indicated by an arrow, a pair of long

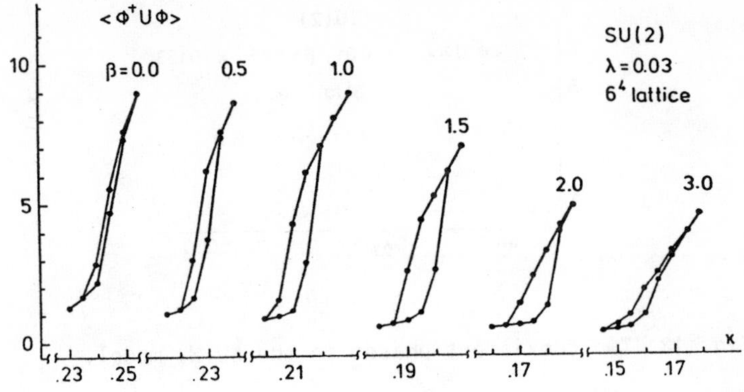

Fig.11. Hystereses on the Higgs PT at λ = 0.03 in the SU(2) model.

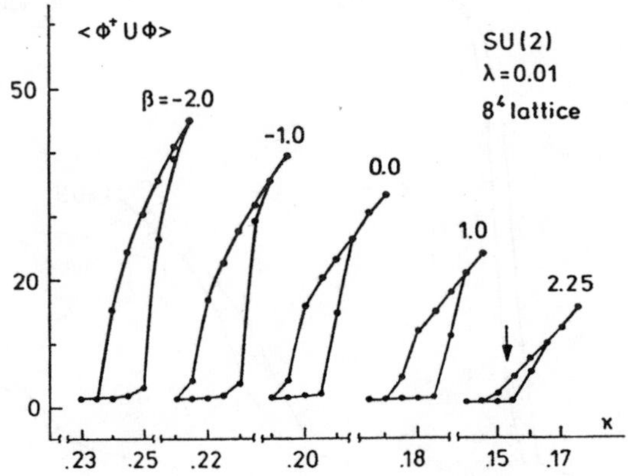

Fig.12. Hystereses on the Higgs PT at λ = 0.01 in the SU(2) model.

iterations with hot and cold starts produces a distinct two-state signal (see Fig.13).

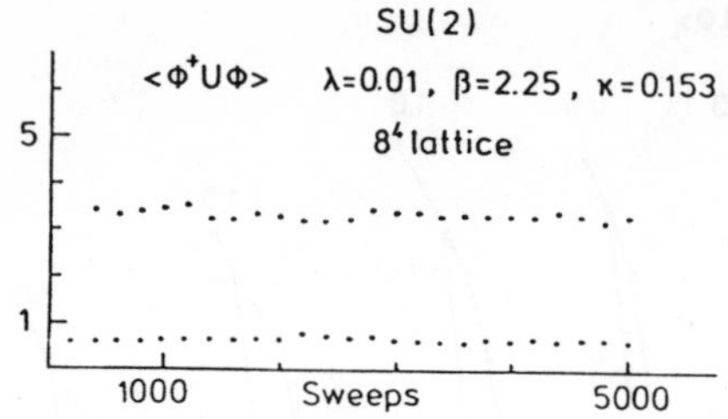

Fig. 13. Two coexistent phases in the SU(2) model.

When β is fixed and λ increases, the Higgs PT weakens, too. This is seen by comparing equivalent thermal cycles at various values of λ, e. g. for the U(1) model in Fig.14. Here we plot the expecta-

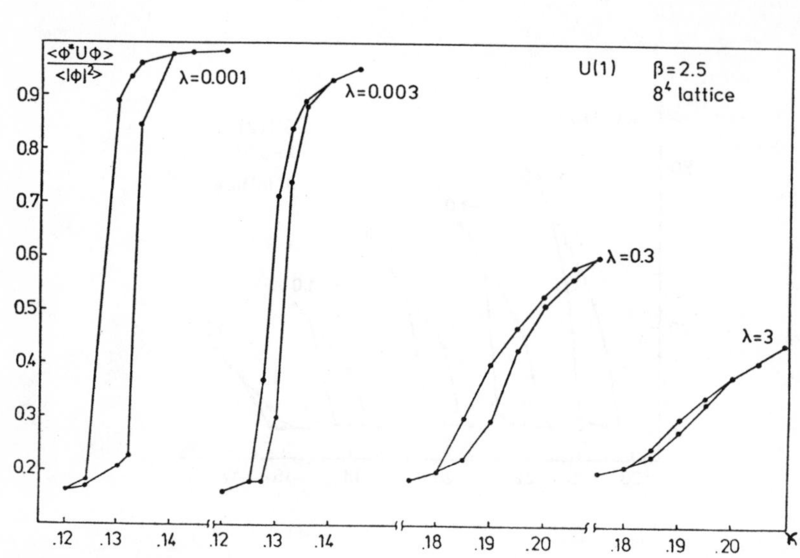

Fig. 14. Comparison of the Higgs PT for various λ in the U(1) model.

tion value of the link product normalized by the squared absolute value of the Higgs field, $\langle \phi^* U \phi \rangle / \langle \phi^* \phi \rangle$, so that the different slopes $4/\lambda$ of the Higgs field dependence on κ (cf. Eq. (4)) are taken into account. One can also observe a decreasing clearness of a two-state signal. For example, for U(1) at λ = 0.001 and β = 2.5 two distinct states are observed in long iterations, see Fig.15. (We have been lucky to observe here also one phase flip, at κ = 0.13.) At λ = 0.3 (Fig.16) a long iteration produces several phase flips. and

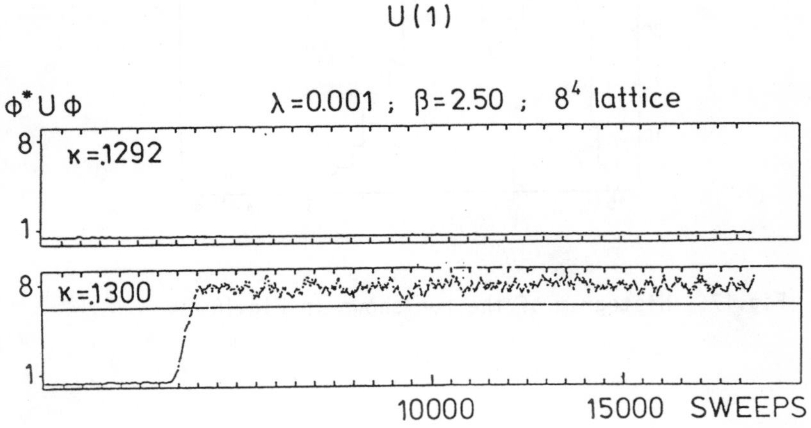

Fig. 15. Two coexistent phases in the U(1) model at λ = 0.001, $\beta = 2.5$.

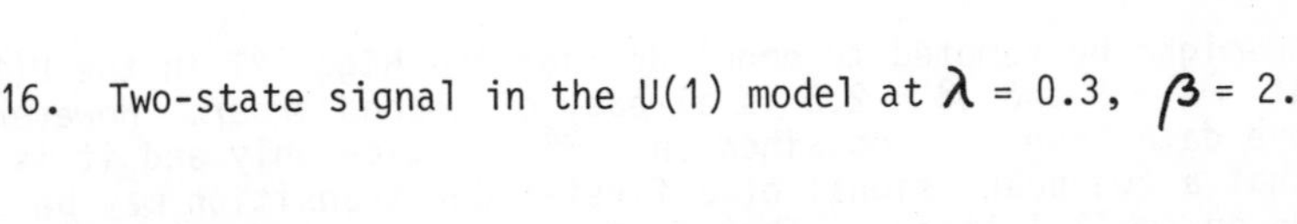

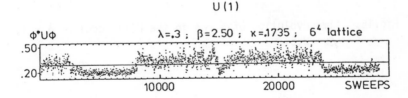

Fig.16. Two-state signal in the U(1) model at λ = 0.3, β = 2.5.

the histogram of this iteration still shows two distinct peaks(Fig.17). At λ = 3 it becomes difficult to distinguish between a two-state signal and long time correlations (Fig.18). No double peak is seen in the histogram (Fig.19).

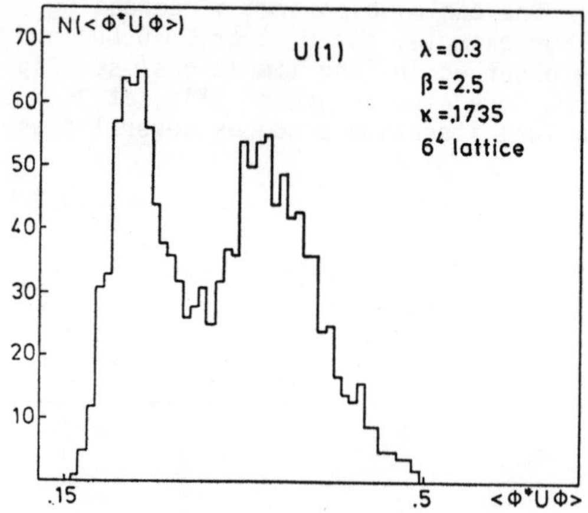

Fig.17. Histogram of the run shown in Fig.16.

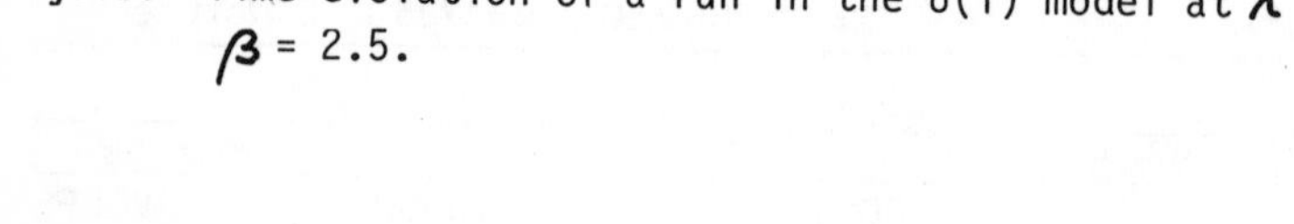

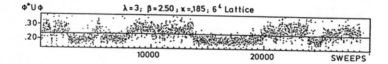

Fig.18. Time evolution of a run in the U(1) model at λ = 3, β = 2.5.

One might be tempted to conclude that the Higgs PT in the U(1) model at λ = 3 and β = 2.5 is already of second order. However, the above data have been obtained on a 6^4 lattice only and it is known that a two peak signal of a first-order transition may be quite obscured on small lattices. Therefore we only conclude that the point

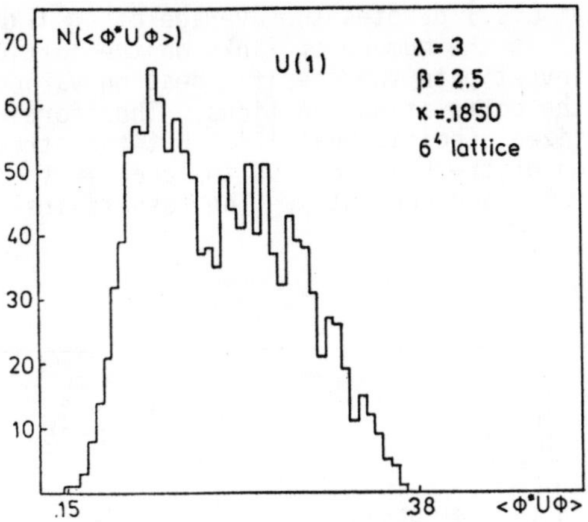

Fig.19. Histogram of the run shown in Fig.18.

is a promissing candidate for a second-order PT and that we should investigate it on lattices as large as possible by methods most suitable for detection of the critical behaviour.

5. SIGNS OF CRITICALITY AT LARGE λ.

At present we are searching for the critical behaviour on the Higgs PT sheet at large λ by calculating the correlation lengths on lattices as large as practically available to us in Bochum ($8^3 \cdot 16$). The inverse correlation lengths are , of course, the Higgs boson and gauge boson masses. Since such calculations are tedious, we can perform them only at a few selected points on the Higgs PT sheet. Let me describe our recent results at two points.

5.1 Point $\lambda = 3$, $\beta = 2.5$ in the U(1) Model

This is the point on the Higgs PT sheet in the vicinity of which we found on a 6^4 lattice a sign of long correlations (Figs. 18, 19). In order to localize the critical region we calculated the link product contribution to the specific heat

$$\partial_\kappa \langle \phi^* U \phi \rangle = N_\ell \left(\langle [\phi^* U \phi]^2 \rangle - \langle \phi^* U \phi \rangle^2 \right). \quad (5)$$

Here the bracket [...] denotes the average of $\phi^* U \phi$ on a given configuration and N_l is the number of links on the lattice. Originally we did not plan to investigate the specific heat on various lattices, aiming directly at the correlation functions. Therefore we did not calculate the normalized specific heat [15]. But the strong size dependence of the position of the PT forced us to localize the PT on several lattices (4^4, 6^4, 8^4) independently. The results are shown in Fig. 20.

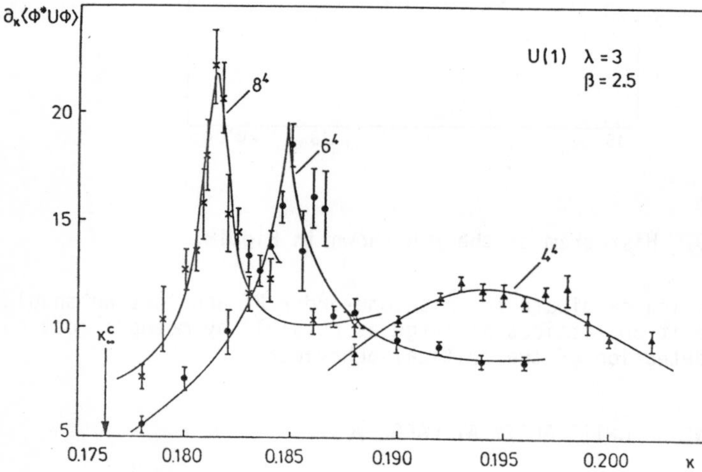

Fig. 20. Specific heat in the U(1) model at $\lambda = 3$, $\beta = 2.5$.

This strong size dependence is a sign of criticality. Fitting the results on each lattice by the function

$$a + b\kappa + \frac{c}{1 + e\,|\kappa - \kappa_L^c|^f} \tag{6}$$

we have extrapolated the position of the peaks κ_L^c on lattices of size L to $L = \infty$ by means of the finite size scaling formula

$$|\kappa_L^c - \kappa_\infty^c| = k\, L^{-1/\nu} \tag{7}$$

and obtained

$$\kappa_\infty^c = 0.17631\,(4)$$
$$\nu = 0.56478\,(13). \qquad (8)$$

Unfortunately, it is not possible to conclude from these results that the PT is of second order. A strong size dependence and consistency with the finite size scaling theory have been observed also in the pure U(1) lattice gauge theory[15], although further investigations [14] demonstrated later that the PT is of first order, albeit close to a tricritical point.

5.2 Point $\lambda = 0.5$, $\beta = 2.25$ in the SU(2) Model

At this point we have found a dramatic size dependence of the width of the specific heat peak. Whereas not much of the PT is seen on a 4^4 lattice, there is a sharp peak, only about $\Delta\kappa \simeq 0.001$ wide, on a $8^3 \cdot 16$ lattice (Fig.21). Such a peak can be easily overlooked in

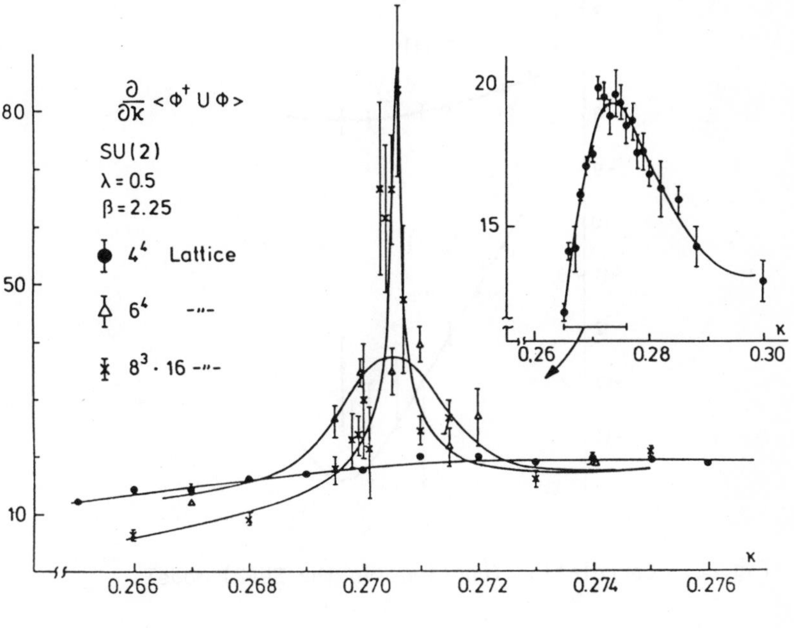

Fig. 21. Specific heat in the SU(2) model at $\lambda = 0.5$, $\beta = 2.25$.

Monte Carlo calculations. It is also very expensive to find such a peak, since one has to look for it on a large lattice and by means of long iteration runs. The points on the $8^3 \cdot 16$ lattice are results of 15 000 - 30 000 sweeps each. We found it therefore more economical to use the asymmetric lattice and to calculate simultaneously to the search for PT also the correlation functions.

I would like to show one pair of correlation functions at $\lambda = 0.5$, $\beta = 2.25$ and $\kappa = 0.2703$, inside of the narrow specific heat peak (Fig.22). The correlation functions are defined

$$\Gamma(t) = \langle \sum_{\vec{x}} \sigma_p(\vec{x},0) \sum_{\vec{y}} \sigma_p(\vec{y},0) \rangle - \langle \sum_{\vec{x}} \sigma_p(\vec{x},0) \rangle^2 , \qquad (9)$$

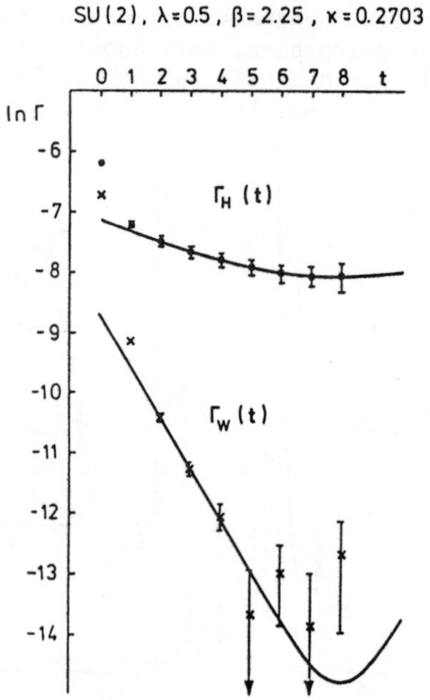

Fig.22. Correlation functions in the SU(2) model.

for the operators $\left(\phi_x = \rho_x \sigma_x \binom{1}{0},\ \rho_x \geq 0,\ \sigma_x \in SU(2) \right)$

$$O_P = \rho_x^2 \quad \text{and} \quad \sum_{\mu=1}^{3} \rho_x \rho_{x+\mu} \, \mathrm{Tr}(\sigma_x^+ U_{x\mu} \sigma_{x+\mu}) \qquad (\underline{1}\ 0^{++}) \tag{10}$$

$$O_P = \rho_x \rho_{x+\mu} \, \mathrm{Tr}(\vec{\tau} \sigma_x^+ U_{x\mu} \sigma_{x+\mu}) \qquad (\underline{3}\ 1^{--}) \tag{11}$$

where $\vec{\tau}$ denotes the Pauli-matrices. These operators contribute to the bound state objects with quantum numbers 0^{++} (isoscalar) and 1^{--} (isovector) respectively. The masses obtained in a fit to the exponential decay

$$\Gamma(t) = A \left[\exp(-mt) + \exp(-m(16-t)) \right] \tag{12}$$

correspond therefore to the Higgs boson and the gauge vector boson particles realized as bound states of the system. Fig.22 displays both the correlation functions and the fits to the points t = 2,...8. In the Higgs boson channel we find a very slow decay and we are able to identify the signal up to the symmetry point (t=8). The dimensionless mass values are

$$\begin{aligned} a\, m_H &= 0.20 \pm 0.03 \\ a\, m_W &= 0.85 \pm 0.08 . \end{aligned} \tag{13}$$

The magnitude of the correlation length in the Higgs channel, $1/m_H \propto 5a$ is about as large as might be expected on an $8^3 \cdot 16$ lattice for a critical system. But, again, what will happen on larger lattices...?

Finally, let me show our preliminary results for the Higgs boson and the gauge boson masses in the vicinity of the Higgs PT point $\lambda = 0.5$, $\beta = 2.25$ (Fig.23). We see a dip in the Higgs boson mass in the same interval in which the specific heat on the $8^3 \cdot 16$ lattice has the sharp peak (Fig.21). The gauge boson mass does not follow this behaviour. We are at present performing similar calculations at the points $\lambda = 3$, $\beta = 2.25$ and $\lambda = 0.5$, $\beta = 2.4$ in order to find out how the masses depend on λ and on β[2].

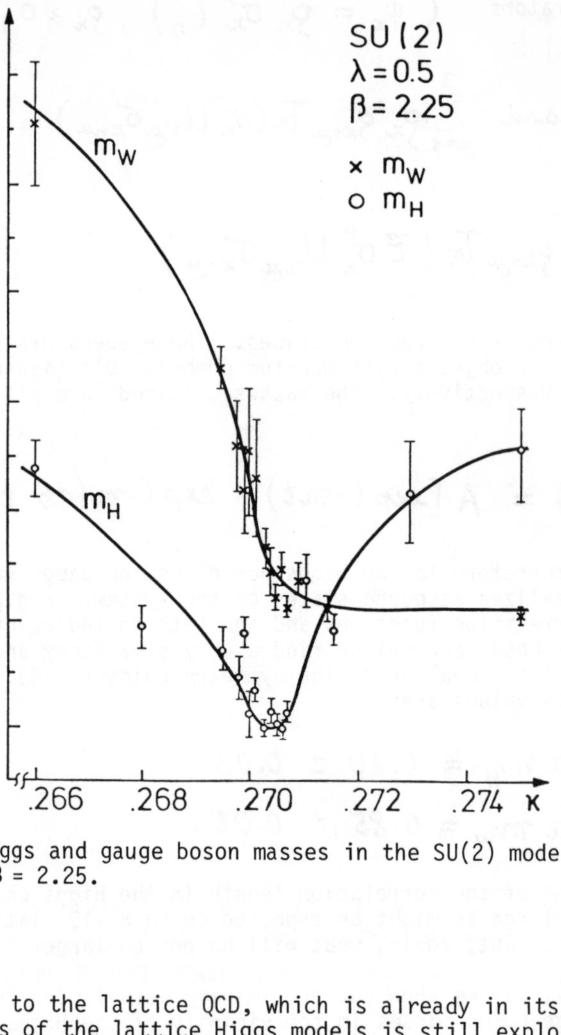

Fig.23. Higgs and gauge boson masses in the SU(2) model at $\lambda = 0.5$, $\beta = 2.25$.

6. CONCLUSION

In contrast to the lattice QCD, which is already in its productive phase, the status of the lattice Higgs models is still exploratory. We still have to learn where and how to perform the continuum limit. The difficulties are caused in part by the large number (3) of the coupling parameters, but also by certain difficulties to determine the order of the Higgs phase transition. We have found some promising signs of criticality at rather low values of β , $\beta = 2.5$. But much work on larger lattices will be required before we shall know reliably the critical points on the sheet of Higgs phase transitions.

ACKNOWLEDGEMENT

I thank C. B. Lang and P. M. Zerwas for stimulating discussions.

REFERENCES

1) K. Jansen, J. Jersák, C. B. Lang, T. Neuhaus and G. Vones, Phys. Lett. 155B, 268(1985);
K. Jansen, J. Jersák, C. B. Lang, T. Neuhaus and G. Vones, Aachen preprint PITHA 85/04;
J. Jersák, C. B. Lang, T. Neuhaus and G. Vones, Aachen preprint PITHA 85/05.

2) H. G. Evertz, J. Jersák, C. B. Lang and T. Neuhaus (in preparation).

3) C. B. Lang, Phys. Lett. 155B (1985) 399, and references therein.

4) D. J. E. Callaway, Nucl. Phys. B223, 189(1984).

5) K. Osterwalder and E. Seiler, Ann. Phys. 110, 440(1978);
E. Fradkin and S. Shenker, Phys. Rev. D19, 3682(1979);
T. Banks and E. Rabinovici, Nucl. Phys. B160, 349(1979);
E. Seiler, "Gauge Theories as a Problem of Constructive Quantum Field Theory and Statistical Mechanics", Lecture Notes in Physics 159(Springer, 1982).

6) S. Dimopoulos, S. Raby and L. Susskind, Nucl. Phys. B173, 208(1980).

7) M. Creutz, Phys. Rev. D21, 1006(1980);
K. C. Bowler, G. S. Pawley, B. J. Pendleton, D. J. Wallace and G. W. Thomas, Phys. Lett. 104B, 481(1981);
D. J. E. Callaway and L. J. Carson, Phys. Rev. D25, 531(1982);
J. Ranft, J. Kripfganz and G. Ranft, Phys. Rev. D28, 360(1983);
V. P. Gerdt, A. S. Ilchev and V. K. Mitrjushkin, Yad. Fiz. 40, 1097(1984);
Y. Munehisa, Phys. Rev. D30, 1310(1984);
G. Koutsoumbas, Phys. Lett. 140B, 379(1984);
D. Espriu and J. F. Wheater, Oxford preprint 3/1985.

8) C. B. Lang, C. Rebbi and M. Virasoro, Phys. Lett. 104B, 294(1981);
V. P. Gerdt, A. S. Ilchev, V. K. Mitrjushkin, I. K. Sobolev and A. M. Zadorozhny, Dubna preprint E2-84-313;
M. Tomiya and T. Hattori, Phys. Lett. 140B, 370(1984);
I. Montvay, Phys. Lett. 150B, 441(1985).

9) H. Kühnelt, C.B.Lang and G.Vones, Nucl. Phys.B230 [FS 10], 16(1984).

10) I. Montvay, DESY preprint 85-005.

11) S. Coleman and E. Weinberg, Phys. Rev. D7, 1888(1973);
M. Peskin, Ann. Phys. 113, 122(1978).

12) H. Kleinert, Phys. Lett. 128B, 69(1983).

13) D. P. Landau and R. H. Swendsen, Phys. Rev.Lett. 46, 1437(1981).

14) J. Jersák, T. Neuhaus and P.M. Zerwas, Phys. Lett. 133B, 103(1983);
H. G. Evertz, J. Jersák, T. Neuhaus and P. M. Zerwas, Nucl. Phys. B251 [FS13], 279(1985).

15) B. Lautrup and M. Nauenberg, Phys. Lett. 95B, 63(1980).

THE STANDARD HIGGS-MODEL ON THE LATTICE

I. Montvay
Deutsches Elektronen-Synchrotron DESY, Hamburg

ABSTRACT

Some recent Monte Carlo calculations in the SU(2) Higgs-model with a scalar doublet field are reviewed. Questions about the dependence on the scalar self-coupling are discussed in the framework of a strong self-coupling expansion. The numerical results are consistent with an asymptotically free continuum limit at vanishing bare gauge coupling.

1. INTRODUCTION

Most of the recent efforts in Monte Carlo simulations of lattice gauge field theories [1,2] are concentrated on the study of pure SU(N) gauge theories and QCD-like theories with SU(N) gauge fields and a number of spin-$\frac{1}{2}$ fermion fields. The numerical study of quantum field theories containing scalar matter fields received up to now only a relatively limited amount of interest. The detailed investigation of pure gauge field theories is certainly the basis for any future understanding of the physical theories with matter fields, therefore it is unavoidable. The extraordinary numerical efforts invested in the latest simulations of QCD with quarks is motivated by the great challenge represented by the complex and experimentally well measured hadron spectrum. The fermionic matter fields are, however, notoriously difficult for numerical studies. Scalar matter fields are, from the numerical point of view, much simpler.

In the standard SU(3) $\otimes$ SU(2) $\otimes$ U(1) theory of strong-electroweak interactions scalar fields play a very important rôle, because they are responsible, via the "Higgs-mechanism", for the masses of all the particles. Experimentally the "Higgs-sector" of the standard model is unknown, the Higgs-particle and its couplings are not yet directly observed. Therefore, the numerical study of field theories with scalar matter fields, in particular the Higgs-sector of the standard model, is both interesting and important. Since from the technical point of view scalar fields are simpler, it is conceivable that the study of quantum field theories containing scalar matter fields can contribute rather substantially to our understanding of the gauge-matter interactions.

One of the possible reasons for the limited interest in performing Monte Carlo simulations with scalar fields is, perhaps, the almost rigorously proven triviality of the simplest renormalizable scalar field theory. Namely, as a result of almost 15 years of hard work [3)], we almost definitely know, that the single-component ϕ^4-theory in the four-dimensional continuum is trivial, i.e. equivalent to a free field theory. This fact, and the large number of apparently free parameters, discredited the Higgs-sector, too, because neglecting fermions and electromagnetism, the Higgs-sector of the standard model is the SU(2)-gauged version of a four-component (doublet) scalar field theory with ϕ^4 self-interaction. It is, however, a priori not clear what is the consequence of the triviality of ϕ^4 interaction for the "standard" SU(2) Higgs-model with doublet scalar field. There are, in principle, several possibilities:

i) the standard Higgs-model, too, is trivial in the four-dimensional continuum;

ii) its continuum limit is non-trivial but λ-independent (the ϕ^4 self-coupling λ is "irrelevant");

iii) it is non-trivial and λ-dependent.

In the first case, the ϕ^4 self-coupling "spoils" the otherwise nice gauge-interaction too, and the only possible continuum limit is a free theory of massive, spin-1 vector bosons ("W-bosons") and a single massive scalar ("Higgs-boson"). In the other two cases the ϕ^4 coupling has to be, presumably, asymptotically free, because the gauge interaction is asymptotically free. At high energies the gauge coupling is negligible and we are left with the "asymptotically trivial" (i.e. asymptotically free) ϕ^4-theory. In the interesting case ii) the continuum theory has one independent parameter less than the bare theory. Therefore, the Higgs-boson mass, for instance, is a function of the W-boson mass and of the (renormalized) gauge coupling constant.

2. STRONG SELF-COUPLING EXPANSION

2.1 Motivation

The standard Higgs-model has three bare coupling parameters: $\beta \equiv 4/g^2$ is the SU(2) gauge coupling, $\varkappa$ the hopping parameter of the scalar doublet field and λ is the ϕ^4 self-coupling (the precise definitions see below in the lattice action). In renormalized perturbation theory the renormalized ϕ^4 coupling λ_{ren} is a free parameter which can be traded, for instance, against the value of the mass of physical Higgs-particle m_H. At tree level the relation is

$$m_H = m_W \frac{\sqrt{8\lambda_{ren}}}{g_{ren}} . \tag{1}$$

Here m_W is the W-boson mass and g_{ren} denotes the renormalized SU(2) coupling constant. The low energy phenomenology (below $\sim$100 GeV) is rather insensitive to λ_{ren} (or to the Higgs-mass m_H). If, however, m_H is very large ($m_H \sim 1$ TeV), then the ϕ^4 self-coupling of the Higgs-field implies a strongly interacting Higgs-sector, which can produce rich, non-perturbative phenomena in the few-hundred GeV range [4,5].

According to the tree-level formula (1), for $\lambda \to \infty$ the Higgs-mass goes to infinity. Of course, due to the strong interaction, the tree level relation is not valid and therefore m_H can stay finite. In fact, there are large-N expansion arguments [6] implying that a "Higgs-remnant" with the quantum numbers of the physical Higgs-particle remains in the spectrum even for $\lambda \to \infty$. A first rough Monte Carlo investigation of the correlations showed [7], that $m_H/m_W = O(1)$ is possible even at infinitely strong bare self-coupling $\lambda = \infty$. A more detailed numerical study of the λ-dependence was also carried out recently [8], showing a remarkable universal behaviour of the mass gaps for different λ-values. Plotting, for instance, the W-mass in lattice units (am_W) for fixed $\beta = 2.3$ and $\lambda = 0.1$, 0.5, 1.0 and ∞ , as a function of an appropriately chosen third variable, one obtains Fig. 1. This shows, that the λ-dependence in the given λ-range is surprisingly weak, in fact, it is too weak to be seen by the limited numerical accuracy in Ref. 8). The interesting question is, of course, whether the λ-dependence goes away completely in the continuum limit or not. We shall see below, that a powerful tool for the study of λ -dependence is the strong self-coupling expansion (SSCE).

2.2 Lattice Action

The gauge field is described on the lattice, as usual, by the link variables $U(x,\mu) \in SU(2)$ (x = lattice point, $\mu = \pm 1, \pm 2, \pm 3, \pm 4$ lattice directions). The doublet field on the lattice points can be represented by its length $\rho_x \geq 0$ and by an angular variable $\alpha_x \in SU(2)$. Since α_x is equivalent to the

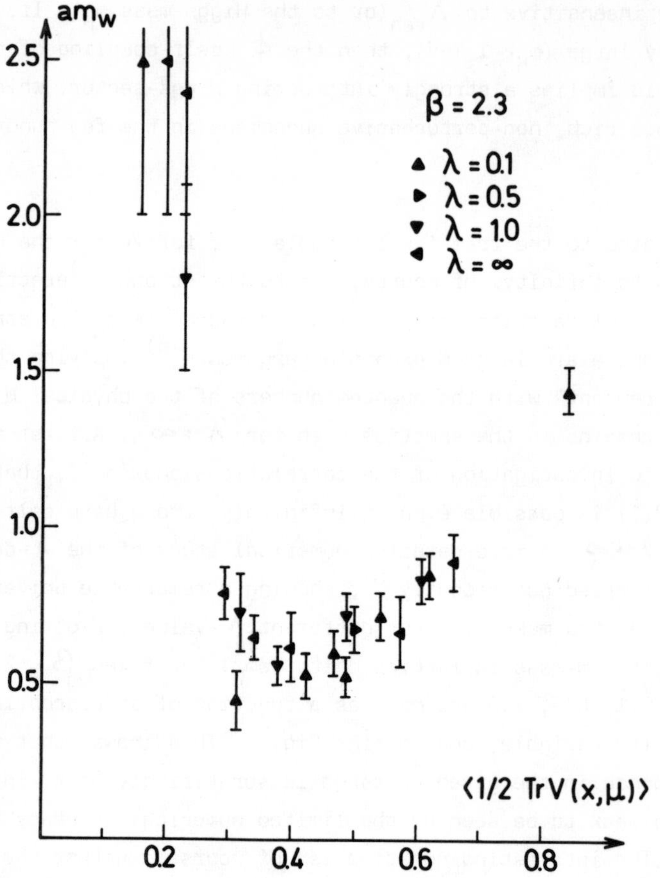

Fig. 1. The W-mass as a function of $\langle Tr\, V(x,\mu)\rangle$ according to Ref. 8).

local gauge freedom, it is possible to introduce the gauge invariant link variables $V(x,\mu) \equiv \alpha^{+}_{x+\hat{\mu}}\, U(x,\mu)\, \alpha_x$ ($\hat{\mu}$ = unit vector in direction μ). In terms of the physical degrees of freedom the lattice action is

$$S^{(s)}_{\lambda\beta\varkappa} = \beta \sum_{P} \left(1 - \frac{1}{2} \mathrm{Tr}\, V_P\right) + \\ + \sum_x \left\{ -3 \ln \rho_x + \lambda s^4 \left[\rho_x^2 - s^{-2}\left(1 - \frac{1}{2\lambda}\right)\right]^2 - \\ - \varkappa s^2 \sum_\mu \rho_{x+\hat\mu}\, \rho_x \, \mathrm{Tr}\, V(x,\mu) \right\} . \tag{2}$$

Here $\sum_P$ means a summation over plaquettes. The arbitrary scale factor $s>0$ in $S^{(s)}_{\lambda\beta\varkappa}$ corresponds to the rescaling $\rho_x \to s\rho_x$ of the integration variable ρ_x. The integration measure for the action (2) is $d\rho_x$ times the SU(2) Haar-measure $d^3V(x,\mu)$. In the limit $\lambda \to \infty$ the length is fixed to the value $\rho_x = s^{-1}$. For s = 1, the $\lambda = \infty$ limit of the action is

$$S^{\lambda=\infty}_{\beta,\varkappa} = \beta \sum_P \left(1 - \frac{1}{2} \mathrm{Tr}\, V_P\right) - \varkappa \sum_{x,\mu} \mathrm{Tr}\, V(x,\mu) . \tag{3}$$

2.3 General Scheme of SSCE

The action in Eq. (2) can be splitted up into its s = 1, $\lambda = \infty$ limit (3) at the hopping parameter value $s^2\varkappa$ plus the rest. The strong self-coupling expansion is obtained by expanding into powers of the coupling term $s^2\varkappa(\rho_{x+\hat\mu}\rho_x - 1)\,\mathrm{Tr}\, V(x,\mu)$ and then performing the integrations $d\rho_x$ over the length of the Higgs-field. The occurring integrals can be expressed by the parabolic cylinder function $D_p(z)$ like

$$J_k(\lambda) = \int_{-\sqrt{\lambda} + (2\sqrt{\lambda})^{-1}}^{\infty} dx\, e^{-x^2} \left(1 + \frac{x}{\sqrt{\lambda}} - \frac{1}{2\lambda}\right)^{1+\frac{k}{2}} = \\ = \exp\left[-\frac{1}{4}\left(\frac{1}{\sqrt{2\lambda}} - \sqrt{2\lambda}\right)^2\right] \frac{\Gamma\left(2 + \frac{k}{2}\right)}{\sqrt{2}\,(2\lambda)^{1/2 + k/4}} D_{-2-\frac{k}{2}}\left(\frac{1}{\sqrt{2\lambda}} - \sqrt{2\lambda}\right). \tag{4}$$

In the expectation values only the ratios (k = 1,2,...)

$$i_k = \frac{J_k(\lambda)}{J_0(\lambda)} = \frac{\Gamma(2+\frac{k}{2})\, D_{-2-\frac{k}{2}}\left(\frac{1}{\sqrt{2\lambda}} - \sqrt{2\lambda}\right)}{(2\lambda)^{k/4}\, D_{-2}\left(\frac{1}{\sqrt{2\lambda}} - \sqrt{2\lambda}\right)} \tag{5}$$

appear. These behave asymptotically like

$$i_k \to \begin{cases} (\lambda\to\infty) & 1 + \frac{k(k-2)}{16\lambda} + o(\lambda^{-2}) \; ; \\ (\lambda\to 0) & \Gamma(2+\frac{k}{2})\left(1 - \lambda\frac{k(k+6)}{4} + o(\lambda^2)\right) . \end{cases} \tag{6}$$

From the first line it can be seen, that for $\lambda\to\infty$ the expansion is similar to a series expansion into the inverse powers of λ. For small λ, however, every term in the series remains finite.

In general, the terms of the SSCE series are given by some correlation functions at $\lambda=\infty$. In particular, as we shall see below, in many important quantities the expectation value of $\mathrm{Tr}\, V(x,\mu)$ appears. Therefore, it is natural to choose the freedom in the scale factor s by requiring

$$\langle \mathrm{Tr}\, V(x,\mu)\rangle_{\lambda\beta\varkappa} = \langle \mathrm{Tr}\, V(x,\mu)\rangle_{\lambda=\infty,\beta,s^2\varkappa} . \tag{7}$$

This means that SSCE is done (for fixed β) along the curves with constant link expectation value $\langle \mathrm{Tr}\, V(x,\mu)\rangle$. Since the phase transition between the confinement-like and Higgs-like phases (for fixed β and different λ) occurs at nearly the same value of $\langle \mathrm{Tr}\, V(x,\mu)\rangle$ [8], the choice in Eq. (7) implies that the expansion is done along curves which do not cross the phase transition surface. This is, of course, important for an optimal convergence radius.

2.4 Examples: Average ρ and Two-Link Correlations

The second order expansion for the average Higgs-field length is, to a good approximation,

$$\langle \rho_y \rangle_{\lambda\beta\kappa} = i_1 + 8\kappa i_1 (i_2 - i_1^2) \langle \mathrm{Tr} V(x,\mu) \rangle_{\lambda=\infty} + \tag{8}$$
$$+ 4\kappa^2 \left(\langle \mathrm{Tr} V(x,\mu) \rangle_{\lambda=\infty} \right)^2 \left(i_3 i_2 + 7 i_3 i_1^2 + 13 i_1 i_2^2 + 30 i_1^5 - 51 i_2 i_1^3 \right) + \cdots$$

The complete second order contains, in addition, an s-dependent piece which is proportional to some specific combinations of the connected 2-link correlations $\langle \mathrm{Tr} V(x_1\mu_1) \mathrm{Tr} V(x_2\mu_2) \rangle^c_{\lambda=\infty}$.
For the choice of s in Eq. (7), however, this is only a small correction to the second order contribution given above. As another example, the lowest non-trivial order of the connected two-link correlation function is

$$\langle \mathrm{Tr} V(y_1\nu_1) \mathrm{Tr} V(y_2\nu_2) \rangle^c_{\lambda\beta\kappa} = \langle \mathrm{Tr} V(y_1\nu_1) \mathrm{Tr} V(y_2\nu_2) \rangle^c_{\lambda=\infty} \tag{9}$$
$$+ \kappa (i_1^2 - s^2) \sum_{x,\mu>0} \langle \mathrm{Tr} V(y_1\nu_1) \mathrm{Tr} V(y_2\nu_2) \mathrm{Tr} V(x,\mu) \rangle^c_{\lambda=\infty} + \cdots$$

Such expressions can be used, for instance, for the SSCE of the mass gaps.

The comparison of Eq. (8) with the Monte Carlo data [8] at λ = 1.0 and 0.1 is shown by Table I. As it can be seen, the agreement of this low order SSCE and the Monte Carlo results is impressive. As a preliminary numerical study has shown, Eq. (8) qualitatively describes the average Higgs-field length $\langle \rho_y \rangle$ even at λ = 0.01. Higher order expansions for the I_W = 0 and I_W = 1 mass gaps, and a detailed comparison to numerical data will be published elsewhere [9]. The present conclusion concerning SSCE is:

i) $\lambda \gtrsim 0.1$ seems to be the region of strong self-coupling, where low orders of SSCE are sufficient;

Table I.

The second order SSCE for the average Higgs-field length $\langle\rho\rangle_{SC}$ compared to the Monte Carlo data $\langle\rho\rangle_{MC}$ [8] at $\lambda = 1.0; 0.1$ and for $\beta = 2.3; \infty$.

$\lambda = 1.0$					
$\varkappa(\beta=2.3)$	$\langle\rho\rangle_{SC}$	$\langle\rho\rangle_{MC}$	$\varkappa(\beta=\infty)$	$\langle\rho\rangle_{SC}$	$\langle\rho\rangle_{MC}$
0.2	1.059	1.060	0.22	1.069	1.070
0.3	1.114	1.115	0.24	1.080	1.081
0.31	1.151	1.152	0.25	1.096	1.092
0.32	1.178	1.178	0.26	1.122	1.124
0.35	1.246	1.241	0.27	1.148	1.149
0.4	1.347	1.331	0.28	1.173	1.173
			0.3	1.219	1.216
			0.32	1.289	1.256
$\lambda = 0.1$					
$\varkappa(\beta=2.3)$	$\langle\rho\rangle_{SC}$	$\langle\rho\rangle_{MC}$	$\varkappa(\beta=\infty)$	$\langle\rho\rangle_{SC}$	$\langle\rho\rangle_{MC}$
0.19	1.336	1.353	0.155	1.280	1.290
0.195	1.423	1.450	0.16	1.291	1.302
0.2	1.523	1.558	0.163	1.306	1.319
0.205	1.594	1.636	0.165	1.330	1.346
0.21	1.663	1.712	0.167	1.360	1.379
0.22	1.780	1.845	0.17	1.408	1.430
0.3	2,535	2.628	0.175	1.484	1.514
			0.18	1.554	1.589

ii) SSCE may well be a convergent expansion for all $\lambda > 0$ values;
iii) a natural variable, instead of the hopping parameter $\varkappa$ (for fixed β and fixed lattice size) is $L \equiv \langle \frac{1}{2} \mathrm{Tr} V(x,\mu) \rangle$.

The practical advantage of SSCE is, that it allows to concentrate the numerical study to $\lambda = \infty$. The time consuming coverage of the whole 3-parameter space $(\lambda, \beta, \varkappa)$ with the measured points is not necessary. Combined with the usual lattice perturbation theory in the gauge coupling g (at $g = 0$), the SSCE may also help to pin down the question of λ -dependence in the continuum limit. A completely analytic study is also possible if, in addition, a hopping parameter expansion is done in the remaining $(\lambda = \infty, \beta = \infty)$ non-linear σ -model. This latter procedure is equivalent to the combination of the usual "high temperature expansion" [10] in the $\beta = \infty$ ϕ^4-model with the g = 0 perturbation theory.

3. SOME RECENT NUMERICAL RESULTS

In two recent papers [7,8] the correlations in W-boson and Higgs-boson channels and the static energy of an external colour charge pair were investigated in the standard Higgs-model by numerical Monte Carlo simulation. (For references to earlier numerical studies in the standard Higgs-model see these papers. For some new results see also the lecture of J. Jersâk in these Proceedings). The numerical study of the correlations is useful for the understanding of the phase structure and of the continuum limit. In general, it is relatively easy to determine the correlations (much easier than e.g. the plaquette-plaquette correlations in pure gauge theory). In the vicinity of the phase transition between the confinement-like and Higgs-like regions, however, the long range correlations require large lattices and the critical slowing down (or metastability) is very dangerous. Sometimes surprisingly long runs are needed in order to be reasonably sure that the results refer to the equilibrium situation.

As examples of some new, high statistics results on 12^4 lattices with the full SU(2) group [11] let us consider the W-boson and Higgs-boson masses in lattice units in two points. At $(\lambda = \infty$, $\beta = 2.3$, $\varkappa = 0.41)$, which is above the phase transition surface, one obtains $am_W = 0.507(14)$ and $am_H = 0.79(3)$. Lorentz-invariance is well satisfied in this point. In particular, one obtains from the zero momentum correlations $am_W = 0.505(15)$ and from the p=1 (in lattice units) correlations $am_W = 0.510(48)$. In another point ($\lambda = \infty$, $\beta = 2.3$, $\varkappa = 0.39$), below the phase transition surface, the result is: $am_W = 1.27(8)$ and $am_H = 0.39(2)$.

In Ref. 8) the static energy of an external colour charge pair (in short, "potential") was investigated in detail on 12^4 lattice and using the icosahedral approximation for SU(2). The aim was to obtain information about the renormalization group trajectories (RGT's), since along a RGT the potential can be rescaled to a common, physical curve. (This method was used for the study of scaling in pure SU(2) gauge theory in Ref. 12).) It turns out 8), that the potential shape sensitively depends on the value of m_H/m_W, therefore the RGT's can be determined quite well. An important point is, that the rescaling of the potential is possible, to a good accuracy, along curves in the λ = const. planes. This and the weak dependence of the masses on λ (if plotted like in Fig. 1) suggests, that the continuum limit can be λ-independent. Of course, this question has to be considered in more detail in future studied. In particular, the crucial point is, whether comparing the appropriate λ = const. RGT's, the λ -dependence weakens for growing β . For fixed λ the scaling properties of the standard lattice Higgs-model are qualitatively similar to the situation in QCD with a single quark mass parameter, i.e. there is an asymptotically free fixed point at $\beta = \infty$ and a (λ -dependent) critical hopping parameter $\varkappa = \varkappa_{crit}(\lambda)$. The different RGT's are parametrized by the mass parameter in the Higgs-potential: above the phase transition line there are the trajectories with spontaneous symmetry breaking, below the phase transition line the trajectories describing a confining theory with scalar matter fields (see Fig. 2).

Once the questions concerning the continuum limit are cleared, it becomes possible to calculate such phenomenologically interesting quantities like the ratio of the Higgs-boson mass to the W-boson mass m_H/m_W. Since the renormalized SU(2) coupling is weak phenomenologically ($g^2_{ren} \cong 0.5$ at the energy scale of m_W), one has to perform the Monte Carlo calculation at high β. This makes the calculation difficult. In a first attempt one can simplify the shape of the RGT's: as a zeroth approximation one can assume that the RGT goes nearly parallel to the phase transition surface between $\beta = \infty$ and the considered large β, and then it departs almost perpendicularly towards $\varkappa = \infty$.

In the Monte Carlo calculation I took $\lambda = 1.0$; $\beta = 8.0$ and $\varkappa = 0.30$ on 10^4 lattice with the full SU(2) group. The measured value of the renormalized gauge coupling, as determined by the Coulomb-potential at lattice distances 1-5, is $\alpha = 0.034(3)$. This roughly corresponds to the expected value $3g^2_{ren}/(16\pi) \cong 0.03$. From 10000 measured sweeps, I obtained for the masses in lattice units $am_H = 1.4(2)$ and $am_W = 0.23(3)$, therefore

$$\frac{m_H}{m_W} \sim 6 \tag{10}$$

I also checked in similar runs, that this ratio does not change appreciably between $\varkappa = 0.28$ and $\varkappa = 0.32$. Eq. (10) gives for the physical Higgs-boson mass $m_H \sim 500$-600 GeV, but this has to be considered only as a first estimate. Completely neglected are here the virtual fermion loops and the electromagnetic U(1)-coupling. Furthermore, there is an unknown (presumably large) error due to finite lattice size effects. One has to study, in the future, also the scaling in this β-range (the precise shape of the physical RGT) and the question of λ-dependence.

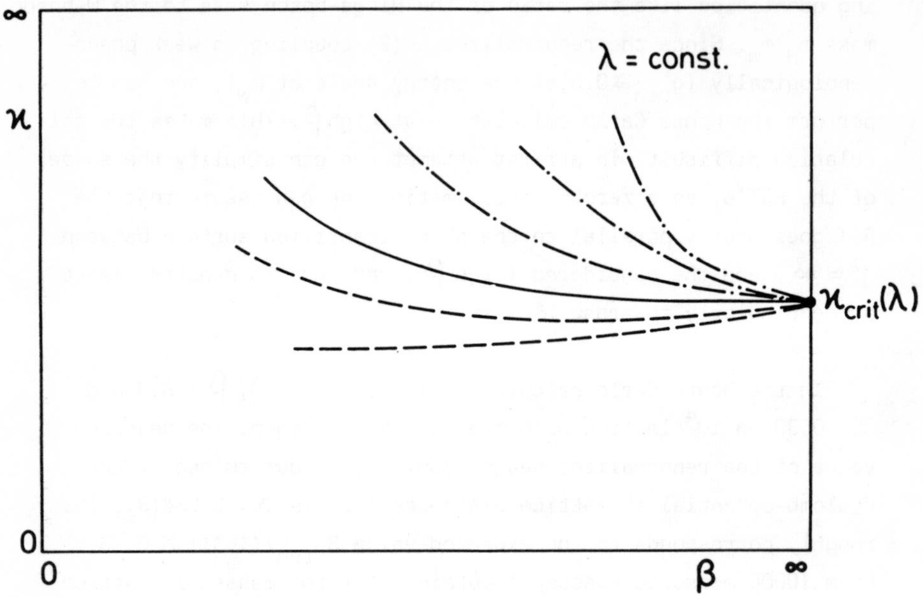

Fig. 2

Fig. 2. The schematic shape of the RGT's in a λ = const. plane. The full line gives the position of the phase transition. The dashed-dotted lines are the RGT's in the Higgs-phase, the dashed ones the RGT's in the confinement-like phase.

4. DISCUSSION

An important parameter in the standard Higgs-model is the ratio of the Higgs-boson mass to the W-boson mass $\xi \equiv m_H/m_W$. According to the Monte Carlo simulations [7,8,11] ξ is greater than 1 in the Higgs phase and smaller than 1 in the confinement phase. For the phenomenological value of the weak SU(2) coupling a first numerical estimate gives $\xi \sim 6$. If in the continuum limit (λ= const., $\beta \to \infty$, $\varkappa \to \varkappa_{crit}(\lambda)$) the second alternative mentioned in the introduction is realized (i.e. if λ is irrelevant) then, besides the Λ-parameter for the SU(2) gauge coupling, ξ is the only free physical parameter of the theory. Every other quantity like for instance $m_W/\Lambda_{SU(2)}$ or the value of the ϕ^4 coupling at some specific point (λ_{ren}), is a function of $\Lambda_{SU(2)}$ and ξ. The renormalization group equations for the renormalized Green's functions are valid with only two coupling parameters. In the usual weak coupling perturbation theory of the standard Higgs-model there are three free parameters. In order to take into account the non-perturbative constraint implied by the requirement of the existence of a mathematically well defined continuum theory, one has to impose on the three parameters of renormalized perturbation theory some external constraint. This is similar to the situation in pure ϕ^4 theory, where $\lambda_{ren} = 0$ is such an external constraint.

REFERENCES

1) K.G.Wilson, Phys. Rev. D10, 2445 (1974)

2) for references and a collection of early papers see "Lattice gauge theories and Monte Carlo simulations", edited by C. Rebbi, World Scientific, 1983

3) an incomplete list of references is:
K.G. Wilson, Phys. Rev. B4, 3184 (1971).
K.G. Wilson, J. Kogut, Phys. Rep. 12C, 75 (1974)
G.A. Baker, J. Kincaid, Phys. Rev. Letters 42, 1431 (1979) and Journ. Stat. Phys. 24, 469 (1981).
B. Freedman, P. Smolensky, D. Weingarten, Phys. Lett. 113B, 481 (1982).
M. Aizenman, Phys. Rev. Letters 47, 1 (1981) and Commun. Math. Phys. 86, 1 (1982).
J. Fröhlich, Nucl. Phys. B200[FS4], 281 (1982) and in "Progress in Gauge Field Theory", Cargèse 1983, ed. by G. 't Hooft et al., Plenum 1984.

4) M. Veltman, Acta Phys. Polonica B8, 475 (1977).
B.W. Lee, C. Quigg, H.B. Thacker, Phys. Rev. D16, 1519 (1977).
T. Appelquist, C. Bernard, Phys. Rev. D22, 200 (1980).
A. Longhitano, Phys. Rev. D22, 1166 (1980).

5) for a more recent work see, for instance, R. Casalbuoni, S. De Curtis, D. Dominici, R. Gatto, Phys. Letters 155B, 95 (1985).

6) M.B. Einhorn, Nucl. Phys. B246, 75 (1984)

7) I. Montvay, Phys. Lett. 150B, 441 (1985)

8) I. Montvay, DESY preprint 85-005 (1985), submitted to Nucl. Phys. B.

9) K. Decker, I. Montvay, in preparation

10) see, for instance, G.A. Baker, J. Kincaid, Journ. Stat. Phys. 24, 469 (1981)

11) W. Langguth, I. Montvay, in preparation

12) F. Gutbrod, I. Montvay, Phys. Lett. 136B , 411 (1984).

SOME RECENT RESULTS ON LATTICE YANG-MILLS THEORIES WITH MATTER FIELDS

Ion Olimpiu Stamatescu

- Institut für Theorie der Elementarteilchen -
Freie Universität, Berlin, W-Germany

ABSTRACT

Monte Carlo results on the phase structure of an SU(3) Higgs model at finite temperature are presented in the first part. Finite step, Markow algorithms for QCD with dynamical quarks, which allow us to obtain unbiased results are presented in the second part. They are applied here to study the deconfining transition in QCD on a $4^3 2$ lattice (SU(3) , Wilson fermions). Only a cross over is found.

1. INTRODUCTION

The work I would like to present here belongs to a larger project on lattice gauge theories with matter fields. People, who at various moments were or are contributing to these analyses are:

Ph.de Forcrand (Cray Research, Chippewa Falls, USA)
F.Karsch (Urbana, Illinois, USA)
E.Seiler (Max-Planck-Inst., Munich, W-Germany)
H.Rothe (Heidelberg, W-Germany)
G.Feuer, C.Hege, V.Linke, A.Nakamura and myself (FU Berlin, W-Germany)

The analyses concern both bosonic matter (Higgs Models) and fermionic matter (QCD). The results which are presented here concern:

- the phase structure of an SU(3)-fundamental Higgs Model: Higgs transition and finite temperature transition,
- tests of a method allowing for realistic calculations in QCD with Wilson fermions on intermediate size lattices, and its application to study the finite temperature transition.

Previous and parallel work includes analyses of the phase structure of the SU(2)-adjoint Higgs Model, study of the properties of the Wilson fermions and the axial anomaly, developement of methods for fermionic calculations, study of 2-dimensional QED and QCD, etc. This of course only covers the part of the activity of these people which more or less belongs to this project of collaborations.

The present calculations in lattice gauge theory seem to have reached the level of precision where in the further comparison with the experiment we need to consider realistic models like the standard model or the QCD. This means that we must introduce matter fields, both bosonic and fermionic. One of the main effects of the coupling to matter is pair production. In the following we shall restrict the discussion to matter fields in the fundamental representation. Among the typical phenomena associated with virtual fundamental charges we have:

- Vanishing of the string tension (loss of the area law) - to be interpreted as hadronization. Notice that this effect appears both for bosonic and fermionic matter.

- For bosonic matter: Higgs mechanismus ("symmetry breaking") - appearence of massive (bleached) W-bosons.

- Weakening of the finite temperature deconfining transition. The shielding of the fractional charges in the hadronic phase tends to wash out the abrupt change in behaviour from this phase to a quark-gluon plasma characterized by Debye-type screening effects.

Only the effects which will be discussed below were mentioned. The next section contains numerical results about the phase structure of an SU(3) Higgs model with complete symmetry breaking and radial degree of freedom on periodic, asymmetric lattices. The character of both the Higgs and the deconfining transitions is found to depend on the Higgs selfcoupling and on a parameter which, in a somehow evasive way, could be related to the number of flavours. In particular, an increase in the latter leads to the disappearence of the deconfining transition for small Higgs mass.

Section 3 contains the presentation and discussion of algorithms which allow for realistic calculations and yield unbiased results for QCD on lattices of intermediate size. These algorithms are then applied to the study of the SU(3) Y M theory with Wilson fermions at finite temperature. On a 4x4x4x2 lattice , no unambiguous signal for a deconfining phase transition is found above $\beta = 3.2$, with 3 flavours and $\kappa = .12$.

2. PHASE STRUCTURE OF AN SU(3) HIGGS MODEL AT FINITE TEMPERATURE

2.1. Motivation

Due to the so-called duality between confinement and Higgs mechanism [1,2] the phase structure of the Higgs models with complete symmetry breaking is not yet well understood. One would like to know, among other things, whether there is a first order phase transition between the "broken" and "unbroken" phases, at least at finite temperature - needed in the "inflationary scenarios" for the early universe [3,4]. Recent results for SU(2) - both adjoint and fundamental - are found in Refs.[5-9]. Here we attack the case of SU(3), which might be interesting: a) as a step toward Grand Unification Groups, b) as a kind of scalar QCD, c) as possibly showing a different deconfining transition - compared with SU(2), and hopefully for further reasons. The work presented here has been done together with F.Karsch and E.Seiler [10].

To "break the symmetry completely" (in the traditional perturbative language) we would need a flavour multiplet of fundamental Higgs fields. In analogy with the SU(2) case, we tried to achieve it by introducing Higgs fields which are SU(3) matrices with a radial degree of freedom. This is not directly related to a conventional 3-flavours model, but we thought that it will not differ in an essential way from the latter. We found, however, quite surprising results, in particular many phase transitions whose order strongly depends on the parameters of the model. We cannot be sure therefore that the properties we have found are shared by more conventional models, and we are currently running checks in that sense with such a 3-multiplets model [11].

2.2. The model

The variables are:

- the gauge fields - link variables $\{U_\mu(x)\}$, $U(x) \in SU(3)$
- The Higgs fields - site variables, consisting of radial and internal degrees of freedom $\{R(x),V(x)\}$, $V(x) \in SU(3)$, $R(x) \in [0, \infty)$.

The action we choose is:

$$S = -\frac{2\varkappa}{3} \sum_{x,\mu} \mathrm{Re\,Tr}\, R_x V_x^{+} U_{x,\mu} V_{x+\mu} R_{x+\mu} + \sum_x R_x^2 + \lambda \sum_x (R_x^2 - 1)^2 - \frac{\beta}{3} \sum_{Plaq.} \mathrm{Re\,Tr}\, U_{\partial P} \tag{2.1}$$

The parameters are $\varkappa, \lambda, \beta$, representing more or less the inverse Higgs mass, the Higgs selfcoupling and the inverse gauge coupling. The first two parameters are chosen such that for $\lambda = 0$ vanishing bare mass means $\varkappa = \varkappa_0 \equiv 1/8$, and for $\lambda \to \infty$ the length of the Higgs field, R(x), is frozen to 1. The integration measure is:

$$Z^{-1} e^{-S} \prod_{x,\mu} dU_{x,\mu} \prod_{x} dV_x \prod_{x} R_x^f \, dR_x \tag{2.2}$$

with dU_μ, dV Haar measures on SU(3), and dR the Lebesgue measure on $[0, \infty)$. Notice the appearence of the additional parameter f in eq.(2.2). The problem is, that the multiples of SU(3) matrices do not form a linear space, as is the case for SU(2). Therefore there is no "natural" choice for f, like $f = 3$ in SU(2) - obtained by going to polar coordinates from a 2-isodoublets model. In fact, one can think of our action eq.(2.1) as arrising from a very complicated Higgs potential, and of f as being a residual arbitrariness left by the choice of such a potential. In the following we took the following values for f : 3 (just a small value), 8 (indicated by the dimension of su(3) and in analogy with SU(2)), 17 (from the dimension of the space of complex 3x3 matrices) and 50 (a large value).

To have a rough suggestion on what the effect of varying f might be, consider there were such a natural choice, say $f = f_0$, leading thus to an "effective potential" without logarithmic terms. Then using the measure eq.(2.2) with $f \neq f_0$ would induce a supplementary term in the potential:

$$-(f - f_0) \ln R_x \tag{2.3}$$

The logarithmic singularity at 0 will be smeared by the quantum effects, leaving a peak for $f > f_0$ which might simulate Higgs mechanism also for positive mass squared. For $f < f_0$, on the other hand, a standard Higgs potential will develop a second deep at origin, which can induce a first order transition. See Fig. 1 . See also Ref.[9].

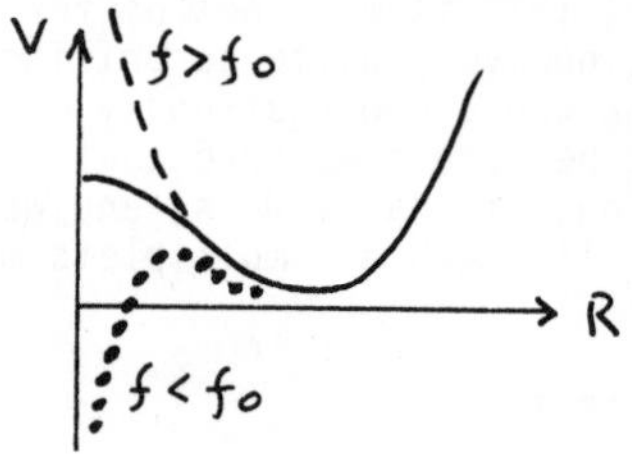

Fig. 1
A heuristic potential for this model

2.3. The calculations

We used the Metropolis algorithm with many hits and updated the compact variables by multiplying with random matrices from a reservoir containing as many as N_V (lattice volume) random matrices and their inverses. This reservoir is renewed each some sweeps and the random matrices are chosen in a neighborhood of 1 , such that the acceptance per hit stays around 1/2. The noncompact variables are updated as

$$R_x \to | R_x + (\text{step size}) \times (\text{rand. nr.} - .5) | \tag{2.4}$$

This a priori probability is symmetric and we can use the Metropolis check

$$\min(1, \varrho) = \min(1, \exp(-\Delta S)) \tag{2.5}$$

(where ρ gives the acceptance rule, after the prechoice eq.(2.4)), instead of the more general

$$\rho = \exp(-\Delta S) \cdot P^{*}_{new \to old} / P^{*}_{old \to new} \qquad (2.6)$$

to be used if the prechoice $P^{*}_{o \to n}$ is not symmetric. See also Ref. [12].

The calculations have been done on the Cray-1 computers of IPP-Garching and of ZIB-Berlin. We used vectorised programs, as described in Ref. [13]. The link variables are distributed in 8 "bushes" and the site variables in 2 "bushes", all the variables in a "bush" being independent and hence succeptible of being simultaneously updated [13] . We typically took some thousands of sweeps on lattices of sizes $N_t \times N_s^3$, with $N_t = 2$ and $N_s = \{4,6,8\}$

2.4. The results

As in the QCD-case [12 - 14] the first order finite temperature transition seems to persist for large Higgs masses (small $\varkappa$). Also the Higgs transition seems to become first order for $\beta < \infty$. Generally this transition become weaker with increasing f and/or λ . The limiting case $\lambda \to \infty$ (corresponding to R = 1) seems to have at most second order transition.

At $\lambda = 0$ the model is unstable for $\varkappa$ above 1/8 . But we find a phase transition for $\beta = \infty$, f = 17 at a $\varkappa < 1/8$, which persists for $\beta < \infty$. We didn´t study the behaviour for very small, nonvanishing λ . In Figs. 2,3 we show the situation for f = 8 and $\lambda = 0.1$.

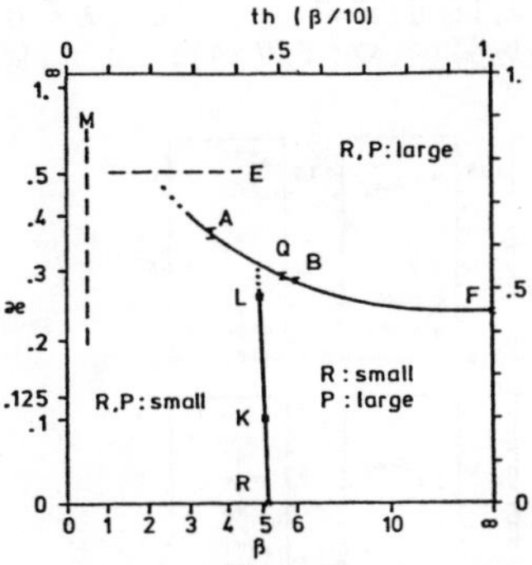

Fig. 2

Phase structure for f=8, λ=0.1 .

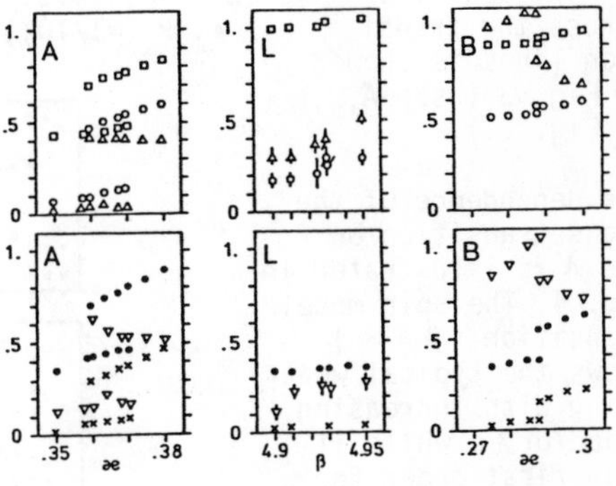

Fig. 3

Behaviour across transitions at A,B,L
o : P; □: A (x1/2 - A,L, x1/4 - B); △: ϵ_G(x4); ●: R (x1/10); ×: H (x1/10); ▽: ϵ_H(x4)

We used the values of $R \equiv \langle R^2 \rangle$ and of the Polyakov loop $P \equiv \langle \mathrm{ReTr} \prod_t U_{\hat{t}}(x,t) \rangle$ to distinguish:

- the Higgs region : R , P large
- the deconfined region : R small, P large
- the confinement region : R , P small

Of course, R and P are not order parameters, however they are sensitive to the various transitions. Further observables are:

$$A = \beta/3 \ \langle \mathrm{Re\,Tr}\,(U_{\partial P}) \rangle$$

$$H = 2\varkappa/3 \ \langle \mathrm{Re\,Tr}\, R(x)\, V(x)^{+}\, U_{\mu}(x)\, V(x+\mu)\, R(x+\mu) \rangle$$

$$\epsilon_G = A(\text{"time like" Plaq.}) - A(\text{"space like" Plaq.})$$

$$\epsilon_H = H(\text{"time like" Link}) - H(\text{"space like" Link})$$

$$\mathcal{M} = \langle |\sum_x V(x)| \rangle / \mathrm{Vol}$$

As one can see from Fig. 2 , the Higgs transition, which is second order in the spin model ($\beta = \infty$), becomes first order for $\beta < \infty$ and then disappears again at very small β. No transition was found on the lines M and E in the diagram, but merely a cross over. The typical behaviour across the transition lines is showed in Fig. 3 (see A,B,L in Fig. 2).

The dependence of the Higgs transition on f and λ is illustrated in Fig. 4. The spin model transition ($\beta = \infty$) shows the typical weakening with increasing f and/or λ, while it gets first order for f = 3 ! For $\beta < \infty$ the transition is steeper, it becomes first order for f = 8 (see Fig. 3) but seems to remain second order for f = 17.

Finally, we look for the deconfining

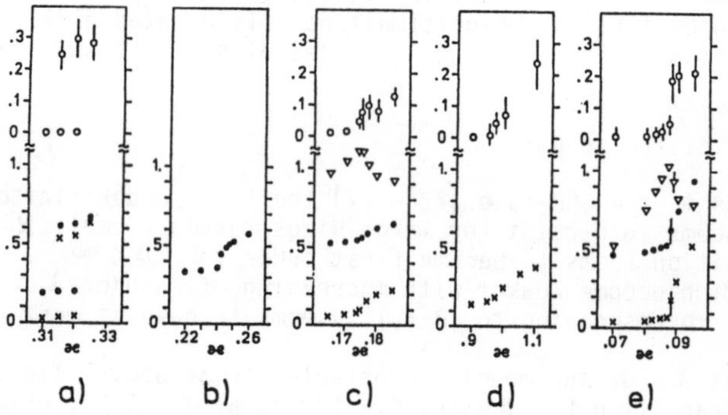

Fig. 4

The spin model transition a):f=3, λ=.1; b):f=8, λ=.1; c):f=17, λ=.1; d):R=1; e):f=17, λ = 0
•: R (x1/10); x : H (x1/5); ▽: ϵ_H (x4); o : $\mathcal{M}$

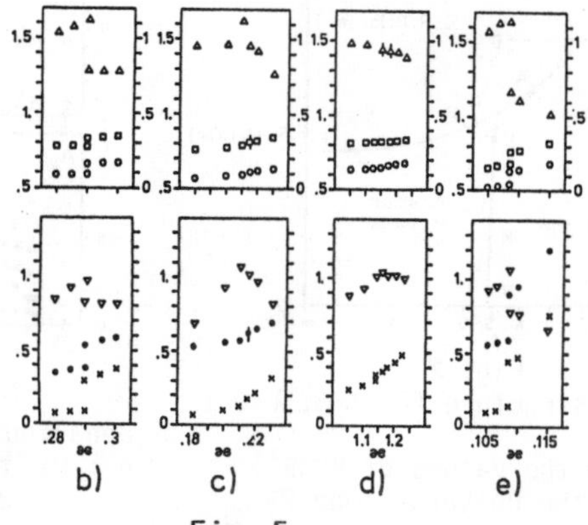

Fig. 5

Higgs transitions b):f=8, λ=.1, β=5.6; c):f=17, λ=.1, β=5.6; d):R=1, β=6; e):f=17, λ=0, β=5.2
o: P; □: A (x1/4); △: ϵ_G(x4; right hand scale); •: R (x1/10, x1/20 - e)); x: H (x1/5); ▽: ϵ_H(x4)

transition. This is first order for small $\varkappa$, weakening with increasing $\varkappa$ (see Fig. 3). In trying to make some connection with QCD we put $\lambda = 0$ and fix f such as to reproduce in lowest order hopping parameter the effect from three flavours of Wilson fermions. The latter lead to an effective action with an additional Polyakov loop term (for $N_t < 4$ and N_f flavours):

$$(2\varkappa)^{N_t}\, 2N_f \left(Tr \prod_{t=1}^{N_t} U_{\hat{t}}(\vec{x},t) + h.c. \right) \tag{2.7}$$

(see Ref. [12]). To the same order in $\varkappa$ our Higgs model produces as additional term in the effective action:

$$(2\varkappa)^{N_t} (\overline{R^2}/6)^{N_t} \frac{1}{3} \left(Tr \prod_{t=1}^{N_t} U_{\hat{t}}(\vec{x},t) + h.c. \right) \tag{2.8}$$

where

$$\overline{R^2} = \int dr\, r^{f+2} e^{-r^2} \Big/ \int dr\, r^f e^{-r^2} = (f+1)/2 \tag{2.9}$$

The two models give identical contributions (to this order in $\varkappa$) if

$$f = 12\,(6N_f)^{1/N_t} - 1 \tag{2.10}$$

which gives f = 50 for $N_f = 3$ and $N_t = 2$. Increasing f we simulate thus more Higgs flavours.

The results in Fig. 6 (f = 50, $\lambda = 0$) show that the deconfining transition remains first order for small $\varkappa$ while moving left, deep into the strong coupling region. With increasing $\varkappa$ the transition weakens and eventually disappears: we could find no signal of it on the lines X and Z′ , and only vague indication of a cross over on the line Z. This means that the behaviour obtained in lowest order hopping parameter [12] is not drastically modified

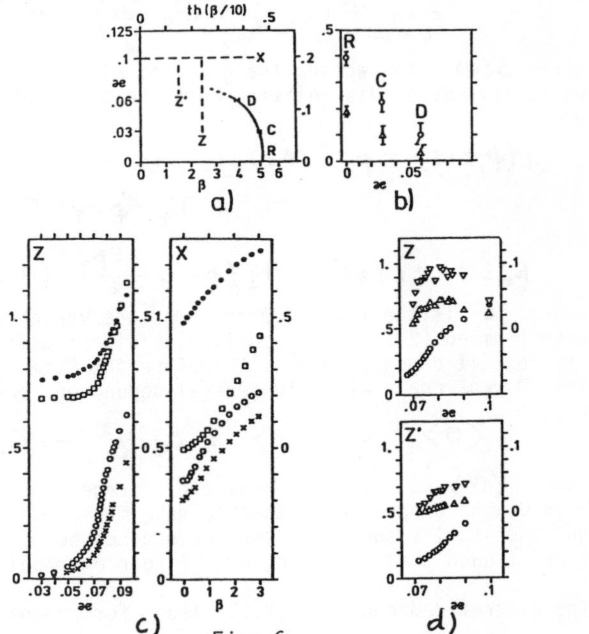

Fig. 6
Deconfining transition: a)transition line; b)discontinuity vs. $\varkappa$ c),d)behaviour on X,Z,Z′ o:P(LH scale); □:A(x1/2, RH scale); x:H(x1/20, LHS); •:R(x1/50, RHS); △:ϵ_G(LHS); ▽:ϵ_H(RHS)

by higher orders from the scalar determinant. Of course, this has no direct relevance for QCD, but it provides an example of disappearence of the deconfining transition in the presence of fundamental matter.

2.5. Testing a new algorithm: Random Walk in the group space.

Under circumstances it may be advantageous to have an alternative algorithm, where a possible dependence on the size of the change for the updating can be systematically analysed. This could help geting rid of metastabilities which are sometimes a nuisance in the Metropolis procedure.

The Random Walk algorithm has been applied successfully to lattice models for noncompact variables [16]. It relies on the diffusion equation

$$\frac{d}{d\tau} P(\phi,\tau) = \frac{\partial}{\partial\phi}\left(\frac{\partial}{\partial\phi} - K(\phi)\right) P(\phi,\tau) \qquad (2.11)$$

whose solution $P(\phi,\tau)$ relaxes, under some general conditions, to an equilibrium distribution

$$\lim_{\tau\to\infty} P(\phi,\tau) = P_E(\phi) = e^{-S(\phi)}/Z \qquad (2.12)$$

where $S(\phi)$ is the action the gradient of which is the driving force K in eq.(2.11). After discretizing both the time and the variable eq.(2.11) reads:

$$P(\phi,\tau+\varepsilon) = p_0 P(\phi,\tau) + T_-(\phi+\eta)P(\phi+\eta,\tau) + T_+(\phi-\eta)P(\phi-\eta,\tau) \qquad (2.13)$$

with

$$p_0 = 1 - 2\varepsilon/\eta^2 \;, \quad T_\pm(\phi) = \frac{1-p_0}{2}\left(1 \pm \frac{1}{2}\eta K(\phi)\right) \qquad (2.14)$$

where ε is the time step and η the variable´s step. As can be easily seen from eq.(2.12), the equilibrium distribution $P_E(\phi)$ does not depend on ε, but it depends on η. Therefore the "time"-averages calculated over this Markow chain will, in general depend on η^2 as given by:

$$\langle O\rangle_\eta = \langle O\rangle + C_1\eta^2 + C_2\eta^4 + \ldots \qquad (2.15)$$

where $\langle O\rangle$ is the exact ensamble average, to be obtained in the limit $\eta\to 0$. The procedure is then to make measurements at 2,3 values of η, not too small (such that the convergence be acceptable) and not too large (such that a simple, e.g. linear extrapolation be possible).

The discretized equation (2.13) leads for noncompact variables to the updating rule

$$\phi \to \phi \ (\text{with } p_0)\ , \quad \phi \to \phi \pm \eta \tag{2.16}$$

with the transition probabilities (see eq.(2.11-13)):

$$T_\pm(\phi) = \frac{1-p_0}{2}\left(1 \pm \frac{1}{2}\eta K(\phi)\right), \quad K = -\frac{\partial S}{\partial \phi} \tag{2.17}$$

For compact variables $U \in SU(N)$ an updating rule could be [12]

$$U \to U \exp(\pm i \eta \lambda^a) \tag{2.18}$$

with

$$T_\pm = \frac{1-p_0}{2}\left(1 \pm \frac{1}{4}\left(S(Ue^{i\eta\lambda^a}) - S(Ue^{-i\eta\lambda^a})\right)\right) \tag{2.19}$$

where λ^a are the generators of the $su(N)$ algebra. Instead of cutting the $T_\pm$ at 0 and 1 (to obtain probabilities) one can use a smooth cut off, e.g. by defining:

$$T_\pm = \frac{1-p_0}{2}\left(1 \pm \tanh\left(\frac{\eta}{2}K\right)\right); \quad \eta K = \frac{1}{2}\left(S(Ue^{i\eta\lambda^a}) - S(Ue^{-i\eta\lambda^a})\right) \tag{2.20}$$

Interestingly enough, "sloppinesses" of this kind, as also the updating of dependent variables as if they were independent (using the old configuration to get the driving force), etc., only affect the coefficients C_i in eq.(2.15). The systematic errors introduced by them are thus eliminated all at the same time. The quality of the extrapolation procedure is illustrated by the following figure ([16] and courtesy of C. Hege).

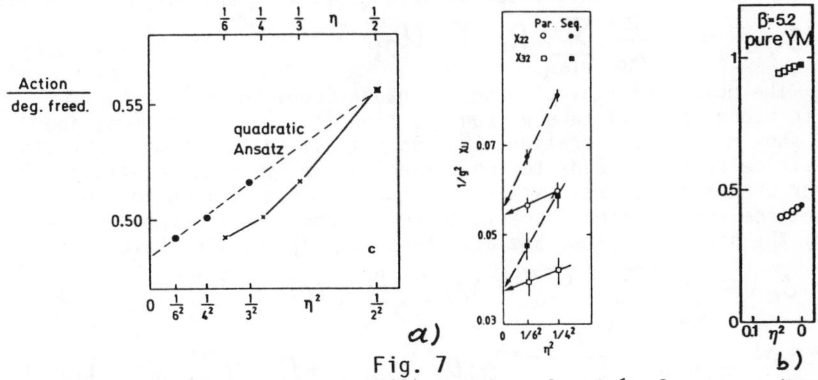

Fig. 7

Random walk a) Noncompact SU(2), action, Creutz´s force vs step size, various updatings of the colours (8 lattice); b) Compact SU(3): Plaquette (□) and Polyakov loop (o) vs step size ($4^3 2$ latt.).

3. MONTE CARLO CALCULATIONS WITH DYNAMICAL FERMIONS

3.1. Approach

With improving quality of the Monte Carlo calculations it becomes evident that for further tests against the experimental data the effects of dynamical quarks must be taken into account. The quenched approximation [17], to which we owe the first impressive results on the chiral symmetry breaking, hadronic spectrum, etc., not only cannot attack such specific effects as hadronization, but seems to have reached its limits also concerning, e.g. baryonic masses calculations, where the agreement with experiment remains systematically poor despite improved statistics.

The present day supercomputers allow computations with dynamical quarks, at least on lattices of intermediate size, and we shall present here fast Monte Carlo algorithms by which interesting problems can be studied. They are first examples of unbiased algorithms, based on Markow chains and which allow for realistic calculations with dynamical Wilson fermions [18].

This work was done with Ph.de Forcrand [19]. The frame of our approach is that of effective action and of pseudofermionic integration [20,21], the latter used in the sense of Ref. [22], after an essential improvement of the pseudofermionic Monte Carlo due to Ph.de Forcrand (Section 3.2.).

We thus proceed from the lattice QCD described by the partition function:

$$Z = \int \prod_{\langle xy \rangle} dU_{\langle xy \rangle}\, e^{-S_{YM}} \int \prod_{x} d\bar{\Psi}_x\, d\Psi_x\, e^{-S_F} \tag{3.1}$$

with $dU_{\langle\rangle}$ the Haar measure for the group variables on the links $\ell = \langle xy \rangle = (x,\mu)$ (we shall use the most convenient notation each time) and Berezin integration for the fermionic site variables [23]. The Yang Mills action is:

$$S_{YM} = -\frac{\beta}{N_c} \sum_{Plaq.} \mathrm{Re}\, \mathrm{Tr}\, U_{\partial P} \tag{3.2}$$

with N_c the number of flavours and the gauge coupling $g^2 = 2\, N_c/\beta$. For the fermions we shall take Wilson´s action [18], which is local and does not show modes proliferation. It does not dilute the fermionic field over a lattice cell and leads to the correct anomally in the continuum. It however breaks the chiral symmetry of the massless fermions on the finite lattice and therefore care is needed in checking for the installation of the scaling regime. See also Refs. [24-26]. We thus have:

$$S_F = \sum_{x,y,\alpha,\beta,a,b} \bar{\Psi}_x^{\alpha,a}\, W_{x,y}^{\alpha,a;\beta,b}(U)\, \Psi_y^{\beta,b} \tag{3.3}$$

$$W_{x;y}^{\alpha a;\beta b} = \mathbb{1} - \varkappa \sum_{\mu=1}^{d} \left(\Gamma_{+\mu}^{\alpha\beta}(r)\, U_{\langle xy \rangle}^{ab}\, \delta_{y,x+\mu} + \Gamma_{+\mu}^{\alpha\beta}\, U_{\langle yx \rangle}^{ab}\, \delta_{y,x-\mu} \right) \tag{3.4}$$

$$\mathcal{U}_{\langle yx\rangle} = \mathcal{U}^{+}_{\langle xy\rangle} = \mathcal{U}^{+}_{x,\mu} \; ; \; \Gamma^{\mu}_{\pm}(r) = r \pm \gamma^{\mu} \; ; \; \{\gamma^{\mu},\gamma^{\nu}\} = 2\delta_{\mu\nu} \tag{3.5}$$

Here the dimension d will be taken 4, $r \in (0,1]$ and the hopping parameter $\varkappa$ is related with the dimensionless bare fermion mass by:

$$\varkappa = 1/(2M + 2dr) \tag{3.6}$$

With this normalization of the fermionic fields the propagator is:

$$G = 2\varkappa \langle \Psi \bar{\Psi} \rangle \tag{3.7}$$

Integrating out the fermions leaves an effective action (for N_f flavours):

$$S_{eff} = S_{YM} - N_f \ln \det W \tag{3.8}$$

and a Markow algorithm for the Yang Mills problem with this action needs:

$$e^{-\Delta S_{YM}} \cdot \varrho_N (\{u\} \to \{u'\}) \tag{3.9}$$

with

$$\varrho_N (\{u\} \to \{u'\}) = (\det W(\{u'\}) / \det W(\{u\}))^{N_f} \tag{3.10}$$

Here N denotes the number of the links which are changed in going from the old configuration $\{u\}$ to the new one $\{u'\}$. Det W is a nonlocal expression, although W is local (nearest neighbour coupling), and this makes the calculation of ϱ_N very time-intensiv.

Processing one link at a time (N = 1) is therefore inefficient. The alternative is to keep N large and induce a seizable change on the lattice, so as to obtain a statistically relevant new configuration with fewer calculations of det W . In this case, however, the large change may be rarely accepted, meaning that we again have to recalculate ϱ_N many times - possibly as many as in the one-link-at-a-time scheme.

To break this vicious circle, two possibilities seem to present themselves:

a) Try to factorize ϱ_N into the contributions from each link (or few links) and then use a link-by-link updating. Such a factorization can only be approximative, of course, due to the nonlocal character of Det W - what we have to ensure is that no bias is introduced in this way and that in fact we can have a control on the error, both in the sense of being able to estimate it and to tune the algorithm such as to limit it. We shall call such algorithms "local". The so-called "pseudofermionic method" [20], is a local algorithm which, however, is biased toward underestimation of the fermionic effects. We shall present in Section 3.3. two alternative local algorithms which fullfill the previous requirements and use them (Section 3.4.) to study the deconfining transition.

b) Try to construct pseudo-heat-bath-type algorithms using the full ρ_N (large N), which will not suffer of a high rejection rate implied by the large changes intervening in the updating. We shall call such algorithms "global" and we shall discuss a possibility in Section 3.5..

Some general comments will be made in Section 3.6.. In the next Section we present our method of using pseudofermionic Monte Carlo and its improvement. In all the following formulae and calculations we shall use Wilson fermions with $r = 1$ and antiperiodic boundary conditions.

3.2. The pseudofermionic integration and its improvement

Fucito, Marinari, Parisi & Rebbi [20] and Weingarten & Petcher [21] observed that one can calculate Det W by a bosonic integration, to wit:

$$(\det W)^{-2} = \int \prod_x d\phi_x^* d\phi_x \exp(-S_{PF}) \tag{3.11}$$

with

$$S_{PF} = \sum_{x,y} \phi_x^* (W^+W)_{xy} \phi_y \tag{3.12}$$

Here $\phi_x^{a,\alpha}$ are commuting Dirac spinors and isospinors - that is, pseudofermions. In the following we shall use the pseudofermionic integration to directly calculate ρ_N , by means of the following series of identities[22]:

$$\begin{aligned}\rho_N(\{U\}\to\{U'\}) &= \left(\int[d\phi]\, e^{-S'_{PF}} \Big/ \int[d\phi]\, e^{-S_{PF}}\right)^{-N_f/2} \\ &= \left(\langle \exp(-\Delta S_{PF})\rangle_{S_{PF}}\right)^{-N_f/2} = \left(\langle \exp(\Delta S_{PF})\rangle_{S'_{PF}}\right)^{N_f/2} \\ &= \left(\langle \exp(-\tfrac{1}{2}\Delta S_{PF})\rangle_{\frac{1}{2}(S_{PF}+S'_{PF})} \Big/ \langle \exp(\tfrac{1}{2}\Delta S_{PF})\rangle_{\frac{1}{2}(S_{PF}+S'_{PF})}\right)^{-N_f/2}\end{aligned} \tag{3.13}$$

where:

$$S_{PF} = S_{PF}(\{U\}),\quad S'_{PF} = S_{PF}(\{U'\}),\quad \Delta S_{PF} = S'_{PF} - S_{PF} \tag{3.14}$$

Here $\langle\ \rangle_S$ means average over the pseudofermionic ensamble described by the action S. While these relations are exact, the fluctuation in S_{PF} is exponentiated, and therefore the variance may get very large. A very efficient way to reduce the variance of the expectation values is to integrate per hand some of the variables [27,28] which ammounts to chose new observables having identical expectations with the old ones [29].

Assume the link $\langle x_0 y_0\rangle$ changes from U to U´ . One can convince himself, that the two observables

$$O_{x_0y_0} = \exp\left(-\left(S_{PF}(U') - S_{PF}(U)\right)\right) \tag{3.15}$$

and

$$\overline{O_{x_0y_0}} = \frac{\int[d\phi_{x_0} d\phi_{y_0}] \exp(-S_{PF}(U'))}{\int[d\phi_{x_0} d\phi_{y_0}] \exp(-S_{PF}(U))} \tag{3.16}$$

have identical expectations. Explicitely we have for the first one:

$$\theta_{x_0 y_0} = \exp\left(\mathrm{Re}\,\mathrm{Tr}\,\left(C.(U'-U)\right)\right)$$
$$C = \varkappa\left(\sum_{z\neq y_0} \phi_z^* W_{z x_0}^+ \Gamma_{\langle x_0 y_0\rangle} \phi_{y_0} + \sum_{z\neq x_0} \phi_{x_0}^* \Gamma_{\langle y_0 x_0\rangle} W_{y_0 z} \phi_z\right) \tag{3.17}$$

The second one can be evaluated by a local Monte Carlo [27] or analytically [28] . We consider it advantageous to use the last method, leading here to:

$$\overline{\theta_{x_0 y_0}} = \exp\left(\mathrm{Re}\,\mathrm{Tr}\,\left(\overline{C}.(U'-U)\right)\right)$$
$$\overline{C} = c_1\left(A_{x_0}^* A_{y_0} + B_{y_0}^* B_{x_0}\right) + c_2\left(A_{x_0}^* B_{x_0} + B_{y_0}^* A_{y_0}\right) \tag{3.18}$$
$$A_{x_0} = \sum_{z\neq x_0, y_0; y\neq y_0} W_{x_0 y}^+ W_{yz} \phi_z \; ; \; B_{x_0} = \sum_{z\neq x_0, y_0} \varkappa\, U_{\langle x_0 y_0\rangle} \Gamma_{\langle x_0 y_0\rangle} W_{y_0 z} \phi_z$$

with ....

$$c_1 = 2/(c_0^2 - 4\varkappa^2),\; c_2 = 4\varkappa/(c_0(c_0^2 - 4\varkappa^2)),\; c_0 = 1 + 16\varkappa^2 \tag{3.19}$$

These formulae are trivially extended to the case of more links changed, if they do not touch. The improvement in the signal to noise ratio obtained by using eq.(3.16) instead of eq.(3.15) is by more than a factor 5 , which means an increase in the algorithm´s efficiency by a factor of 10 - 100. In Fig. 8 we compare the convergence of eq.(3.15) and eq.(3.16) for small lattices where exact results can be obtained with standard routines.

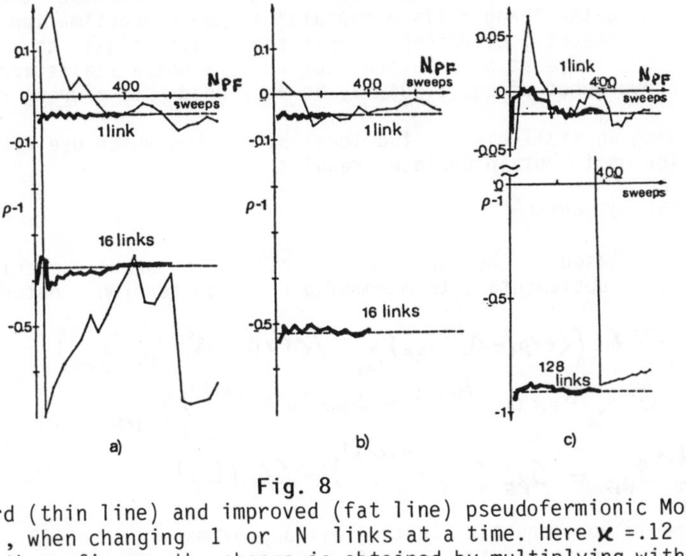

Fig. 8

Standard (thin line) and improved (fat line) pseudofermionic Monte Carlo for ρ_N , when changing 1 or N links at a time. Here $\varkappa$ =.12 and N_f=2. In all these figures the change is obtained by multiplying with $\exp(i\eta\lambda^a)$ - see eqs.(2.18),(2.19). "Background" refers to the old configuration.
a): 4×2^3 lattice, η=.4, N=16, hot background (β =4); b): as a), η =.8;
c): 4^4 lattice, η=.4, N=128, cold background ($\beta = \infty$).

The improvement of the convergence obtained by using eq.(3.16) is striking. We found out that with $N \sim 100$, to obtain ρ_N within 20% errors we need of the order of 500 standard or of the order of 50 improved pseudofermionic sweeps. This applies typically to calculations at $\beta \sim 4 - 5$ and with $\varkappa \simeq .12$, i.e. far beyond the validity of the approximation given by the lower orders of the hopping parameter expansion. Of course, one can use improved formulae also for other observables, like W^{-1}, etc.[30].

3.3. Local algorithms

An immediate factorization of ρ_N is given by the direct linearization ρ'_N:

$$\rho'_N(\{U\} \to \{U'\}) = \exp\left(\tfrac{1}{2} N_f \langle \Delta S_{PF} \rangle_{S_{PF}}\right) \tag{3.20}$$

This is the pseudofermionic method, with $N = 4 \times N_V$ (all the links of the lattice). However, due to Jensen´s inequality:

$$\begin{aligned} \rho'_N(\{U\} \to \{U'\}) &= \exp\left(\tfrac{1}{2} N_f \langle \Delta S_{PF} \rangle_{S_{PF}}\right) \\ &\geqslant \langle \exp(-\Delta S_{PF}) \rangle_{S_{PF}}^{-N_f/2} = \rho_N(\{U\} \to \{U'\}) \end{aligned} \tag{3.21}$$

ρ'_N represents always an overestimation of the correct ρ_N. Now the typical procedure is to make a prechoice, U´, using only the Yang Mills action, and then accept it with probability min (1 , ρ_N). The fermions then act by selecting among the "Yang Mills proposals" and an overestimation of ρ_N leads to an (erroneous) overacceptance at the fermionic gate as if the fermions were not there at all. Notice that this is not a statistical effect: the convergence of eq.(3.20) may be very good - to a wrong result.

In the following we shall discuss two local algorithms which use finite steps and allow us to obtain unbiased results.

A) A Random Walk algorithm

The algorithm is based on the formulae in Section 2.5. where the driving force contains a supplementary term comming from the fermionic determinant

$$\begin{aligned} \eta K_F^{\ell,s} &= -\frac{N_f}{4} \ln\left(\langle \exp(-\Delta_+^{\ell,s} S_{PF}) \rangle_{S_{PF}} / \langle \exp(-\Delta_-^{\ell,s} S_{PF}) \rangle_{S_{PF}}\right) \\ &\simeq \frac{N_f}{4} \langle S_{PF}(U_\ell e^{i\eta\lambda^s}) - S_{PF}(U_\ell e^{-i\eta\lambda^s}) \rangle_{S_{PF}} \end{aligned} \tag{3.22}$$

with

$$\Delta_\pm^{\ell,s} S_{PF} = S_{PF}(U_\ell e^{\pm i\eta\lambda^s}) - S_{PF}(U_\ell) \tag{3.23}$$

Although there is no inequality in the last approximation in eq.(3.22) one does not need to use it, as the exact form is just as easy to calculate.

The algorithm reads:

- For a given configuration we calculate K_{YM} and K_F for a bush of links to be changed by steps in the 8 colour directions. For K_F we use an improved pseudofermionic Monte Carlo (see eqs.(3.13),(3.16)).
- A sweep is achieved in N_B such steps, N_B being the number of bushes.
- After the algorithm converges at a given step size η we continue with a smaller value for η . We need convergence at 2-3 values for η.
- The final results are obtained by extrapolation to $\eta = 0$.

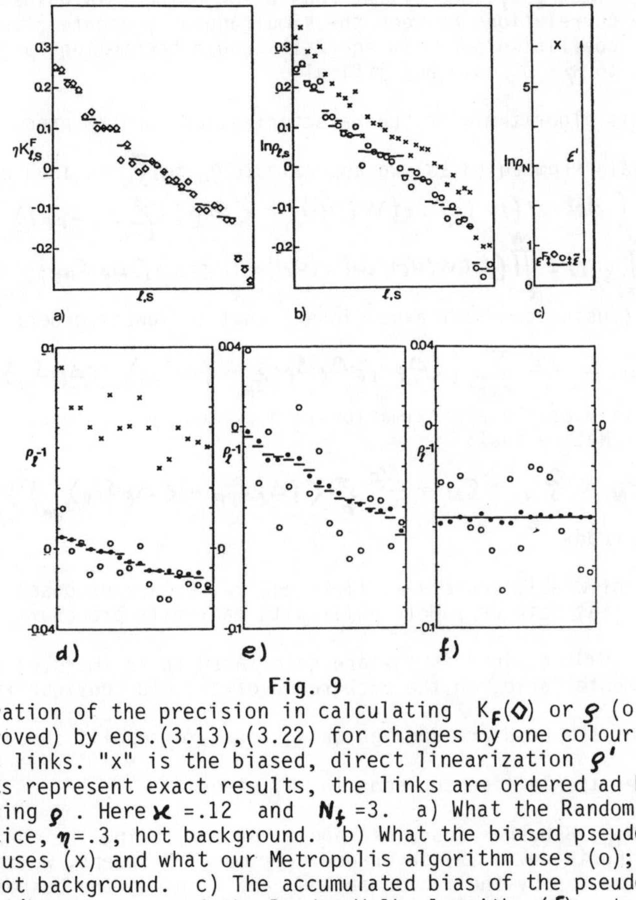

Fig. 9

Illustration of the precision in calculating K_F(◇) or ϱ (o: standard, •: improved) by eqs.(3.13),(3.22) for changes by one colour step in various links. "x" is the biased, direct linearization ϱ' . The line segments represent exact results, the links are ordered ad hoc with decreasing ϱ . Here $\varkappa$ =.12 and N_f =3. a) What the Random Walk uses; 2^4 lattice, η=.3, hot background. b) What the biased pseudofermionic method uses (x) and what our Metropolis algorithm uses (o); 2^4 lattice, η=.3, hot background. c) The accumulated bias of the pseudofermionic method (ε'), the error of the Random Walk algorithm (ε^F) - to be eliminated by extrapolation - and the error $\tilde{\varepsilon}$ of our Metropolis algorithm; 2^4 lattice, η=.3, hot background. d) Same as b), 4×2^3 lattice, η =.4, hot background. e) Same as d), η =.8. f) Same as d), cold background.

In Fig. 9 we present illustrations of the quality of the pseudofermionic Monte Carlo for $K^F_{\ell,s}$ and show the difference to a biased pseudofermionic method which uses the direct linearization eq.(3.20). Of course, the pseudofermionic method was not supposed to work for such large changes (large η) as our algorithm can afford. However, using very small steps makes the convergence extremely slow, by itself a very unpleasant feature. Moreover, even a small updating bias will accumulate in the long - just because it is a bias, i.e. it drives always in the same direction.
In all these calculations we use a partition of the links in $N_B = 8$ bushes, the links in a bush having all the same direction but no site or plaquette in common [31]. One of the reason for using this bushes here is to weaken the correlations between the simultaneously updated links - otherwise the coefficients C_i in eq.(2.15) could become too large and the extrapolation to $\eta = 0$ could be difficult.

B) A Metropolis algorithm with "bush-factorization" and YM prechoice.

Proceeding again from eq.(3.13) we approximate ρ_N by $\tilde{\rho}_N$ defined by:

$$\begin{aligned} \rho_N &= \left(\det W(\{u'\}) / \det W(\{u\}) \right) = \langle \exp(-\sum_{\ell=1}^{N} \Delta_\ell S_{PF}) \rangle_{S_{PF}}^{-N_F/2} \\ \simeq \tilde{\rho}_N &= \prod_{\ell=1}^{N} \rho(\ell) = \prod_{\ell=1}^{N} (\det W(u'_\ell) / \det W(u_\ell)) = \prod_{\ell} \langle \exp(-\Delta_\ell S_{PF}) \rangle_{S_{PF}}^{-N_F/2} \end{aligned} \tag{3.24}$$

One can show (using cumulant expansions) that to lowest order:

$$\tilde{\rho}_N - \rho_N \simeq \frac{N_F}{2} \sum_{\ell \neq \ell'} \left(\langle \Delta_\ell S_{PF} \cdot \Delta_{\ell'} S_{PF} \rangle_{S_{PF}} - \langle \Delta_\ell S_{PF} \rangle_{S_{PF}} \langle \Delta_{\ell'} S_{PF} \rangle_{S_{PF}} \right) \tag{3.25}$$

i.e., the quality of the approximation is dictated by the correlations between links. Notice that:

$$\rho'_N - \rho_N \cong \tilde{\rho}_N - \rho_N + \frac{N_F}{4} \sum_{\ell} \langle (\Delta_\ell S_{PF} - \langle \Delta_\ell S_{PF} \rangle_{S_{PF}})^2 \rangle_{S_{PF}} \tag{3.26}$$

The algorithm reads:

- For a bush of weakly correlated links new values are proposed using a Yang Mills heat bath or a Metropolis-with-many-hits procedure.
- The factors $\rho(\ell)$ contained in $\tilde{\rho}_N$ are calculated in an improved pseudofermionic Monte Carlo, in the background of the old configuration (N_{PF} sweeps)
- The links of the bush are updated using local Metropolis checks with $\rho(\ell)$
- A sweep over the lattice succeeds in N_B steps.

We used the $N_B = 8$ bushes described above. By monitoring the correlation matrix appearing in eq.(3.25), averaged over many YM sweeps, we can judge the error introduced by the factorization, and thin out the bushes if necessary. The quality of the approximations used is illustrated in Fig. 9 and Table 1. Here the columns "I" and "II" contains exact ρ_N and $\tilde{\rho}_N = \prod \rho(\ell)$(standard NAG routines), column "III" contains the $\tilde{\rho}_N$ computed with pseudofermionic Monte Carlo (standard on 2^4, improved on 4×2^3 lattice).

Table 1.
Exact and pseudofermionic calculation of ϱ_N, $\tilde{\varrho}_N$

Nr.	Exact: $\ln\varrho_N$ (I)	Exact: $\ln\tilde{\varrho}_N = \ln\prod_l \varrho_l$ (II)	Ps.f. Monte Carlo $\langle e^{\Delta}\rangle$: $\ln\tilde{\varrho}_N = \ln\prod_l\varrho_l$ (III)	Ps.f. Monte Carlo $\langle e^{\Delta}\rangle$: $\ln\varrho_N$ (IV)	Ps.f. Monte Carlo $e^{\langle\Delta\rangle}$: $\ln\varrho'_N$ (V)	Example Description: latt.	back-ground config.	η	N x Nr. of colour steps
1	.53	.74	.68	1.05	6	2^4	hot ($\beta \simeq 4.7$)	.3	8 x 8
2	.43	.33	.34	-1.01	4.2				
3	-.16	-.16	-.14(-.23)	-.14(-.41)	1.1	4×2^3	hot ($\beta \simeq 4.$)	.4	16 x 1
4	-.69	-.70	-.67(-.80)	-.67(- 1.02)	5.4			.8	
5	-.28	-.28	-.28(-.31)	-.28(-.22)	.9		cold ($\beta \simeq \infty$)	.4	
6	-1.10	-1.07	-1.10(-1.03)	-1.10(-.03)	4.9			.8	

3.4. The deconfining transition in the presence of quarks

By analogy with spin models it is expected that the ordering effect of the matter fields will weaken the deconfining transition. We have seen that the diamagnetismus of the Higgs fields can lead to the complete disappearence of the deconfining transition (Section 2.4.). Whether the effect of the fermionic paramagnetismus should also be so drastic seems not yet to be agreed upon, although the tendency is clear.

Because a transition affects bulk properties, it can be studied also on smaller lattices and therefore we choose this problem to illustrate the algorithms developed above. For that we used a $4^3\times2$ lattice with Wilson fermions and SU(3) colour group. We used both the Random Walk and the Metropolis algorithm of Section 3.3. We do 100 improved pseudo-fermionic sweeps in the long directions and 400 standard sweeps in the short one (remember that the links in our bushes have all the same direction) in order to calculate K_F(for RW) and ϱ (for Metropolis). The algorithms run on the Cray-XMP´s at Cray Research and on the Cray-1 of Z I Berlin and take about 3´ per full Yang Mills sweep.

A Random Walk step is a complex of 8 steps done successively in the 8 (positive or negative) colour directions in some random order. For the Metropolis algorithm the prechoice is done by previously running a pure Yang Mills Metropolis with 50 hits (average acceptance per hit: 1/2).

All the runs start by first doing some 1000 pure Yang Mills sweeps at the given β and then turning on the fermions. Then we do one or more groups of 150 sweeps with fermions, one after the other: first Random Walk with the largest step size, followed by further Random Walks with decreasing step sizes, and followed finally by Metropolis runs. For some points no Random Walk runs are done, the first group of sweeps are Metropolis. We use the last 100 sweeps of each group for taking averages.

Results are presented for

- the average Plaquette

$$A = \langle \tfrac{1}{3} \mathrm{Re}\, \mathrm{Tr}\, (U_{\partial P}) \rangle \tag{3.27}$$

- the average Polyakov loop

$$P = \langle (\mathrm{Re}\, (\tfrac{1}{3} \mathrm{Tr}\, U_{Pol})^3)^{\frac{1}{3}} \rangle \underset{\varkappa > 0}{\simeq} \langle \tfrac{1}{3} \mathrm{Re}\, \mathrm{Tr}\, U_{Pol} \rangle \tag{3.28}$$

where $U_{\partial P}$ is the product of links around a plaquette and U_{Pol} the product of N = 2 links on a thermal string, both averaged over the lattice, and

- the physical gluonic energy

$$\epsilon_G = \beta \; (\text{A(Time-like)} - \text{A(Space-like)}) \tag{3.29}$$

We first did runs on a 2^4 lattice, where we could produce also exact results by running in parallel a standard Metropolis algorithm with link by link updating and calculation of the fermionic determinant at each updating via standard NAG routines. The comparison is presented in Table 2 and shows a good agreement between the "exact" results obtained in this way and the "Metropolis" results - obtained with our Metropolis algorithm of Section 3.3., with bush-factorization. The latter uses 400 (after 40 for heating) pseudofermionic sweeps to calculate $\tilde{\varrho}_N$. With either procedure, after ~ 1000 pure YM sweeps at each β the fermions are turned on, the first ~ 50 heating sweeps discarded and the next 100 sweeps or more used for averages. The errors correspond to binning in groups of 10 sweeps. Notice that a small lattice fluctuates more.

Table 2.
Exact and bush-factorized Metropolis results for the 2^4 lattice.

Example	Plaquette (A)		Polyakov loop (P)	
	Exact	Metropolis	Exact	Metropolis
$\beta = 4.7,\ \varkappa = .12,\ N_f = 3$	.530 ± .010	.530 ± .010	.505 ± .020	.450 ± .040
$\beta = 4.0,\ \varkappa = .15,\ N_f = 2$	.390 ± .010	.415 ± .010	.350 ± .020	.370 ± .020

The results for the $4^3\,2$ lattice are presented in Figs. 10 and 11. We use $\varkappa = .11 - .12$, such that we remain in the realm of intermediate masses. We took $N_f = 3$, the number of flavours lighter than, say, 1Gev. In accordance with the discussion in Sections 3.2.,3.3. we tend to trust our algorithms up to some 20% error in dealing with the fermions. The statistical errors measured are much smaller. Correspondingly, our data for A, P and ϵ_G could show errors of ~5%, 10% and 20%, respectively, if we cumulate all the sources. This is only a rough estimation and therefore we didn´t put error bars on the points.

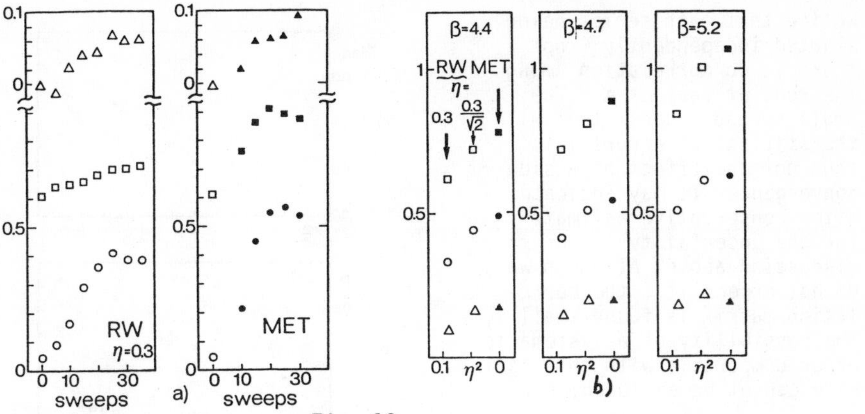

Fig. 10
Random Walk and bush-factorized Metropolis on a 4^3 2 lattice. a) Turning on the fermions, after 1000 pure YM sweeps. Here N_f =3, κ=.11, β=4.7. b) Comparison of the Random Walk and Metropolis data (successive groups of 150 sweeps, after 1000 pure YM). Here N_f=3, κ=.12, various β . □: A(x2); o: P; △: ϵ_G (x2 - b))

The first observation is the swiftness of the response, when we turn on the fermions: in something like 40 sweeps the Polyakov loop values rise from 0 to around .5! (see Fig. 10a). Remarkably enough, this is induced by a contribution from the fermionic determinant to the updating of only 10 - 20% (the fermionic driving force K_F is around 15% of the YM force K_{YM} in the Random Walk algorithm, similarly the rejection rate forced by the fermions in the Metropolis algorithm is about 15%) - see also Ref. [32]. This is how the ordering effect of the fermions works. Thus stating that "the effects of the determinant are small" may be very misleading.

As one can see from Fig. 10b, the Random Walk results extrapolate correctly to our Metropolis results, although for higher β a linear extrapolation seems too rough (too large η). As a finite step size η brings more disorder into the system, the direction of the extrapolation is well defined. Hence the exact value is bounded by the one at the smallest η.

Fig 11 shows our Metropolis data for β between 3.2 and 5.2. Here κ =.12 and N_f=3 were used. Some of the β-points were not run with Random Walk but only with the (bush-factorized) Metropolis, therefore we only show the latter results. In general these agree with the extrapolated Random Walk results, as discussed above, well within the expected precision.

As can be seen on Fig. 11, all the gluonic expectations (A, P, ϵ_G) are much larger then in the pure gauge case, and behave very smoothly. There might be a kind of cross over around β =4.2, and one would need larger lattices to make clear statements - but these data strongly suggest that there is no phase transition for these fermionic parameters.

Notice that each series being started independently (hot), there is no correlation among the runs at various β . The small spread (consistent with the statistical errors) is thus not the effect of a slow convergence. It may indicate a too conservative estimate for the uncertainty in our discussion above. Although we do not expect it (the correlation matrix is found small), the possibility of a systematic error comparable with this estimate cannot be excluded, however.

For comparison, we ran the "pseudofermionic method" (see eq.(3.20)) at β =4.4 with the same fermionic parameters and N_B=8 bushes. We took N_{PF}=50, after 10 heating sweeps, and the prechoice is done with only one Metropolis hit, leading to an acceptance of 85% (these numbers are typical for the applications of this method). The results are roughly 1/2 of our Metropolis or Random Walk values, suggestive for the bias and/or slow convergence of the pseudofermionic method.

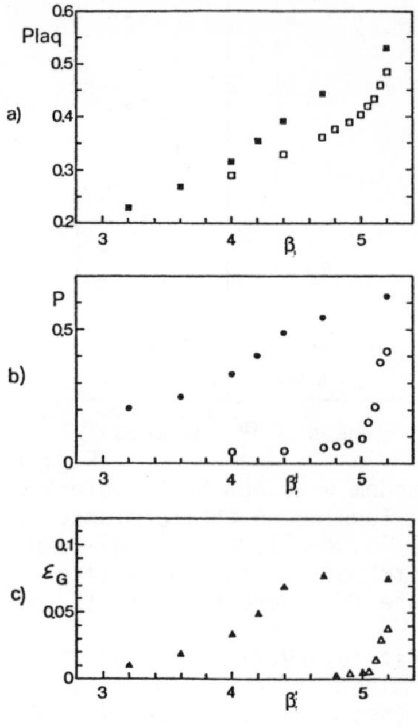

Fig. 11
Metropolis data for: a) Plaquette (A) b) Polyakow loop (P), and c) Gluonic energy (ϵ_G). Empty symbols denote the pure gauge expectations (~2000 runs).

3.5. A global algorithm

In order to avoid any systematic error we can try to do global updating (notice that in our local algorithms there is no accumulated error - the updating is not biased toward a given direction -, but some remnant systematic imprecision may appear, either due to the extrapolation to η =0 of the Random Walk data or due to surviving correlations in the bush-factorized Metropolis algorithm). A global algorithm not suffering from a high rejection rate (which can result from the large variation in the determinant induced by the many simultaneous changes) is:

- In preparing to update a given bush of links perform a pseudofermionic Monte Carlo on the background of a given links configuration. Calculate for each pseudofermionic configuration n the coefficients $c_{n,\ell}$ of the ΔU_ℓ in the bush (see eqs.(3.17),(3.18)) and store them. The needed storage is only: $2\text{x}9\text{x}N\text{x}N_{PF} \cong 200$ kwords for $N \cong 100$, $N_{PF} \cong 100$.

- Implement a pseudo heat bath with the determinant: propose new values for the links in the bush by a pure YM prechoice (pseudo heat bath or Metropolis with very many hits) and compute

$$\rho_N(\{U\} \to \{U'\}) = \left(\frac{1}{N_{PF}} \sum_{n=1}^{N_{PF}} \exp\left(-\sum_{l=1}^{N} \mathrm{Re\,Tr}\left(C_{n,l}\,\Delta U_l\right)\right)\right)^{-\frac{N_f}{2}} \qquad (3.30)$$

using the stored coefficients $\{C_{n,l}\}$; use the ρ_N/ρ_N(Max) as updating probability, where ρ_N(Max) is some bound on ρ_N, estimated e.g., via the paramagnetic inequality $\rho_N(\{U\}\to\{U'\}) \leq \rho_N(\{U\}\to\{1\})$; keep doing all that untill a change (of the whole bush) is accepted.

Notice that the calculation of ρ_N by eq.(3.30) takes practically no time and thus even very high rejection rates can be aforded. The limitation on the size of the bush comes rather from the convergence of the pseudofermionic Monte Carlo (how large N_{PF} are needed). The results up to now indicate that $N_{PF} = \mathcal{O}(100)$ improved sweeps are enough to get ρ_N with 90% precision for $N = \mathcal{O}(100)$ links. Work with this algorithm is in progress [33].

3.6. Discussion

All the algorithms described above perform a full YM sweep on a 4^4 lattice in about 4′ (SU(3), Wilson fermions). For intermediate size lattices and for not too high β - that is, as long as we can hold fixed the number of bushes -, the algorithm is linear in the lattice volume. When we have to limit the number of links in the bush, the algorithm goes quadratically with the lattice volume. However, more refined estimation, similar to those in Ref.[34], should be made.

The necessity of enlarging the lattice is clear already for the question of the finite temperature transition (see e.g. Ref. [35]). Still larger lattices will be needed to study the quark-quark potential and the spectrum in scaling regime. In my opinion, we need both the Gflops of the new super and dedicated computers, and the good ideas which help us to depart more and more from the brutal approach and devise clever algorithms. It appears to me that there should be no intrinsic lack of phantasy related to the Monte Carlo approach - in a generalized sense.

Some new algorithms are studied in the recent literature. An appealing idea, for example, is to use the error of the determinant calculation as stochastic noise [36]. One only needs an unbiased estimate for , as given possibly by eq.(3.13), with very few sweeps [37]. Further works I am aware of are quoted in Ref [38], but this is clearly very incomplete.

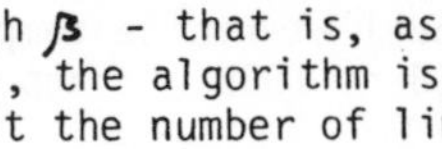

I would like to thank my colleagues in the collaboration mentioned above for participation and discussions, and D. Duke and the S C R I - Florida for inviting me to talk at this meeting. I would like to aknowledge partial travel support from the Freie Universität Berlin.

References

1 K.Osterwalder and E.Seiler, Ann. Physics **110**, 440 (1978)

2 E.Fradkin and D.Shenker, Phy. Rev. **D19**, 3682 (1979)

3 A.Guth, Phys. Rev. **D23**, 347 (1981)

4 A.Albrecht and P.Steinhardt, Phy. Rev. Lett. **48**, 1220 (1982)

5 F.Karsch, E.Seiler and I.O.Stamatescu, Phys. Lett. **131B**, 138 (1983)

6 T.Munehisa and Y.Munehisa, Nucl. Phys. **B215** FS7 , 508 (1983)

7 H.Kühnelt, C.B.Lang and G.Vones, Nucl. Phys. **B230** FS10 , 31 (1984)

8 V.P.Gerdt, A.S.Ilchev, V.K.Mitryushkin, I.K.Sobolev and A.M.Zadorozhny, "Phase structure of the SU(2) lattice gauge-Higgs theory", preprint Dubna E2-84-313 (1984)

9 J.Jersak, contribution to these proceedings
I.Montvay, contribution to these proceedings

10 F.Karsch, E.Seiler and I.O.Stamatescu, "Study of the phase structure of an SU(3) Higgs model at finite temperature", preprint FU-Berlin ,Jan. 1985, to appear in Phys. Lett.

11 C.Hege, V.Linke, E.Seiler and I.O.Stamatescu, work in progress

12 I.O.Stamatescu, "On lattice gauge theory with matter", Proceedings of the NATO Advanced Study Inst., München (1983)

13 I.O.Stamatescu, "Simultaneous processing of variables in lattice gauge theory calculations", preprint München, MPI-PAE/PTh 92/83

14 P.Hasenfratz, F.Karsch and I.O.Stamatescu, Phys. Lett. **133B**, 221 (1983)

15 J.Polonyi, H.W.Wyld, J.B.Kogut, J.Shigemitsu and D.K.Sinclair, Phys. Rev. Lett. **53**, 644 (1984)
T.Celik, J.Engels and H.Satz, Phys. Lett. **133B**, 427 (1984)

16 I.O.Stamatescu, U.Wolff and D.Zwanziger, Nucl. Phys. **B225** FS9 , 377 (1983)
E.Seiler, I.O.Stamatescu and D.Zwanziger, Nucl. Phys. **B239**, 177 (1984)

17 E.Marinari, G.Parisi and C.Rebbi, Nucl. Phys. **B190** FS3 , 734 (1981)

18 K.Wilson in Vol. "New phenomena in subnuclear physics" (Erice 1975) A.Zichichi Ed. (Plenum, New York, 1977)

19 Ph.de Forcrand and I.O.Stamatescu, "Fast algorithms for fermionic Monte Carlo", preprint FU Berlin, June 1985

20 F.Fucito, E.Marianari, G.Parisi and C.Rebbi, Nucl. Phys. **B180** FS2 , 360 (1981)

21 D.H.Weingarten and D.N.Petcher, Phys. Lett. **99B**, 333 (1981)

22 G.Bhanot, U.Heller and I.O.Stamatescu, Phys. Lett. **129B**, 440 (1983)

23 F.A.Berezin, "The method of second quantization" Academic, New York, 1966

24 L.H.Karsten and J.Smit, Nucl. Phys. **B183**, 103 (1981)

25 W.Kerler, Phys. Rev. **D23**, 2384 (1981)

26 E.Seiler and I.O.Stamatescu, Phys. Rev. **D25**, 2177 (1982)

27 G.Parisi, R.Petronzio and F.Rapuano, Phys. Lett. **128B**, 418 (1983)

28 Ph.de Forcrand and C.Roisnel, Phys. Lett. **151B**, 77 (1985)

29 H.B.Callen, Phys. Lett. **4B**, 161 (1983)

30 A.Nakamura, private communication

31 I.O.Stamatescu, "Remarks on Monte Carlo calculations for lattice gauge theory with fermions", preprint München, MPI-PAE/PTh 82/81

32 M.Fischler and R.Roskies, Phys. Lett. **145B**, 99 (1984)

33 Ph.de Forcrand, A.Nakamura and I.O.Stamatescu, work in progress

34 D.Weingarten, "Algorithms for Monte Carlo calculations with fermions", preprint IBM Watson Research Center (1985)

35 U.Heller and F.Karsch, preprint CERN, TH.4078/84
R.V.Gavai and F.Karsch, preprint Illinois, ILL-(TH)-85-#19

36 A.D.Kennedy and J.Kuti, preprint Santa Barbara, NSF-ITP-85-21
G.Bhanot and A.D.Kennedy, preprint Santa Barbara, NSF-ITP-85-24

37 Ph.de Forcrand, work in progress
A.D.Kennedy, J.Kuti et al., private communication

38 e.g.:
A.Nakamura, Phys. Lett. **149B**, 391 (1984)
S.Otto, Phys. Lett. **135B**, 129 (1984)
Ulli Wolff, Phys. Rev. **D30** 2236 (1984)
M.Karowski, R.Schrader and H.J.Thun, preprint Göttingen 1984
G.Parisi, preprint Frascati LNF-84/4 (P)
contributions to these proceedings by: I.Barbour, G.Batrouni, F.Fucito, J.Nesic, M.Ogilvie, S.Otto and others

SPONTANEOUS SYMMETRY BREAKING BY ADJOINT SCALARS

K. Olynyk [1]

Department of Physics, The Ohio State University, 174 W. 18th Ave., Columbus, Ohio 43210

(Address after October 1, 1985)
The Fermi National Accelerator Laboratory
P.O. Box 500, Batavia, Illinois 60510

ABSTRACT

The calculation of gauge invariant signals of spontaneous symmetry breaking in SU(N) scalar models is reviewed. As an application of the ideas presented, a Monte Carlo calculation is performed for SU(5) adjoint scalar matter fields without and with gauge fields. The tree-level scalar potential is chosen to break SU(5) to SU(3)XSU(2)XU(1), but it is found that radiative corrections due to the gauge bosons give rise to the possibility of breaking to a metastable (perhaps even stable) SU(4)XU(1) phase for certain ranges of the bare scalar mass. The transitions between the various phases of the theory are found to be first order. The implications of these results to continuum physics are as yet unclear. As an added bonus some interesting matrix identities are included in an Appendix.

---

[1] Much of this article reports work that I have done in cooperation with with Junko Shigemitsu.

1. INTRODUCTION

Lattice gauge theory, and in particular the Monte Carlo technique, has proved an effective method for the study of non-Abelian gauge theories with and without matter fields. While a large fraction of research in the field has been directed toward the understanding of nonperturbative effects in QCD there has been significant interest in the area of spontaneous symmetry breaking [1,2,3]. At the same time, spontaneous symmetry breaking has become an indispensable ingredient in the understanding of elementary particle interactions. The low energy physics of any model is dependent upon the symmetry of the vacuum and the structure of the vacuum is determined by the dynamics of the model. Coleman and Weinberg [4] pointed out some time ago that radiative corrections can give rise to an effective potential that can be dramatically different from what is seen at the tree-level of the Lagrangian. It is therefore important to have a reliable method of calculating the effect of the interactions upon the vacuum structure in order to have a qualitative and quantitative understanding of the resulting physics.

We have recently shown that for SU(N) gauge theories (N>2) coupled to adjoint scalars, there can be defined simple gauge invariant observables which can detect and characterize the spontaneous breakdown of the vacuum symmetry [2]. Here we wish to report the application of this technique to the study of a model that may be of interest to physicists not working in lattice gauge theory. We have studied SU(5) gauge theory broken by adjoint scalars, namely the Higgs sector responsible for the first stage of symmetry breaking in minimal SU(5) grand unified theory [5].

In order to summarize the main result of this article, we need to first recall an earlier work where we considered a theory of scalar fields transforming under the adjoint representation of a global SU(5) symmetry [3]. The symmetry breaking pattern there was found to be in accordance with standard tree-level analysis [6]. In the present investigation, however, we find that the calculated observables indicate that the dynamics of the theory gives rise to a metastable SU(4)XU(1) phase in addition to the symmetric (SU(5)) and the broken (SU(3)XSU(2)XU(1)) phases that would be expected from a tree-level analysis. However, it is not totally clear whether the SU(4)XU(1) phase is stable or metastable. Once the system settled into any of the three observed phases, subsequent tunneling into another phase was never observed. On the other hand we find that the size of the dynamically generated SU(4)XU(1) phase in parameter space depends strongly upon the starting lattice configuration, thus indicating metastability. Our parameter space consists of the gauge and quartic scalar couplings and the negative bare scalar mass. In all our calculations to date we have only varied the last quantity and kept all other parameters fixed. As we varied the bare scalar mass it was found that the system would make two transitions -- the first transition is from the symmetric state to an SU(4)XU(1) phase; then at a more negative value of the bare mass a second transition to an SU(3)XSU(2)XU(1) phase would occur. For the fixed quartic couplings used throughout the calculation, the transitions were always found to be first order.

It should be noted that results of a similar nature have been obtained in continuum calculations where implications were drawn about possible early universe scenarios [7]. For this investigation (where there is little understanding of the continuum limit -- a generic problem of theories with couplings that are not asymptotically free) such conclusions are not justified. However, we find it interesting that two vastly different calculational schemes (perturbative Coleman-Weinberg versus strong coupling Lattice Gauge Theory) give qualitatively similar results.

2. THEORY

For SU(N) gauge theories with adjoint scalars (N>2) there exists dimensionless, local, and gauge invariant observables which can determine the symmetry structure of the vacuum [2]. These are

$$m_1 = \frac{\mathrm{tr}\Phi^3}{(\mathrm{tr}\Phi^2)^{3/2}}, \; m_2 = \frac{\mathrm{tr}\Phi^4}{(\mathrm{tr}\Phi^2)^2}, \; \ldots\ldots, \; m_{N-2} = \frac{\mathrm{tr}\Phi^N}{(\mathrm{tr}\Phi^2)^{N/2}} \tag{1}$$

where the scalar field is cast in the form of NXN

traceless Hermitian matrices

$$\Phi(n) = \sum_{i=1}^{N^2-1} \phi^i(n) \frac{\Lambda^i}{2} \qquad (2)$$

the Λ's are the NXN analogues of the Gell-Mann matrices. In fact the list given in equation (1) is complete. That is, any other local, dimensionless, and gauge invariant observable must be a function of the terms given in (1). To convince yourself of the generality of this set it is helpful to consider the most general form of an object that is gauge invariant. The following figure shows an example of such an object.

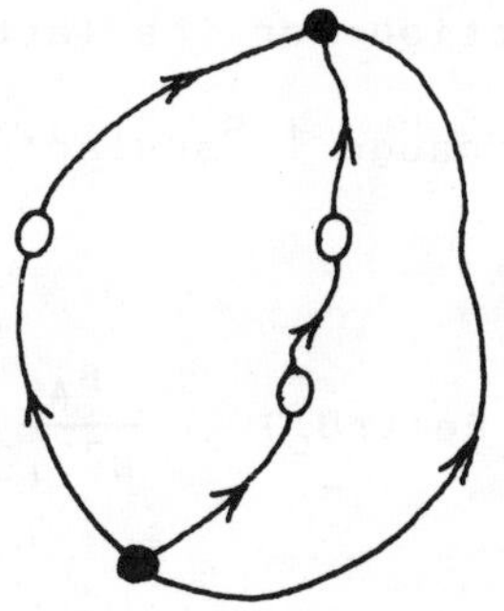

Fig.1 A gauge invariant 'loop' in SU(3) gauge theory. The lines represent SU(3) string segments a la Wilson, the open circles represent the scalar matrices defined in equation (2), and the solid dots represent summations with the completely antisymmetric tensor $\varepsilon_{abc\ldots}$.

In the limit that the string segments are shrunken to zero length, the expression shown in Fig. 1 becomes

$$I(x) = \varepsilon_{abc}\, \Phi(x)_{aa'}\, \Phi^2(x)_{bb'}\, \delta_{cc'}\, \varepsilon_{a'b'c'}$$
$$= \mathrm{Tr}\Phi\mathrm{Tr}\Phi^2 - \mathrm{Tr}\Phi^3 \quad . \qquad (3)$$

The most complicated expression one can think of can always be expressed in terms of the eigenvalues of the Hermitian Φ's. The eigenvalues are in turn functions of the set

$$\{\mathrm{Tr}\Phi, \mathrm{Tr}\Phi^2, \ldots\ldots, \mathrm{Tr}\Phi^N\} \ . \qquad (4)$$

The interested reader is directed to the the Appendix for a proof of this claim. In a direct extension of this idea, it is reasonable that any dimensionless and gauge invariant object must be a function of the eigenvalues of the dimensionless matrix $\Phi' = \Phi/\sqrt{\mathrm{Tr}\Phi^2}$. The eigenvalues of Φ' are functions of the set in (1).

The Euclidean action for the lattice is given by

$$S = S_{gauge} + S_{scalar}, \qquad (5)$$

$$S_{gauge} = -\sum_p \{ \frac{\beta}{N} \mathrm{Re}(\mathrm{tr}U_p) + \frac{\beta_A}{N^2-1} \mathrm{tr}_A\, U_p \} \ , \qquad (6)$$

$$S_{scalar} = 2\sum_{n,\mu} \{ \mathrm{tr}\Phi^2(n) - \mathrm{tr}(\Phi(n)U_\mu(n)\Phi(n+\mu)U_\mu^+(n))) \}$$
$$+ \sum_n \{ -m^2\mathrm{tr}\Phi^2(n) + \lambda_1(\mathrm{tr}\Phi^2(n))^2 + \lambda_2\mathrm{tr}\Phi^4(n) \}$$
$$(7)$$

U_p is the product around the plaquette p of the links which are in the fundamental representation of SU(5), and tr_A is the trace in the adjoint representation. We have excluded cubic terms in the scalar potential by imposing the discrete symmetry $\Phi \to -\Phi$.

Tree level analysis predicts that for $m^2, \lambda_1, \lambda_2 > 0$ the system will undergo spontaneous symmetry breaking according to the pattern SU(5) → H = SU(3)XSU(2)XU(1) [6]. As we shall see, this has been confirmed in a pure scalar theory for $\lambda_1=\lambda_2=0.05$ [3]. To check that this pattern is maintained in the presence of gauge fields we looked at the following signals --

$$VAC_1 = 30\, m_1^2 = 30 \; \frac{(tr\Phi^3)^2}{(tr\Phi^2)^3} \;, \qquad (8)$$

$$VAC_2 = \frac{30}{7} m_2 = \frac{30}{7} \; \frac{tr\Phi^4}{(tr\Phi^2)^2} \;, \qquad (9)$$

$$VAC_3 = \frac{30^3}{13^2} m_3^2 = \frac{30^3}{13^2} \; \frac{(tr\Phi^5)^2}{(tr\Phi^2)^5} \;, \qquad (10)$$

$$\gamma = 2 \left| \frac{(\#\text{ of sites with } tr\Phi^3 > 0)}{\text{total } \#\text{ of sites}} - \frac{1}{2} \right| . \qquad (11)$$

The normalization of the VAC's is such that they would equal unity if the true symmetry of the theory is H = SU(3)XSU(2)XU(1) exactly. On the other hand, if the true symmetry were SU(4)XU(1) exactly then VAC_1= 27/2, VAC_2=39/14, and VAC_3=70227/1352. Of course, the quantum fluctuations will change the results somewhat, but even so, the difference between the two symmetry breaking patterns is dramatic and should be easily distinguished.

A state with symmetry unbroken is characterized by a value of γ close to zero. Any breaking of symmetry also breaks the discrete symmetry $\Phi \to -\Phi$ and produces a nonzero value of γ. It is this changing of γ that signals the onset of symmetry breaking, but it will not distinguish among different symmetry breaking patterns. The symmetric state is also characterized by the values of the VAC's. In principle the VAC's can be calculated analytically -- for example, it has been shown that VAC_1 takes on the value of 27/52 in the symmetric phase [3].

The average plaquette for the SU(5) gauge group is defined by

$$\text{AVG.PLAQ.} = 1 - \frac{1}{5}\langle \mathrm{tr}U_p \rangle \qquad (12)$$

where the average is over all sites and configurations. A large value for the average plaquette is indicative of a disordered gauge configuration. It is reasonable that as the true symmetry group of the vacuum becomes smaller, the average plaquette decreases.

3. RESULTS

For the special case of the pure gauge theory, we plot the average plaquette versus the fundamental gauge coupling for various values of the adjoint gauge coupling in Fig. 2. The bulk transition is clearly seen for the case of zero adjoint coupling which is in agreement with other work [8]. The point circled in Fig. 2 corresponds to $\beta = 22.0$ and $\beta_A/(N^2-1) = -0.4$. For the pure gauge theory, this choice of couplings; (1) avoids the spurious effects of the bulk transition; (2) leaves the system in the strong coupling phase thus minimizing finite size effects and; (3) has a sensible weak coupling limit with $g^2_{eff}>0$ [9]. For these reasons we decided to use these very same couplings throughout the calculation in the full theory of gauged adjoint scalars. Of course, in the presence of dynamic scalar fields the corresponding couplings will not be the same as in the pure gauge case. We did not calculate this shift in couplings so we gambled that the choice motivated by the pure gauge theory would suffice.

Presented first in Figs. 3 and 4 are results for the pure scalar theory with the same quartic couplings used in the calculation in the full theory ($\lambda_1 = \lambda_2 = +0.05$). These results will serve to show the dramatic difference between the physics of the two theories.

Fig. 3 shows $\mathrm{tr}\Phi^2$ versus m^2. It is seen that $\mathrm{tr}\Phi^2$ follows the classical prediction for the symmetry breaking pattern SU(5) → SU(3)XSU(2)XU(1). Fig. 4 shows VAC_1, VAC_2, and VAC_3 for the same values of the quartic couplings. VAC_1 takes on a value close to that predicted for the symmetric state for $m^2 < 0.4$. There is a dip for $0.4 < m^2 < 1.5$ and after that, VAC_1

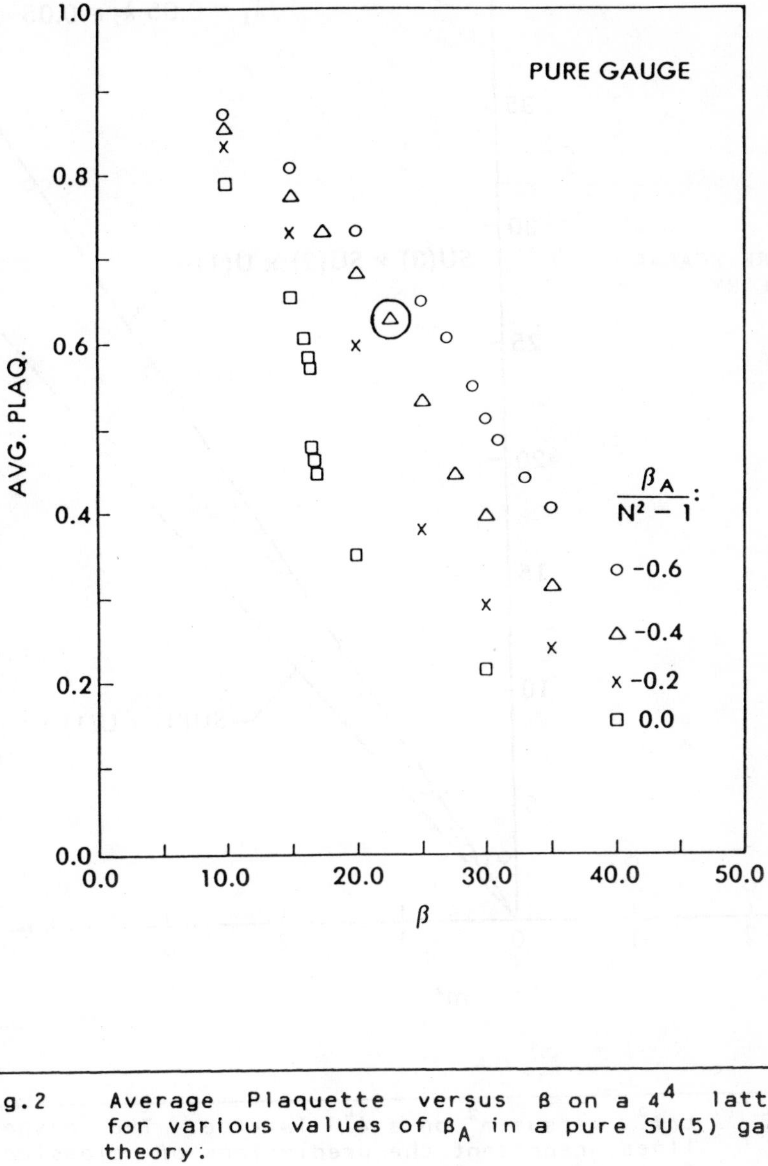

Fig.2 Average Plaquette versus β on a 4^4 lattice for various values of β_A in a pure SU(5) gauge theory.

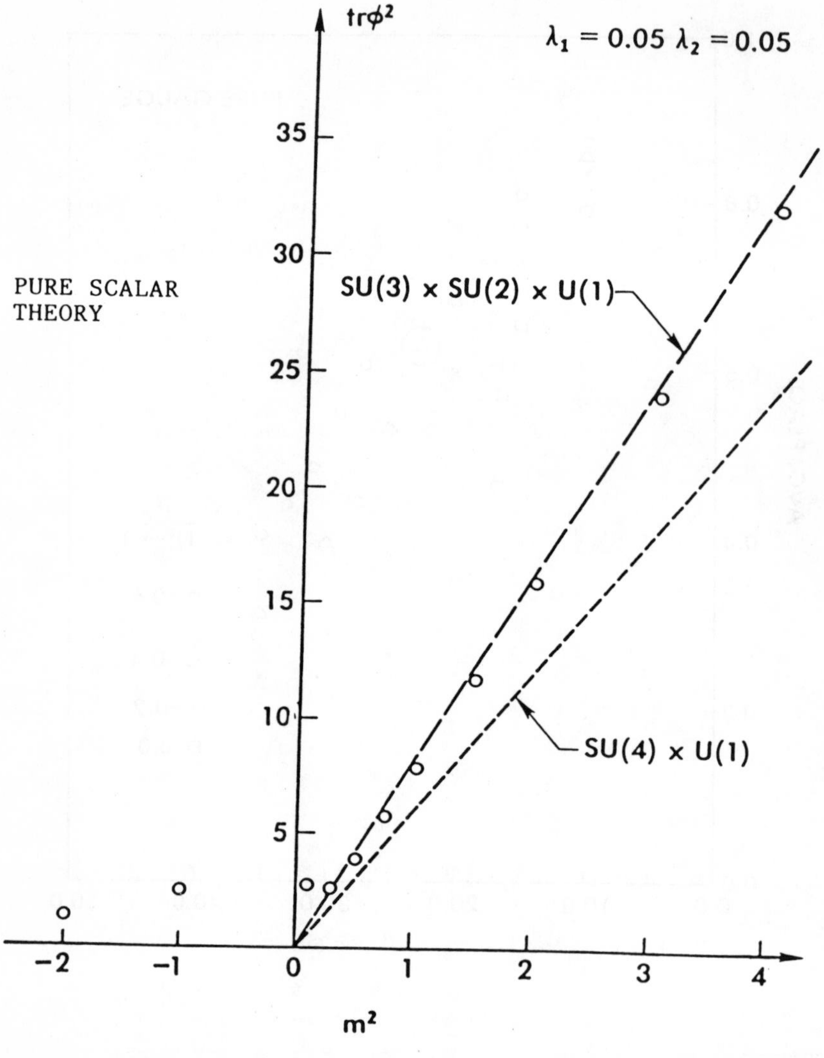

Fig.3 $\mathrm{tr}\phi^2$ versus m^2 on a 4^4 lattice. The dashed lines represent the predictions of classical field theory.

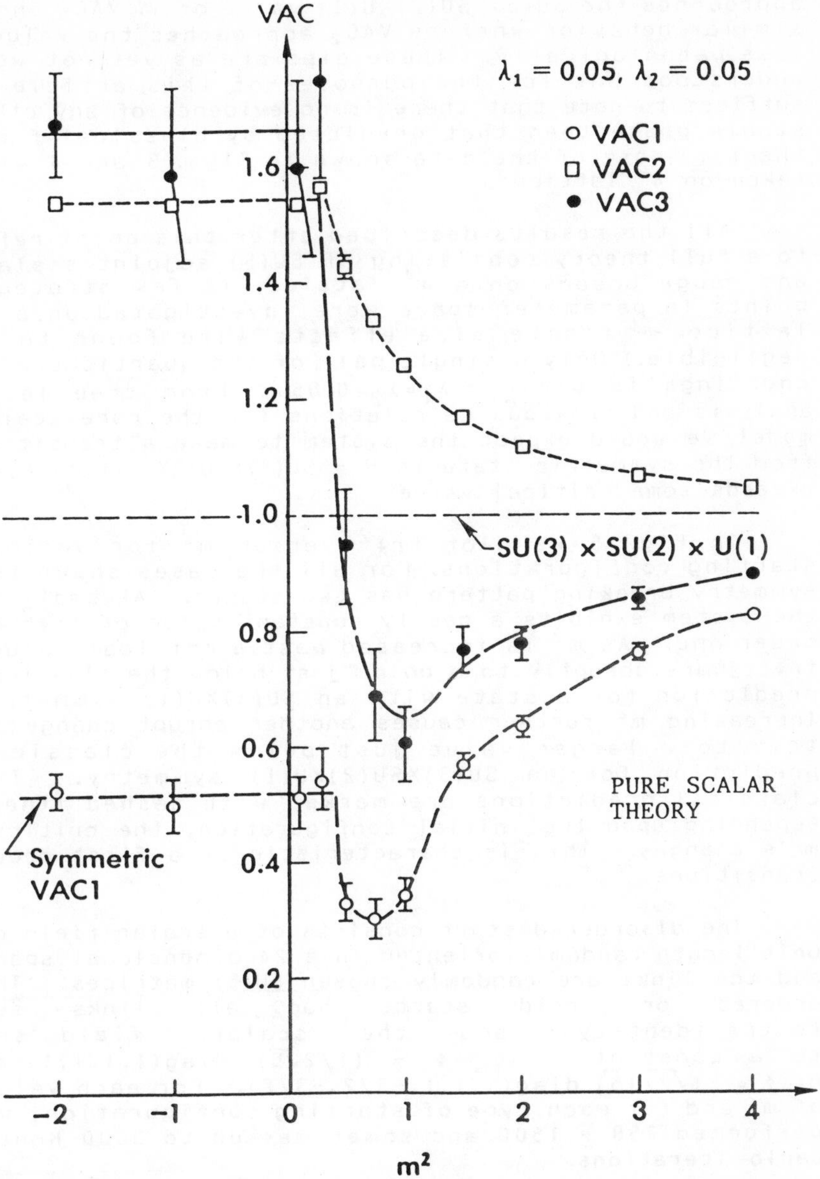

Fig.4 VAC_1, VAC_2, and VAC_3 versus m^2 for $\lambda_1 = \lambda_2 = 0.05$. The classical SU(3)XSU(2)XU(1) value is unity for all three quantities.

approaches the SU(3)XSU(2)XU(1) value of 1. VAC_3 shows similar behavior whereas VAC_2 approaches the value 1 less pathalogically. These dips are as yet not well understood but for the purposes of this article it suffices to note that there is no evidence of any other stable phase than that predicted by classical field theory. Most of the data shown in Figs. 3 and 4 were taken on 4^4 lattices.

All the results described after this point refer to a full theory consisting of SU(5) adjoint scalars and gauge bosons on a 4^4 lattice. (A few strategic points in parameter space were investigated on a 6^4 lattice -- finite size effects were found to be negligible.) Only a single pair of the quartic scalar couplings is used -- $\lambda_1=\lambda_2=0.05$. From tree level analysis and previous calculations for the pure scalar model we would expect the system to make a transition from the symmetric state to H = SU(3)XSU(2XU(1) once m^2 exceeds some critical value.

In Fig. 5 we plot $tr\Phi^2$ versus m^2 for various starting configurations. For all the cases shown the symmetry breaking pattern has two stages. At small m^2 the system exhibits a nearly constant value of $tr\Phi^2$ of order one. As m^2 is increased past a critical value, $tr\Phi^2$ jumps abruptly to a point just below the classical prediction for a state with an SU(4)XU(1) symmetry. Increasing m^2 further causes another abrupt change in $tr\Phi^2$ to a larger value just below the classical prediction for an SU(3)XSU(2)XU(1) symmetry. The classical predictions are marked with dashed lines. Depending upon the initial configuration, the critical m^2's changes. This is characteristic of a first order transitions.

The disordered start consists of a scalar field of unit length randomly oriented in a 24-dimensional space and the links are randomly chosen SU(5) matrices. The ordered or cold starts had all links set to the identity and the scalar field set to a constant $\Phi = (1/2\sqrt{5})\ \mathrm{diag}(1,1,1,1,-4)$ or $\Phi_2 = (\sqrt{2/15})\ \mathrm{diag}(1,1,1,-3/2,-3/2)$. For each value of m^2 and for each type of starting configuration, we performed 750 ~ 1500 and sometimes up to 3000 Monte Carlo iterations.

To guide the readers eye we have connected the points of the SU(4)XU(1) start which shows the two transitions most clearly. For all three starting configurations the transition from the symmetric to the

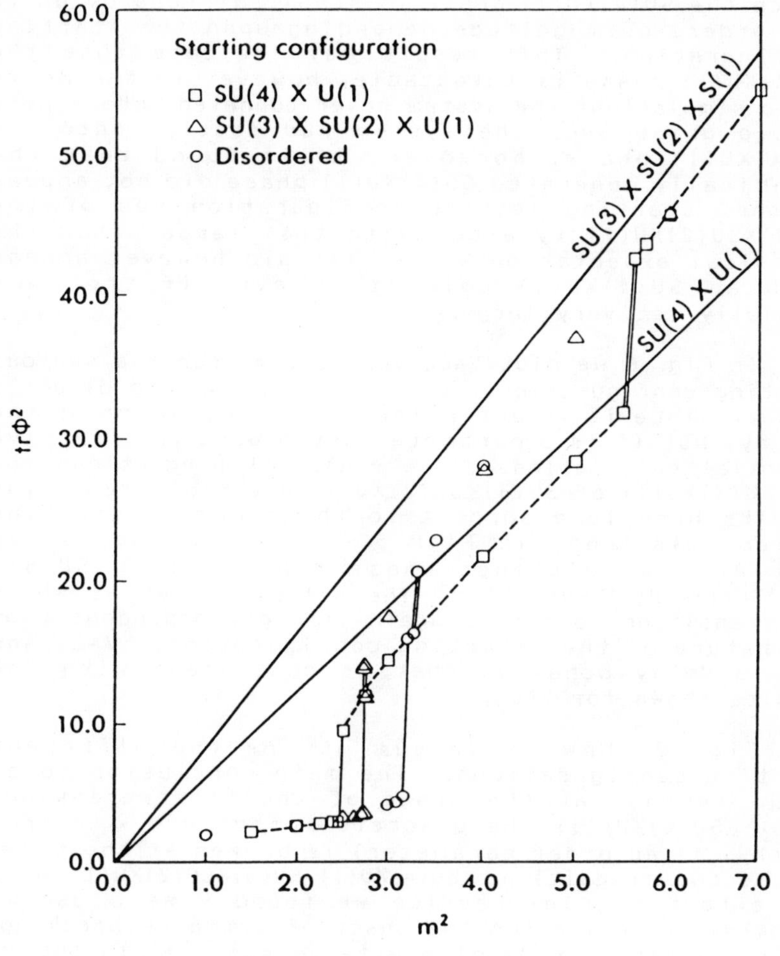

Fig.5 $tr\phi^2$ versus m^2 for the SU(5) gauge-Higgs system with various starting configurations.

SU(4)XU(1) state is marked by a single line and the subsequent transitions to SU(3)XSU(2)XU(1) are marked by a double line. Notice that the range of m^2 over which the SU(4)XU(1) phase is stable differs by up to two orders of magnitude depending upon the starting configuration. This result may indicate that the SU(4)XU(1) phase is metastable, however in the Monte Carlo simulations the system never tunneled into a more stable phase once the system had settled into the SU(4)XU(1) phase. Moreover, it is found that the dynamically generated SU(4)XU(1) phase did not appear if the starting lattice configuration was of the SU(3)XSU(2)XU(1) type but with $\mathrm{tr}\Phi^2$ larger than the classical expectation value. (It did however appear with an SU(4)XU(1) cold start even if $\mathrm{tr}\Phi^2$ was initially set very large.)

In Fig. 6 we plot VAC_2 versus $\mathrm{tr}\Phi^2$ for the various starting configurations. $\mathrm{tr}\Phi^2$ cannot be varied directly as such since it is determined by the dynamics of the theory, but it is a parameter often used in effective potential calculations. The classical predictions for H = SU(4)XU(1) or SU(3)XSU(2)XU(1) are shown too. The results here lend support to the symmetry breaking pattern discerned in Fig. 5 -- namely, the system displays two distinct phase transitions: SU(5)→SU(4)XU(1)→SU(3)XSU(2)XU(1). The value of $\mathrm{tr}\Phi^2$ at which the transitions occur is again strongly dependent upon the nature of the starting configuration. VAC_1 and VAC_3 display behavior that is consistent with the results shown for VAC_2.

Fig. 7 shows γ versus m^2 for the different starting configurations. The main conclusion to be drawn is that at the stage of the first breaking, SU(5)--SU(4)XU(1), the discrete symmetry $\Phi \to -\Phi$ (for which γ is an order parameter) is broken at the time. The second transition, SU(4)XU(1)→SU(3)XSU(2)XU(1) does not effect γ. In practice we found γ as a useful symmetry criterion for the onset of symmetry breaking. In Fig. 8 the average plaquette versus $\mathrm{tr}\Phi^2$ is shown. It clearly shows that each stage of symmetry breaking gives rise to a more ordered gauge configuration in agreement with intuition. The corresponding symmetry group, as deduced from the previous figures, are labeled.

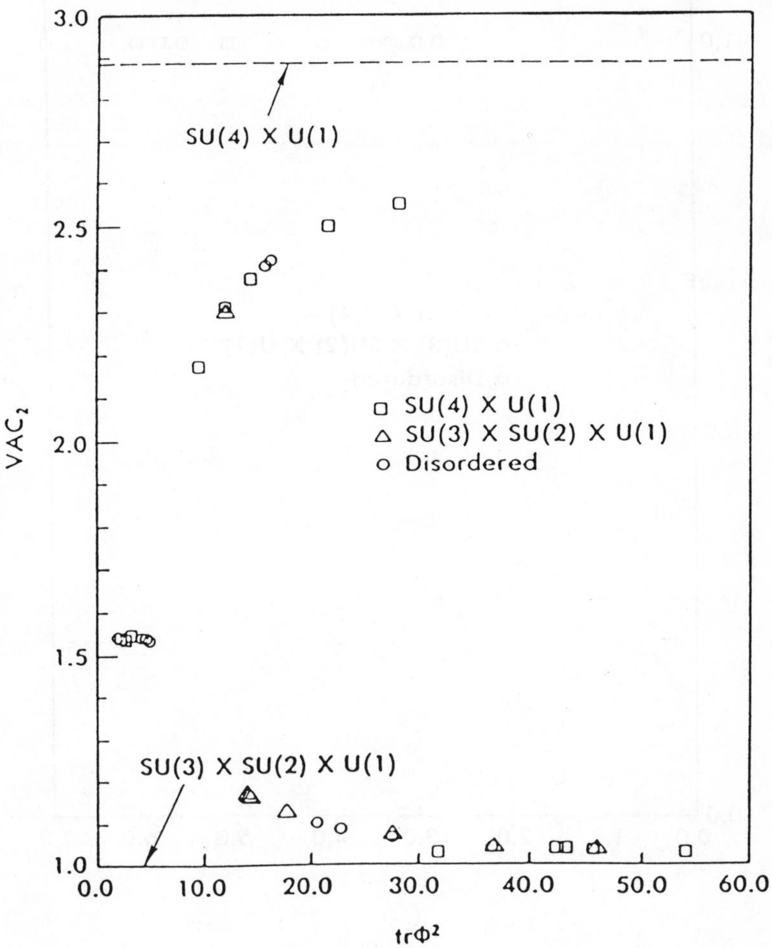

Fig.6 VAC_2 versus $tr\Phi^2$ for various starting configurations of the SU(5) gauge-Higgs system.

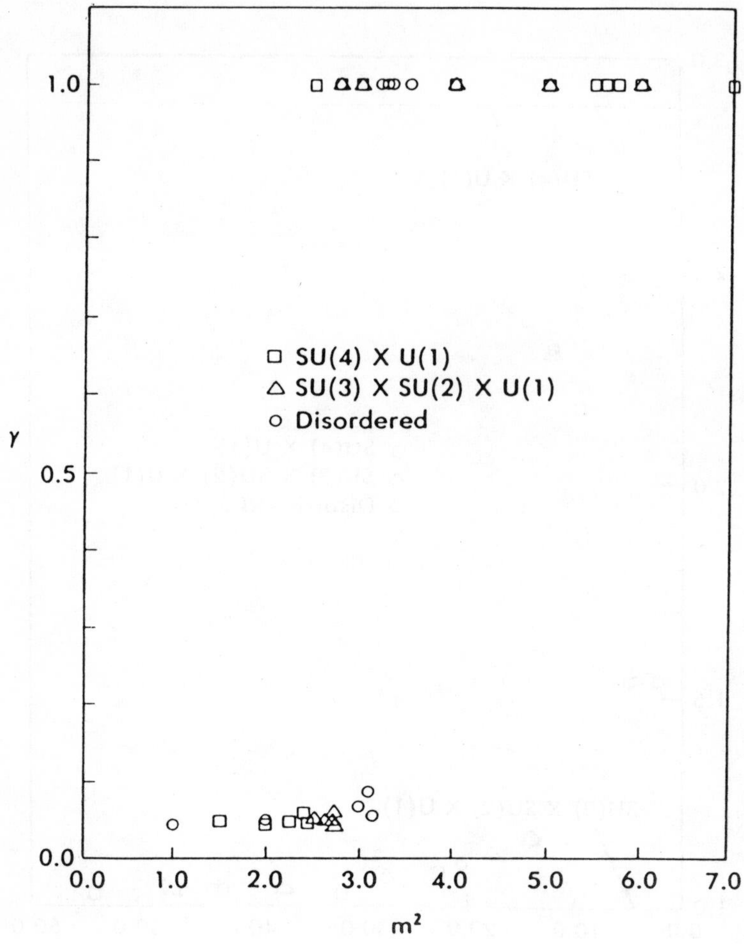

Fig.7 γ versus m^2 for various starting configurations of the SU(5) gauge-Higgs system.

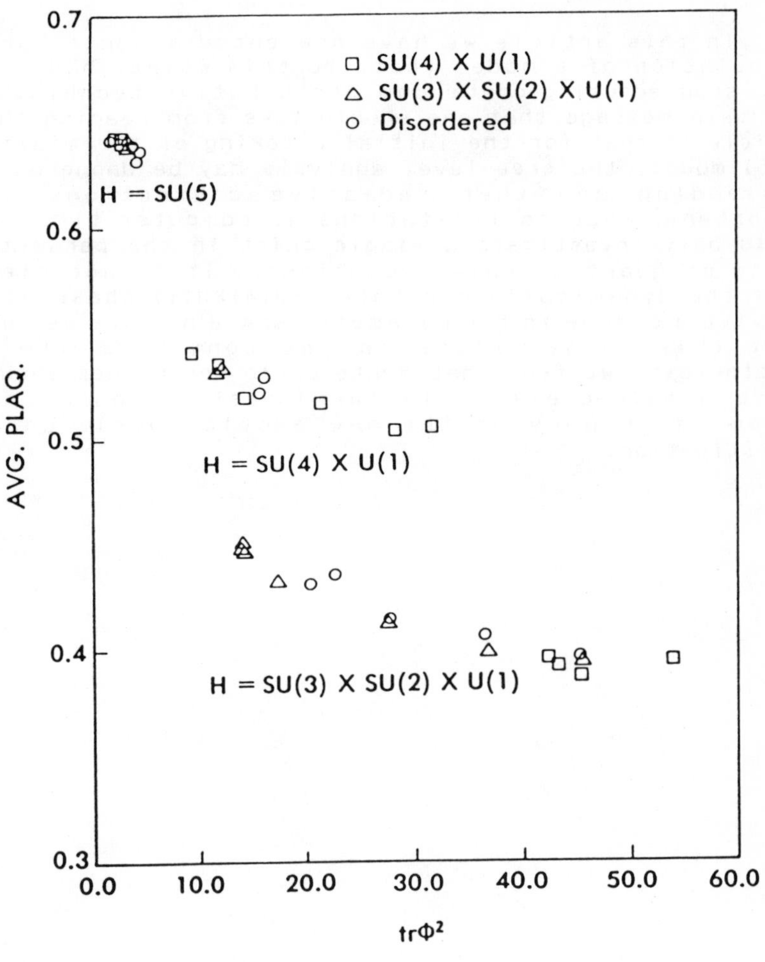

Fig.8 Average Plaquette versus $\mathrm{tr}\Phi^2$ for various starting configurations of the SU(5) gauge-Higgs system.

4. CONCLUSION

In this article we have presented a Monte Carlo calculation of a model that, to this point, had only been studied using continuum perturbative techniques. The main message that one should take from reading this article is that for the initial breaking of the minimal SU(5) model, the tree-level analysis may be dangerously misleading and that radiative corrections are important. Due to limitations in computer time, we could only investigate a single point in the parameter space of quartic scalar couplings. It is not clear that the dynamically generated SU(4)XU(1) phase will survive a change in the parameter space nor are we sure that this phase exists in the continuum limit. Nonetheless, we feel that Monte Carlo techniques should prove a valuable tool in the investigation of the dynamics of many of the new exotic models under investigation.

APPENDIX

In this section I shall present a method for expressing the determinant of any finite dimensional matrix A in terms of traces of various powers of this same matrix. In addition I shall present a number of interesting corollaries.

The standard expression for a determinant is given by

$$\det A \;=\; \varepsilon_{i_1 i_2 i_3 \ldots i_n} A_{1i_1} A_{2i_2} \ldots A_{ni_n} \,. \tag{A.1}$$

Using this same formula it follows that

$$\det(I - A) \;=\; 1 - \mathrm{tr}A + \ldots + (-1)^d \det A \,, \tag{A.2}$$

where d is the dimension of A. It is crucial to notice that (A.2) is a power series in the elements of A terminating with a term of the d-th power. The coefficients of all the higher order terms are identically zero. Another standard way of expressing the left hand side of (A.2) is

$$\det(I - A) \;=\; \exp\{\, \mathrm{tr} \ln(I - A) \,\}$$

$$= \exp\{\, - \sum_{1}^{\infty} \mathrm{tr}(A^n)/n \,\} \,. \tag{A.3}$$

Some simple combinatorics allows us to reexpress the right hand side of (A.3) as

$$\det(I - A) \;=\; \prod_{k=1}^{\infty} \left[\, 1 + \sum_{n_k=1}^{\infty} (-)^{n_k} \frac{(\mathrm{tr}A^k)^{n_k}}{k^{(n_k)}\, n_k!} \,\right] \tag{A.4}$$

Arranging the terms of (A.4) in increasing powers of A gives:

$$\det(I - A) = 1 - \mathrm{tr}A$$

$$+ \frac{1}{2!}\left[(\mathrm{tr}A)^2 - (\mathrm{tr}A^2) \right]$$

$$- \frac{1}{3!}\left[(\mathrm{tr}A)^3 - 3(\mathrm{tr}A)(\mathrm{tr}A^2) + 2(\mathrm{tr}A^3) \right]$$

$$+ \frac{1}{4!}\left[(\mathrm{tr}A)^4 - 6(\mathrm{tr}A)^2(\mathrm{tr}A^2) + 3(\mathrm{tr}A^2)^2 + 8(\mathrm{tr}A)(\mathrm{tr}A^3) - 6(\mathrm{tr}A^4) \right]$$

$$- \frac{1}{5!}\left[(\mathrm{tr}A)^5 - 10(\mathrm{tr}A^3)(\mathrm{tr}A^2) + 15(\mathrm{tr}A)(\mathrm{tr}A^2)^2 + 20(\mathrm{tr}A)^2(\mathrm{tr}A^3) - 20(\mathrm{tr}A^2)(\mathrm{tr}A^3) - 30(\mathrm{tr}A)(\mathrm{tr}A^4) + 24(\mathrm{tr}A^5) \right]$$

$$+ \ldots\ldots$$

$$+ \frac{(-1)^d}{d!}\left[(\mathrm{tr}A)^d + \ldots. + (-1)^{d+1}(d-1)!(\mathrm{tr}A^d) \right]$$

$$+ \ldots.. \qquad \text{(A..5)}$$

Since the series must terminate at the d-th order in the elements of A, each term of power greater than d must vanish order by order. Also, the d-th order term in (A.5) must be equal to $(-1)^d \det A$.

As an example of the usefulness of (A.5) I shall consider a three dimensional matrix M. Immediately, the determinant is given by

$$\dim(M) = 3 \ ,$$

$$\det M = \frac{1}{3!} [\ (trM)^3 - 3(trM)(trM^2) + 2(trM^3)\ ] \ , \tag{A.6}$$

and the first vanishing term in (A.5) tells us that

$$trM^4 = \frac{1}{6} [\ (trM)^4 - 6(trM)^2(trM^2) + 3(trM^2)^2 + 8(trM)(trM^3)\ ] \ . \tag{A.7}$$

Similarly, the higher order terms in equation (A.5) can be used to express traces of higher powers of M in terms of trM, trM^2, and trM^3. The eigenvalue equation for M is obtained by the replacement $M \to M/x$ in (A.5);

$$0 = x^3 - x^2 trM + x[(trM)^2 - (trM^2)]/2 - [(trM)^3 - 3(trM)(trM^2) + 2(trM^3)]/6 \ . \tag{A.8}$$

That is,

$$x = F(trM,\ trM^2,\ trM^3) \ . \tag{A.9}$$

When looking at adjoint scalars, the matrices have been traceless. If I assume that trM = 0 and convert to dimensionless quantities:

$$\underline{x} = x/(trM^2), \quad m_1 = trM^3/(trM^2)^{3/2} \ , \tag{A.10}$$

the dimensionless eigenvalue equation becomes

$$0 = \underline{x}^3 - \underline{x}/2 - m_1/3 . \quad (A.11)$$

These examples that I have show for 3X3 matrices are trivialy extended to larger matrices.

ACKNOWLEDGMENT

This research was supported in part by a grant from U.S. Department of Energy DE-AC02-ER01545. The calculations were carried out on the CRAY-XMP at the National MFE Computer Center. The authors thank the DOE for providing them with time at the center. One of us (K.O.) thanks the Aspen Center for Physics where many of the ideas presented here were formulated.

REFERENCES

1. C. B. Lang, C. Rebbi, and M. Virasoro; Phys. Lett. 104B (1981) 294.

   R. C. Brower, D. A. Kessler, T. Schalk, H. Levine, and M. Nauenberg; Phys. Rev. D25 (1982)3319.

   F. Karsch, E. Seiler and I. O. Stamatescu; Phys. Lett. 131B (1983) 139.

   H. Kuhnelt, C. B. Lang, and G. Vones; Nucl. Phys. B230 [FS10] (1984) 16.

   S. Gupta and U. Heller; Phys. Lett. 138B (1984) 171.

   M. Kikugawa, T. Maehara, T. Minazuki, J. Saito, and H. Tanaka; Prog. Theor. Phys. 69 (1983) 1207.

   I. Montvay; Phys. Lett. 150B (1985) 441; and DESY Preprint 85/005.

2. K. Olynyk and J. Shigemitsu; Nucl. Phys. B251 [FS13] (1985) 472.

3. K. Olynyk and J. Shigemitsu; Phys. Lett. 154B (1985)278.

4. S. Coleman and E. Weinberg; Phys. Rev. D7 (1973) 1888.

5. H. Georgi and S. Glashow; Phys. Rev. Lett. 32 (1974) 438.

6. Ling-Fong Li; Phys. Rev. D9 (1974) 1723.

A. J. Buras, J. Ellis, M. K. Gaillard and D. V. Nanopoulos; Nucl. Phys. B135 (1976) 66.

7. A. J. Breit, S. Gupta and A. Zaks; Phys. Rev. Lett. 51 (1983) 1007.

J. Kodaira and J. Okada; Phys. Lett. 133B (1983) 291.

C. W. Kim and J. S. Kim; Nucl. Phys. B244 (1984) 523.

I. G. Moss; Phys. Lett. 128B (1983) 385.

P. Salomonson. B. -S. Skagerstam, and A. Stern; Phys. Lett. 151B (1985) 243.

8. M. Creutz; Phys. Rev. Lett. 46 (1981) 1441.

K. Moriarty; Phys. Lett. 106B (1981) 130.

9. Constantin P. Bachas and Roger F. Dashen; Nucl. Phys. B210 [FS6] (1982)583.

SPIN GLASS MODELS OF NEURAL NETWORKS ; SIZE DEPENDENCE OF MEMORY PROPERTIES

D.J. Wallace

Department of Physics, University of Edinburgh,
Mayfield Road, Edinburgh, EH9 3JZ, U.K.

ABSTRACT

The main features of a class of neural network models are reviewed. Results are reported of numerical simulations using the ICL Distributed Array Processor, exploring size dependence, scaling properties and some finite size effects, in networks with up to 512 nodes.

1. INTRODUCTION

A talk on neural networks may seem a surprising intrusion in a Conference dedicated to Advances in Lattice Gauge Theory. I am happy it is being included, however, because (a) I believe the subject is of very great intrinsic interest and a fertile area for theoretical and numerical work, (b) the class of models which we consider includes site (neurons) and link (synaptic connections) variables and various neural functions are modelled by dynamics with different quenching procedures, and (c) a straw poll at a lunch table favoured this topic over the review of hadron masses I was originally scheduled to give.

The details of neural function and neural anatomy are myriad and incompletely known[1]. For the purposes and motivation of this talk, the key features are:

- the network is composed of nodes (neurons), which are linked by synaptic connections;
- the neurons fire electrical pulses along the synaptic connections, at a rate which depends upon the electrical potential of the neuron, as shown qualitatively in fig. 1;
- the electrical potential of the neuron is in turn dependent on the totality of electrical signals it is receiving from the neurons which are connected to it.

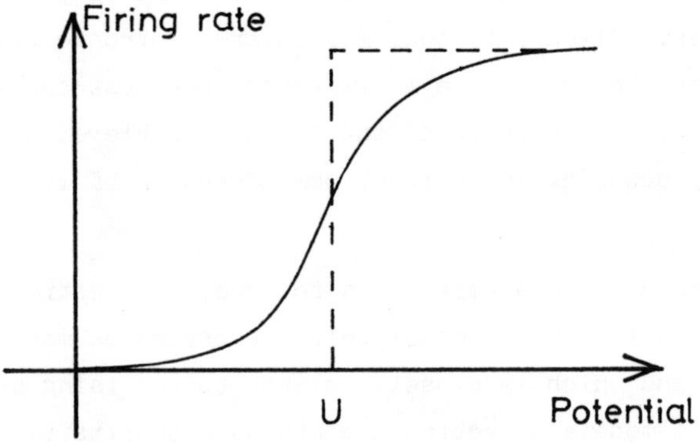

Fig. 1 : The continuous line indicates the qualitative firing-rate response of a neuron as a function of its potential. In the simplest models it is replaced by a step-function at the threshold U.

There are many fundamental questions which this description should stimulate.

- To what extent is the performance of the neural system determined by the non-linear response of the individual neuron, and by the complexity and collective behaviour of the network? Present belief is that both aspects are important, as we shall discuss.

- To what extent are the connections structured and to what extent are they random? Both characteristics have advantages; structured connections can presumably evolve to optimise their architecture for specific neural functions, and randomly connected structures are intrinsically robust to (random) cell deaths.
- A related question concerns the interplay of parallel and serial processing. The neural structure outlined above is clearly capable of parallel processing since in principle each neuron can act independently in response to the stimuli it receives. Given a typical molecular-electronic time constant of the order of a millisecond, one must conclude that the processing power of the brain is achieved by highly parallel processing in at least some fraction of its 10^{10} neurons.

The structure of this talk is as follows. In section 2 we shall review briefly a basic model which incorporates many of the above features and which is closely related to the Ising spin glass and Mattis models. Section 3 contains a qualitative discussion of learning and memory-like properties in the model. In section 4 we present the results of numerical simulations on the ICL Distributed Array Processor (DAP) for networks of 64, 128, 256 and 512 nodes. Our principal aims at this stage are to explore quantitatively the size dependence of memory properties and to expose finite size effects.

2. A MODEL

The particular model which we shall consider is identified with Hopfield[2)], although it has substantial origins in the work of earlier authors. It is only one of the classes of neural network models of current interest; an excellent introduction to the general subject is given in the book edited by Hinton and

Anderson[3]). It should be clear from the following that this Hopfield model captures many of the important features mentioned in the introduction, albeit in a highly simplified form. From a physiological point of view it might model the intensely connected clusters of 100 - 10,000 neurons which are observed in the nervous system.

The starting point is the approximation of the response function in fig. 1 by a step function, as indicated by the dashed lines. Thus a node can be in one of only two states; 1 (firing at maximum rate), or 0 (not firing) according to whether its potential is above or below its threshold, indicated by U in the figure. At any given instant, the state of the entire network of nodes is therefore given by a vector of bits V_i (= 1 or 0), i = 1, 2,..N, where N is the number of nodes.

The second ingredient of the model is the connection strength T_{ij}, which represents the behaviour of the synaptic connection joining "neurons" i and j. In the model it is frequently presumed that T_{ij} is symmetric, $T_{ij} = T_{ji}$, and that in principle all T_{ij}'s may be non-zero, i.e. a connection exists between every pair of nodes (hence the "intensely connected cluster" remark above). The quantitative property of T_{ij} is that it determines the contribution to the potential of node i if node j is firing; if node j is not firing, there is zero contribution. The net resulting potential of node i, ϕ_i from all other neurons is therefore

$$\phi_i = \sum_{j \neq i} T_{ij} V_j$$

$$= \sum_{j=1}^{N} T_{ij} V_j ;$$

if the convention $T_{ii} = 0$ (i = 1,...N) is understood. The T_{ij}'s may be both excitatory ($T_{ij} > 0$) and inhibitory ($T_{ij} < 0$).

To complete the "cellular automaton loop", we now use the simplified response function in fig. 1 to determine the new state V_i' of neuron i:

$$V_i' = \begin{matrix} 1 \\ 0 \end{matrix} \quad \text{if} \quad \sum_j T_{ij} V_j \begin{matrix} > U_i \\ < U_i \end{matrix} \qquad (1)$$

where U_i is the value of the threshold at node i.

It is clear that this model forms a basis to investigate some of the important questions discussed in the introduction. The non-linearity of neural firing response is maintained, in an extreme form. The complexity of the network is in principle as great as we wish, and the question of structured and random connectivity can be explored by appropriate constraints on the T_{ij}'s.

In order to address the question of parallel versus serial processing, and indeed to specify the model completely as a dynamical system, we must prescribe more precisely the updating process which determines the new state of a node. There are many possibilities: the following three correspond very loosely to the broad classification of computer architectures.

- Serial update - pick a node i (at random or in some fixed order), update it, pick another node, update it etc. Only one "Central Processing Unit" is performing at anyone time (but, yes, the sum in $\sum_j T_{ij} V_j$ could, of course, be vectorised!).
- Lockstep parallel - all the nodes update simultaneously, using the network state from the previous step. This procedure would be rather analogous to computation on processor arrays such as the I.C.L. Distributed Array Processor (DAP) or the Goodyear Massively Parallel Processor (MPP), which are examples of the Single Instruction-stream, Multiple Data-stream class of computer arrays.

- Asynchronous parallel - all the nodes update concurrently and independently in time according to the signals which they are receiving at any instant from the other neurons. This is analogous to the Multiple Instruction-stream Multiple Data-stream class of computers such as the Denelcor HEP, and, with small numbers of processors, many mainframes and e.g. CRAY XMP. From the physiological point of view it would seem the most natural dynamics. One suspects that it is also the richest in behaviour since it incorporates intrinsically the possibility of time delay (the time it takes for the firing of neuron i to influence neuron j, which depends on e.g. the length of the synaptic connection); time delay could, of course, also be explored in the two dynamical schemes above.

The serial update algorithm is the one which is actually employed by Hopfield, the node to be updated being chosen at random, with the motivation that this might be more akin to the asynchronous parallel update than would a regular serial update. It is attractive from a theoretical point of view if T_{ij} is symmetric, because any starting state $\{V^{init}\}$ just iterates to a local stationary point of an energy function

$$E = -\frac{1}{2} \sum_{ij} V_i T_{ij} V_j + \sum_i U_i V_i \quad : \qquad (2)$$

in any update E never increases, so generically $\{V^{init}]$ iterates to a local minimum of E.

The analogy with the Ising spin glass should be clear; the node variables V_i correspond to spins and the connection strengths T_{ij} to exchange constants. Since in principle all pairs of nodes may be connected, the closer analogy is with the Sherrington-Kirkpatrick model[4)]. The thresholds U_i correspond to quenched random external fields[5]. To effect a more explicit analogy one should rewrite eq. 2 in terms of the 1, -1 variables $2V_i - 1$, as in the lattice gas to Ising model transformation; the net result is a shift in the thresholds U_i. The deterministic dynamics described above corresponds to the zero temperature limit of the model. Finite temperature effects are equivalent to noise and are important[6]; they play a central role in another class of models called Boltzmann machines[7].

3. MEMORY AND LEARNING

In the context of its memory and learning properties, the simplest interpretation of the model is that stable vectors of the iteration (1) correspond to the concepts or images stored in memory. Behind this interpretation is the idea that we might be "aware" of stable patterns of neural activity, but might be "unaware" of the host of different transient flows towards the stable patterns. In cases where there exists an energy function E such as in eq. (2), the contents of the memory correspond to the local minima of E. It is well known that the positive and negative competing interactions in the Ising spin glass give rise to an energy function with many competing local minima. In fact the number of metastable states in the Sherrington-Kirkpatrick model is estimated[8] to increase with the number N of nodes as ~ exp(0.1992 N); hence there is enormous potential memory capacity in this class of models.

Before we turn to the question of how one can "learn" by adjusting the connection strengths so that a given required set of minima are created, it is worth pointing out that this form of memory storage is very different from typical computer memory. Information stored in a computer is localised on an area of silicon and recalled by means of an address which must be exactly specified; if even one bit in the address is wrong, the wrong data is retrieved. By contrast, the memory in the neural network is not localised in any of the nodes but is represented by collective patterns of activity over the whole network, controlled by the connection strengths T_{ij}. Moreover, provided sufficient external stimulus is given to initiate an activity pattern which is in the basin of attraction of the desired minimum, then the pattern will iterate to that minimum. The model thus has the ability to recall a complete concept or image from partial information about it; this property is called content-addressable memory. We note that a natural measure of the "distance" between two patterns of activity is the Hamming distance, defined as the number of nodes which are not in the same state in the two configurations. In terms of the two vectors $V^{(1)}$ and $V^{(2)}$ describing the configurations,

$$H^{12} = \sum_{i=1}^{N} |V_i^{(1)} - V_i^{(2)}| \quad . \tag{3}$$

The robustness of this memory storage to cell death of some neurons and synaptic connections is an attractive feature of the model.

The "recall" or "identifying" mode of operation of the network should now be clear; an external stimulus activates some firing pattern, the connection strengths are quenched at the values T_{ij}, and the firing pattern evolves to stability. The question naturally arises of how the T_{ij}'s are modified to "learn" specific facts, i.e. how can one systematically adjust the T_{ij}'s to achieve a given pattern of minima? Some prescriptions for learning procedures exist[7,9], although they do not always appear particularly natural from a physiological point of view. There is also a suggestion for a "sleep" mode of activity[10]. The typical feature of these modes is that they involve iterations with the nodes and connection strengths successively quenched and active; it might, therefore, be possible to capture a range of neural functions by different modes of operation of the network.

Rather than dive into such speculations, we restrict ourselves in the rest of the paper to one particular model for the connection strengths T_{ij}, the storage prescription model[11]. We shall also consider only the case where the thresholds V_i are zero. Suppose we have a cluster of N neurons and we wish to store in memory n states $\{V_i^{(s)}: i = 1,2,\ldots N,\ s = 1,2,\ldots n\}$; we shall refer to these as the the nominal vectors. Define

$$T_{ij} = \begin{cases} \sum_{s=1}^{n} (2V_i^{(s)} - 1)(2V_j^{(s)} - 1) & i \neq j \\ 0 & i = j \end{cases} \tag{4}$$

Thus we add or subtract 1 in T_{ij} according to whether $V_i^{(s)}$ is or is not the same as $V_j^{(s)}$.

The qualitative characteristics of the model are easy to understand. Imagine starting from a nominal vector $V^{(s)}$ and consider first the effect of the contribution of that nominal vector in T_{ij}:

$$\sum_j (2V_i^{(s)} - 1)(2V_j^{(s)} - 1)V_j^{(s)} = (2V_i^{(s)} - 1) \sum_j V_j^{(s)}$$

$$\sim \frac{N}{2} (2V_i^{(s)} - 1) , \tag{5}$$

where the N/2 follows because a random vector V will have typically N/2 1-bits and N/2 0-bits. Thus the contribution in T_{ij} of a nominal vector strongly reinforces the stability under the iteration (1) of that nominal vector. The effect of the other nominal vectors in T corresponds crudely to the additive noise of (n-1) N/2 numbers each of which is ± 1. If n is not too large then the signal term (5) will dominate and the nominal vectors will be stable; as n increases, more and more bits in the nominal vectors will be incorrectly stored, as the noise begins to overwhelm the signal.

One can ask many specific questions of this model. What fraction of the nominal states is perfectly stored, as a function of n and N? How close must an initial state be for the entire nominal state to be recalled? If this memory is truly a collective property of the network to what extent is it universal, i.e. independent of the details of the prescription for the T_{ij}'s and the U_i's? The first two of these questions, and others, are addressed in Hopfield's paper[2]), with rather low statistics, for networks of 30 and 100 nodes. We now turn to the results of higher statistics simulations on the ICL DAPs at Edinburgh, for networks of 64, 128, 256 and 512 nodes.

4. DAP SIMULATIONS

The DAP consists of 4096 bit-serial processing elements, hardwired in a 64 × 64 square array, hosted on an ICL mainframe. In the two machines at Edinburgh, each processing element has 4K bits of local memory and data can be communicated amongst the processing elements by global North, South, East and West shifts. The array is one example of SIMD architecture; 4096 identical operations are performed simultaneously on the local data. An additional powerful feature is an activity register with which any subset of processing elements can be masked out. For 32-bit floating point arithmetic the machine performs at roughly 20 M flops. For short word or bit-manipulation problems it is extremely powerful: a 3-d Ising model runs at 200 million update attempts per second[12]). The language used is a convenient parallel extension of Fortran called DAP Fortran. The Edinburgh machines have been used to study a range of problems in condensed matter and particle physics; for a review and other references see Bowler and Pawley[13]).

In the results reported here we follow exactly the random serial update algorithm as in Hopfield[2]). The parallelism of the machines is fairly efficiently exploited by the parallel masking and sum involved in each iteration (1), and by running several independent simulations simultaneously to increase statistics.

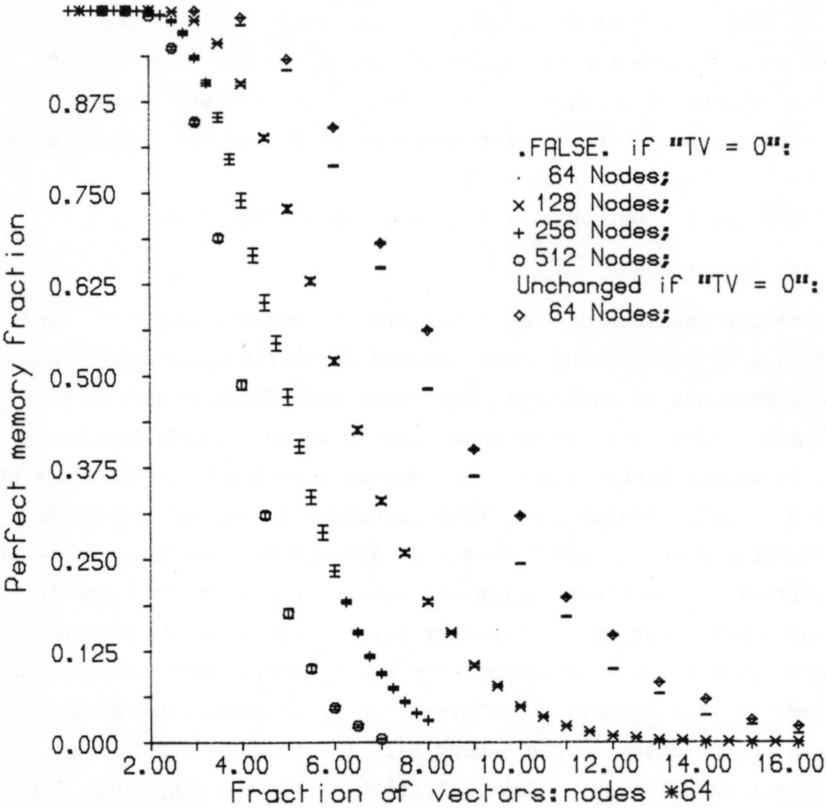

Fig. 2 Perfect storage fraction as a function of the number of nominal vectors (n) and nodes (N); the error bars on some data points are smaller than the resolution of the graph.

We start by showing in fig. 2 the raw data for the fraction of nominal vectors which are perfectly stored i.e. for which all nodes i are stable under the iteration (1). (Note that this is independent of whether serial or parallel updating is done.) It is plotted as a funtion of n/N, the ratio of nominal vectors to nodes, for 64, 128, 256 and 512 nodes. The thresholds U_i are chosen to be zero, as in Hopfield[2]; our 64-node results are in qualitative agreement with an interpolation of his 30 and 100 node data, but with much higher statistics. In order to begin exploring finite size effects, we show the results of two 64-node runs which differ according to the prescription adopted for the new state in the marginal case where $\sum T_{ij}V_j$ is zero. In one, the new state is unchanged, (i.e. the neuron continues in its previous state), in the other the new state is 0, i.e. the neuron fires only if it is strictly above threshold. The effect is significant with the "unchanged" data higher, as anticipated. Note that the discrepancy changes by a factor of roughly two in going from odd to even n; for odd n, zero can be obtained only if the number of 1-bits in the nominal vector is even. We have verified that the magnitude of the discrepancy decreases as the number of nodes increases.

Let us turn now to the extent to which we can describe theoretically the data for various N. There is a simple model [see e.g. ref.2)] based on the "signal + noise" discussion in eq. (5) et sequi. In it, one estimates the probability p that any node in any nominal vector will flip in terms of the probability that (n-1)N/2 numbers each randomly ± 1 will add up to a number greater than the signal N/2. For large n - 1 and N, the result is

$$p = \frac{1}{\sqrt{2\pi}\ \sigma} \int_{N/2}^{\infty} \exp(-\ x^2/2\sigma^2)dx$$

where $\sigma^2 = (n-1)N/2 \approx nN/2$ is the variance of the noise. Scaling x, p is trivially obtained in terms of the complementary error function.

$$p = \frac{1}{2}\ \mathrm{erfc}\ (\tfrac{1}{2}(N/n)^{\frac{1}{2}}) \qquad (6)$$

Notice that this function scales, i.e. depends only on the ratio n/N. If none of the nodes flips in a nominal vector of length , we obtain the perfect storage fraction P :

$$P = (1 - p)^N \quad . \qquad (7)$$

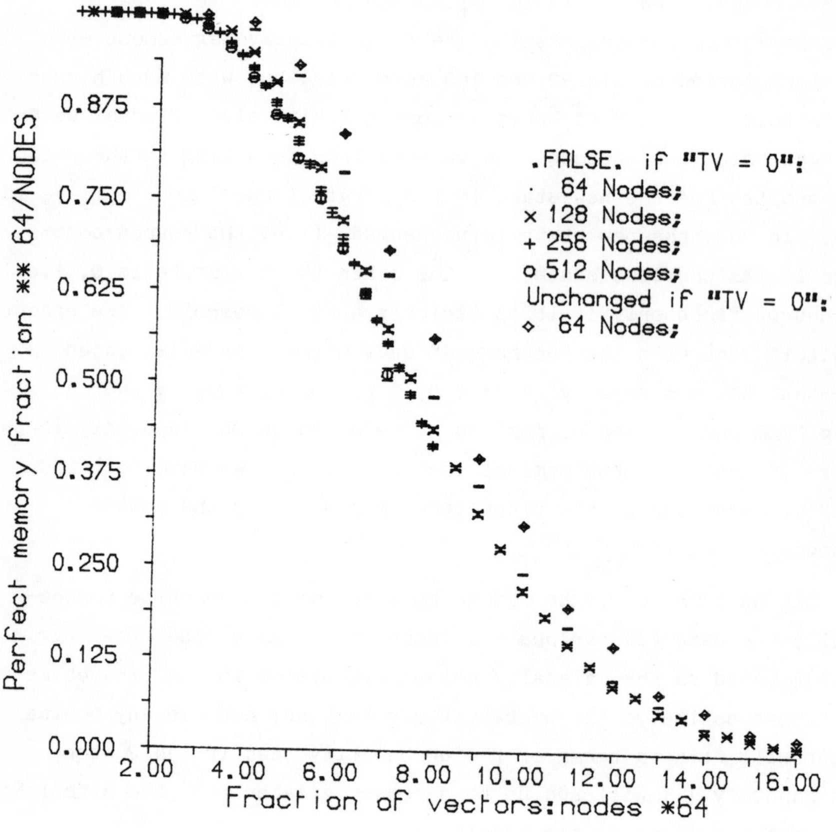

Fig. 3. The data of fig. 2, raised to the power of 64/N, to expose the scaling property (8); the error bars on some data points are smaller than the resolution of the graph.

This "signal + noise" model is certainly wrong, because there are correlations in the T_{ij}'s of which the most obvious is symmetry: $T_{ij} = T_{ji}$. Nevertheless it has some useful qualitative features which help to analyse the data. In particular it suggests a scaling form

$$P = [f(n/N)]^N \quad . \qquad (8)$$

To test this form we show in fig. 3 the same data as in fig. 2, but taking the square root of the 128 data, the fourth root for 256 nodes and the eighth root for 512. If eq. (8) is valid, all the data should, up to finite size effects, fall on a single scaling curve. The evidence from fig. 3 that this indeed happens is fairly convincing. (The error in the final data point for 512 nodes may be underestimated because of the incorrect application of gaussian statistics to small numbers.)

However the data also establish that the "signal + noise" model is quantitatively wrong. In fig. 4, we show the same data analysed according to eq. (7); the numerical probability for a single node to flip divided by the naive theoretical estimate (6) is shown as a function of n and N. We note here two features. First, finite size effects are very large for small n. This can be anticipated from the signal + noise model, where the asymptotic form of the error function is, with n replaced by the more accurate n - 1,

$$\left(\frac{n - 1}{\pi N}\right)^{\frac{1}{2}} \exp\left(- \frac{1}{4} N/(n - 1)\right) \approx \exp\left(- N/(4n^2)\right)p(n/N);$$

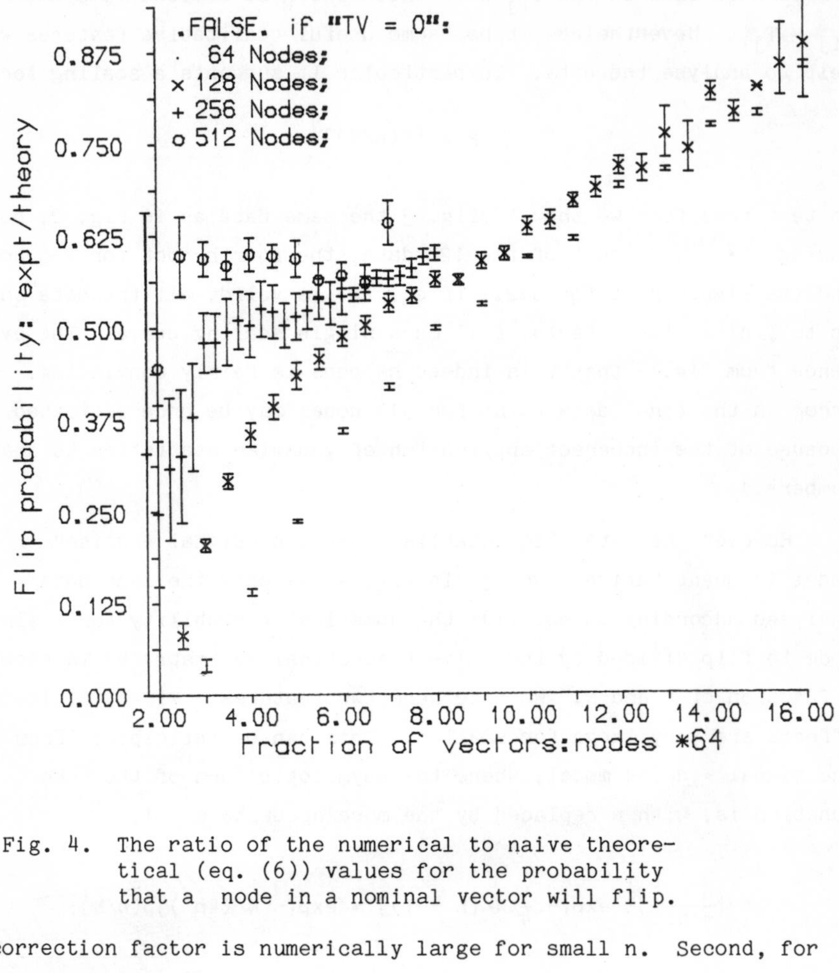

Fig. 4. The ratio of the numerical to naive theoretical (eq. (6)) values for the probability that a node in a nominal vector will flip.

The correction factor is numerically large for small n. Second, for larger n a scaling form does appear to be emerging which is significantly different from that of the signal + noise model.

We note here that figs. 2 and 3 and the related discussion in the text qualify Hopfield's result that in the storage prescription roughly 0.15 N nominal vectors can be stored; for a given value of n/N, the mean number of bits which are wrong increases unboundedly with N.

All of the above is concerned with the stability property of nominal vectors, and can be accumulated by the equivalent of a single sweep through the network. We turn now to the probability distribution for the Hamming distance between an initial nominal vector and the final vector to which it iterates. In fig. 5 we show the 64 node data in a two-dimensional plot, for Hamming distances 0 to 64, and with 1 to 16 nominal vectors in T_{ij}; the square root

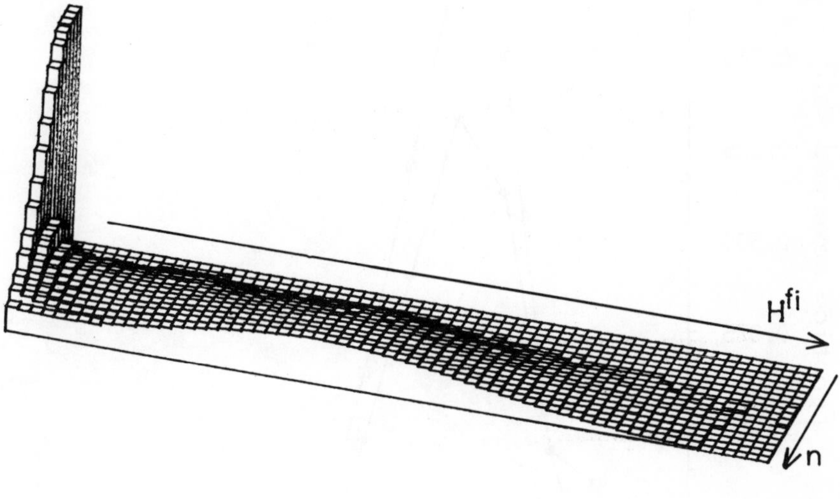

Fig. 5. Histogram of Hamming distance H^{fi} between initial nominal and final vector for 64 nodes, for n = 1 to 16 nominal vectors.

of the distribution is shown, in order to enhance the interesting features. The section corresponding to Hamming distance zero is the perfect storage fraction of fig. 2 (64 nodes: state is 0 if $\sum T_{ij}V_j = 0$). The histogram illustrates rather clearly the "leakage" away from zero Hamming distance as the number of nominal vectors increases. For 16 vectors there is clear evidence of the crossover towards a binomial distribution which would correspond to no correlation between initial and final vectors.

There is also just visible in this plot a non-zero value for Hamming distance 64. This is shown in expanded form in fig. 6. This effect, in which a nominal vector iterates to its complement

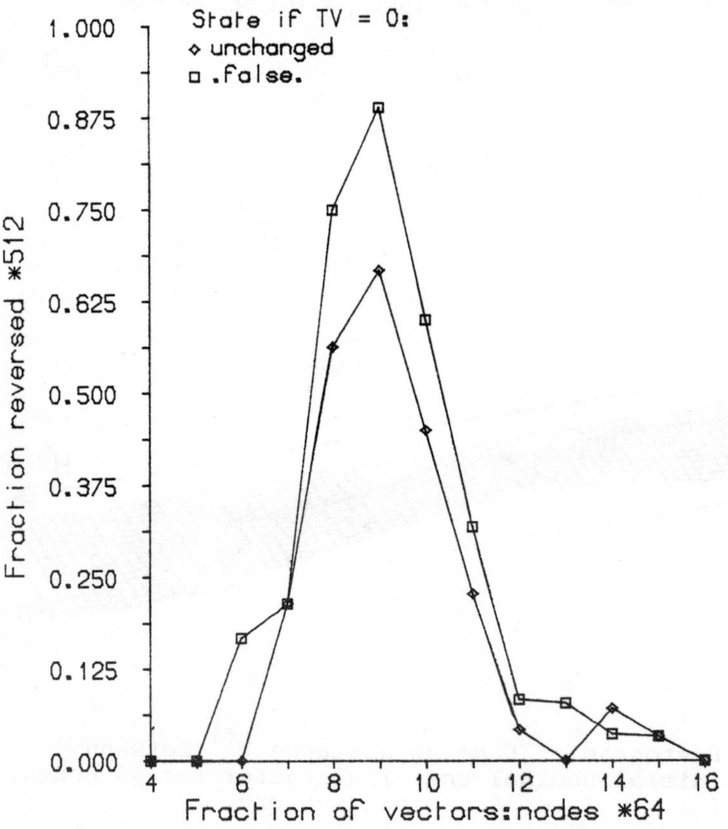

Fig. 6. Probability for final vectors to be the complement of the initial nominal vector.

occurs only in an intermediate range of n. The reason is as follows. It is easy to see on the basis of the signal + noise analysis that the complement of a nominal vector is also a potential memory state. For small n, however, the nominal vector is likely to be stable, and if it is unstable will still be typically highly correlated with the final vector. For large n, both nominal vectors and their complements are unlikely to be stable. Only in the intermediate regime is there a finite probability that a nominal vector will be unstable, and in the basin of attraction of its stable complement. This is a classic finite-size effect. (Only in the infinite network limit, $N \to \infty$, is there a symmetry under complement for the "$T_{ij}V_j$" model; the "$T_{ij}(2V_j - 1)$" model has complement as an exact symmetry and would not exhibit this effect.)

Finally, we remark briefly on the potential universality of the results. It is trivial to apply the signal + noise model to other schemes. For example, in the model touched upon in the lines above, in which the new state of neuron i is determined by the sign of $\sum T_{ij}(2V_j - 1)$, one readily calculates the node flip probability to be

$$p = \frac{1}{\sqrt{2\pi}\,\sigma} \int_N^\infty \exp(-\,x^2/2\sigma^2)dx, \qquad \sigma^2 = nN$$

$$= \frac{1}{2}\,\mathrm{erfc}\,\left((N/2n)^{1/2}\right)$$

Thus p(n/N) has the same functional form as in (6), up to a scale factor of 2 in the argument, i.e. we have a one-scale-factor universality in this simple theoretical model. Whether this property persists for the exact solutions of a wider range of models remains to be established.

In conclusion, I hope you agree that this is a fascinating topic with great scope for theoretical and numerical work. It is also likely to have many important application areas.

ACKNOWLEDGEMENTS

I thank Rex Whitehead for introducing me to neural networks, Nick Stroud for introducing me to ERCC graphics and Tony Hey for copies of the references by Hopfield and Hinton et al. I have benefitted from useful discussions with Alastair Bruce and David Willshaw. I am grateful to Dennis Duke for the opportunity to present this work here. The DAPs at Edinburgh are supported in part by SERC grants NG11849 and NG15908.

REFERENCES

1. See for example, Jack, J.J.B., Noble, D. and Tsien, R.W., "Electric current flow in excitable cells", Clarendon Press (Oxford, 1975).

2. Hopfield, J.J., "Neural networks and physical systems with emergent computational abilities", Proc. Natl. Acad. Sc. USA, 79, 2554 (1982).

3. Hinton, G.E. and Anderson, J.A., eds., "Parallel models of associative memory", Lawrence Erlbaum (Hillsdale, New Jersey, 1981).

4. Sherrington, D. and Kirkpatrick S., "Solvable model of a spin-glass", Phys. Rev. Lett. 35, 1792 (1975).

5. See Imry, Y., "Random external fields", in Proceedings of the Fifteenth IUPAP Int. Conf. on Thermodyn. and Stat. Mech., J. Statist. Phys. 34, 849 (1984).

6. Amit, D.J., Gutfreund, H. and Sompolinsky, H. "Spin glass models of neural networks", Racah Inst., Hebrew Univ. preprint (1985), discuss the finite temperature behaviour of the model in the limit n fixed, N → ∞; they find a phase transition from a low temperature phase with 2n ground states of the Mattis type. (Mattis, D.C., "Solvable spin systems with random interactions", Phys. Lett. 56A, 421 (1976). In this paper we are concerned with the limit N → ∞, n/N fixed.

7. Hinton, G.E. and Sejnowski, T.J., "Optimal perceptual inference", in Proc. IEEE Comp. Soc. Conf. on Computer Vision and Pattern Recognition, Washington D.C. (1983).
Ackley, D.H., Hinton, G.E. and Sejnowski, T.J., "A learning algorithm for Boltzmann machines", Cog. Sc. 9, 147 (1985).

8. A review is given by Moore, M.A., in "Statistical and particle physics: common problems and techniques", eds. Bowler, K.C. and McKane, A.J., Proc. 26th Scottish Univ. Summer School in Physics, SUSSP Publications (Edinburgh 1984).

9. See for example, Kohonen, T., Oja, E. and Lehtiö, P., "Storage and proccessing of information in distributed associative memory systems", in ref. 3 above, p. 105; Aleksander, I., "Emergent intelligent properties of progressivley structured pattern recognition nets", Pattern Recog. Letts. 1, 375 (1983).

10. Crick, F.C. and Mitchison, G., "The function of dream sleep", Nature 304, 111 (1983); Hopfield, J.J., Feinstein, D.I. and Palmer, R.G., "'Unlearning' has a stabilising effect in collective memories", Nature 304, 158 (1983).

11. Cooper, L.N., in Proceedings of the Nobel Symp. on Collective Properties of Physical Systems, eds. Lundqvist, B. and Lundqvist, S., Academic Press (New York, 1973).

12. Reddaway, S.F., Scott, D.M. and Smith, K., "A very high speed Monte Carlo simulation on DAP", Proc. VAPP II Conf., Comp. Phys. Comm. 37 (1985).

13. Bowler, K.C. and Pawley, G.S., "Molecular dynamics and Monte Carlo simulations in solid-state and elementary particle physics", Proc. IEEE 72, 42 (1984).

SIMULATION OF DISCRETE EUCLIDEAN QUANTUM GRAVITY

by

BERND A. BERG*
Supercomputer Computations Research Institute**
Florida State University
Tallahassee, Florida 32306
and
II. Institut fur Theoretische Physik, Hamburg

ABSTRACT

A recent exploratory numerical simulation of 4d discrete Euclidean quantum gravity is reported. Relying on Regge calculus |1| and the Regge-Einstein action models suitable for such simulations are investigated. Monte Carlo calculations support the following results: In the volume → ∞ limit canonical dimensions are realized and a finite action density is obtained.

*Supported in part by DFG Contract BE 915
**Permanent address after September 1, 1985

It is well-known that classical gravity is inconsistent with quantum principles. At a length scale of the order of the Planck length l_p classical space-time has to break down and we expect quantum fluctuations in space and time. The best analogy -if there is any- to quantizing gravity might be the quantization of electromagnetism. Gravity and electromagnetism have in common that they are both classically well understood and valid over many orders of magnitude in length scale. For electromagnetism the "straightforward" quantization of the classical action was tremendously successful. On the other hand quantum gravity is not even qualitatively understood. A main reason is that the gravitational coupling is dimensionful and straightforward quantization leads to a non-renormalizable quantum field theory. Therefore the perturbative calculus provides little insight. Furthermore the classical action is unbounded, what seems to be a major obstacle also to any non-perturbative quantization. We may of course like to modify the classical action, but without experimental guidance we have no intuition about the physics at a scale of the order of the Planck length. It seems to be hopeless to make a serious choice from the variety of possible actions. In particular there is no a-priori reason why concepts successful in the description of strong interactions, as for instance asymptotic freedom, should be meaningful for gravity. Even more radical: The whole concept of interactions mediated by particles may be questionable for gravity.

It is, however, a possibility that the difficulties concerning the quantization of the classical action are due to inadequate methods. In this talk I like to report on my recent attempts [2] of using Monte Carlo (MC) methods for gaining some insight into the problems of quantizing the classical Einstein action. My main guiding principle is simplicity. As experimental guidance is missing this seems to be a good choice, which leads to some systematics of exploratory numerical studies. Furthermore is has the advantage to keep the computational effort reasonable.

A natural framework for numerical studies is the Regge skeleton space [1]. Amazingly the Regge skeleton space was already introduced more than 20 years ago. On a Regge skeleton space the Einstein action becomes [1,3]

$$S = \sum_t \alpha_t A_t . \tag{1}$$

The sum goes over all triangles of the skeleton space. A_t is the area of triangle t and α_t the associated deficit angle. Following Ref. [4] the theory is formulated in the Euclidean. Minkowskian results are conjectured to be obtained by analytic continuation.

We would like to calculate vacuum expectation values with respect to the partition function

$$Z = \int_V D\,[\text{space}]\; e^{m_p^2 S} . \qquad (2)$$

Here m_p is the Planck mass and the sign of the action is choosen to give negative modes and zero modes for small fluctuations around flat space |5,3|. The measure D [space] is choosen to be

$$D\,[\text{space}] = \prod_l d \ln (x_l) \prod_p F_\Theta (x_{p_1}, \ldots, x_{p_{10}}) . \qquad (3)$$

Here x_l is the length of link l, $\prod_l$ extends over all links l and $\prod_p$ over all pentahedra p (= 4-simplexes) of the Regge skeleton. The function $F_\Theta(x_1, \ldots, x_{10})$ takes care of the pentahedra constraints: $F_\Theta = 1$ if a flat pentahedron can be constructed from the ordered links of length $x_1, \ldots, x_{10}$, and otherwise $F_\Theta = 0$. The measure is the simplest |2|, allowing simulations within a fixed finite volume. The length scale is set by defining the expectation value of one length to be constant. For instance:

$$l_o = \sqrt[4]{v_o} \text{ with } v_0 = \langle v_p \rangle , \quad V = N_p v_o . \qquad (4)$$

v_p is the volume of pentahedron p, and l_o is the analog of the lattice spacing "a" on a rigid lattice.

Numerical simulations are done under the following two aspects:

1.) Are canonical dimensions realized? 2.) Do we obtain a finite action density ($m_p^2 \neq 0$) ?

1.) By equation (4) the length scale is introduced in a rather arbitrary way. This is satisfactory if other length scales are equivalent. For instance:

$$l_0' = \langle x_l \rangle \quad \text{and} \quad l_0'' = \sqrt{\langle A_t \rangle} \quad \text{etc.} \tag{5}$$

Equivalent means $l_0' = c' l_0$ and $l_0'' = c'' l_0$ in the limit of an infinite number of links. In other words: we have canonical dimensions.

2.) The Regge-Einstein action S is unbounded, but a finite action <u>density</u> is still possible due to entropy of D [space]. A numerical study can decide these questions.

My MC calculation is carried out for the hypercubic model [5] and a simplicial variant of it [2]. By theoretical reasons [6] one would prefer to use the random lattice of Christ, Friedberg and Lee [7], but unfortunately the computational effort would increase drastically. For this and no other reasons simple models were chosen in the exploratory study [2].

I have carried out MC calculations for $m_p^2 = 0, \pm 0.3$. Systems of size $N=2^4$ and $N=3^4$ sites were used and measurements were performed after each sweep.

$m_p^2=0$ results are entirely due to the entropy of the measure (3). These results are collected in Table 1. The approach to equilibrium is (for the hypercubic model and 3^4 sites) depicted in the Figure. $\langle x_1^i \rangle$ (i=1,2,3,4) are restrictions of the link length to subclasses of links, which differ in the initial confuguration. After about 60 sweeps equilibrium is reached and the system has completely lost any memory of the original configuration in flat space. Final link averages depend slightly on the numbers of pentahedra sharing the link. Interesting is the negative curvature $\langle S \rangle$. Finite size effects are small in both models. This strongly supports canonical dimensions. In other words: Setting the scale by means of equation (4) or one of equations (5) is identical for $N_p \to \infty$ (N_p= number of pentahedra).

Of major interest is analyzing $\langle S \rangle$ as a function of m_p^2. Unfortunately $m_p^2 \neq 0$ slows down the computational speed by a factor of order ≈40. So far I have only carried out calculations for $m_p^2 = \pm 0.3$, which are collected in Table 2. Naively the correct sign is $m_p^2 > 0$, but the connection to Minkowskian gravity is not well understood. Therefore one should first of all investigate Euclidean gravity in itself and consider the whole range $-\infty < m_p^2 < \infty$. Finite size effect ($2^4 \to 3^4$) are now strong, but clearly indicate finite ratios in the limit $N_p \to \infty$. We conclude again in favor of canonical dimensions. Most strikingly the curvature expectation value $\langle S \rangle$ follows this pattern, implying a finite action density $\langle S \rangle$ for $N_p \to \infty$.

Guided by the achieved qualitative numerical understanding, I like to discuss questions concerning the continuum limit and universality. The region of physical interest is

$$l_p \ll L = \sqrt[4]{V} . \tag{6}$$

Here $l_p = |m_p^{-1}|$ is the Planck length and L the edge length of the finite system.

Let me consider two relevant scenarios. The conventional picture is to send $l^0(l_p) \to 0$ in units $[l_p]$ of the Planck length. This means $m_p^2 \to 0$ in system units $[m_0 = l_0^{-1}]$ |4|. In physical units $[l_p^2]$ a finite action density can only be obtained, if in system units $[l_0]$ $\langle S \rangle \to 0$ for $m_p^2 \to 0$. In our two models this is not the case.

T.D. Lee |6| advocates a fundamental length, which may provide a natural cut-off for ultraviolet divergencies. This means $l_0 \propto l_p$ and this is very attractive in view of the canonical dimensions. By fine tuning m_p^2 one could fix $\langle S \rangle$ to any requested value, for instance $\langle S \rangle = 0$. Carrying out the limit $N_p \to \infty$ leads to $L \propto l_0 (Np)^{1/4} \to \infty$. The cosmological constant is exactly zero, because $\langle S \rangle$ is by construction volume independent.

Final results have to be universal: They are not allowed to reflect short-distance artifacts of the used models. Non-universal features of our models are the prescriptions of glueing links at sites together. But the situation is very different from lattice gauge theory, where the O(4) invariance is broken down to the hypercubic group. In lattice gravity the Regge-Einstein action is invariant under general coordinate transformations |3| and a lattice does not really exist: The link length are dynamical variables. Further clarification of universality is desirable. In particular the random lattice could be a fundamental concept in itself and no universality with respect to other models would be needed. According to Ref. |6| unitarity of the S - matrix can be proven, if the links are appropriately re-linked Unfortunately this is difficult to implement in computer simulations. Our considered models can, however, be regarded as qualitative approximations of the random lattice.

In conclusion my investigation |2| is a starting point for numerical work on discrete Eudlidean quantum gravity. Presently the main results are canonical dimensions between dimensionful quantities, including the finite action density. An obvious next step is to investigate in more detail the m_p^2 dependence of ⟨S⟩. Future work may concentrate on coupling with an asymptotically free field theory and on using the random lattice. The aim is to understand qualitatively Euclidean quantum gravity with the Regge Einstein action.

Finally I like to mention independent work by Hamber and Williams |8|, which has some overlap with my work |2|. In particular also the measure D [space] as given by equation (3) is used and one should be able to compare $m_p^2 =$ 0 results. The general philosophy of the author of Ref. |8| is, however, rather different. Their aim is mainly to construct a renormalizable, asymptotically free theory of quantum gravity and to analyze this theory numerically.

Acknowledgements

I benefitted from discussions with D. Gross, H. Hamber, N. Christ and T.D. Lee. Finally I would like to thank D. Duke for organizing a very pleasant conference in Tallahassee.

References

|1| T. Regge, Nuovo Cim. 19 (1961) 558

|2| B. Berg, revised and extended version of Preprint, DESY 84-119

|3| R. Friedberg and T.D. Lee, Nucl. Phys. B242 (1984) 145; G. Feinberg, R. Friedberg, T.D. Lee and H.C. Ren, Nucl. Phys. B245 (1984) 343

|4| S.W. Hawking in S.W. Hawking and W. Israel (eds.): "General Relativity", Cambridge University Press 1979.

|5| M. Rocek and R.M. Williams, Z. Phys. C21 (1984) 171

|6| T.D. Lee, Erice lectures 1982 and Preprint CU-TP-297

|7| N.H. Christ, R. Friedberg and T.D. Lee, Nucl. Phys. B202 (1982) 89; Nucl. Phys. B270 [FS6] (1982) 337

|8| H. Hamber and R.M. Williams, in preparation; H. Hamber, Les Houches lectures 1984

	Model 1, 2^4	Model 1, 3^4	Model 2, 2^4	Model 2, 3^4
Statistics	6 x 2000	5 x 400	8 x 1000	10 x 200
$\langle x_1 \rangle$	2.8779 (17)	2.8651 (16)	2.8605 (16)	2.8557 (18)
$\langle A \rangle$	3.3814 (23)	3.3609 (23)	3.3571 (08)	3.3475 (11)
$\langle \alpha \rangle$	-.0142 (06)	-.0148 (04)	-.0136 (06)	-.0147 (06)
$\langle S \rangle$	-.283 (04)	-.280 (03)	-.338 (08)	-.326 (10)
$\langle x_1^1 \rangle$	2.9278 (10)	2.9133 (19)	2.9214 (51)	2.9142 (42)
$\langle x_1^2 \rangle$	2.8503 (34)	2.8417 (27)	2.8623 (59)	2.8570 (51)
$\langle x_1^3 \rangle$	2.8578 (28)	2.8396 (35)	2.8126 (71)	2.8049 (83)
$\langle x_1^4 \rangle$	2.9244 (56)	2.9148 (44)		
$\langle x_1^5 \rangle$			2.9306 (94)	2.9210 (82)
$\langle x_1^6 \rangle$			2.8831 (51)	2.8746 (70)
$\langle x_1^7 \rangle$			2.7965 (120)	2.8026 (95)

Table 1

Numerical results with $m_p^2 = 0$. The statistics are given in sweeps. Error bars are calculated with respect to the indicated number of bins. (The numbers in parenthesis are statistical errors in the last digits.)

	$m_p^2=+0.3,\ 2^4$	$m_p^2=+0.3,\ 3^4$	$m_p^2=-0.3,\ 2^4$	$m_p^2=-0.3,\ 3^4$
Equilibrium	300	700	600	1 100
Statistics	5 x 1 200	8 x 400	11 x 600	11 x 400
$\langle x_1 \rangle$	4.33 (05)	3.86 (02)	4.29 (04)	3.99 (06)
$\langle A \rangle$	6.43 (12)	5.22 (04)	6.62 (09)	5.74 (02)
$\langle \alpha \rangle$	-1.066 (05)	-1.015 (04)	0.129 (03)	0.111 (02)
$\langle S \rangle$	5.01 (11)	3.42 (04)	-5.70 (18)	-5.15 (03)

Table 2

Model 1: MC results for $m_p^2= 0.3$. Error bars are calculated as in Table 1. They may, however, be unreliable due to metastable states and limited statistics. "Equilibrium" gives the sweeps as omitted for reaching equilibrium.

MODEL 1 (3^4 SYSTEM)

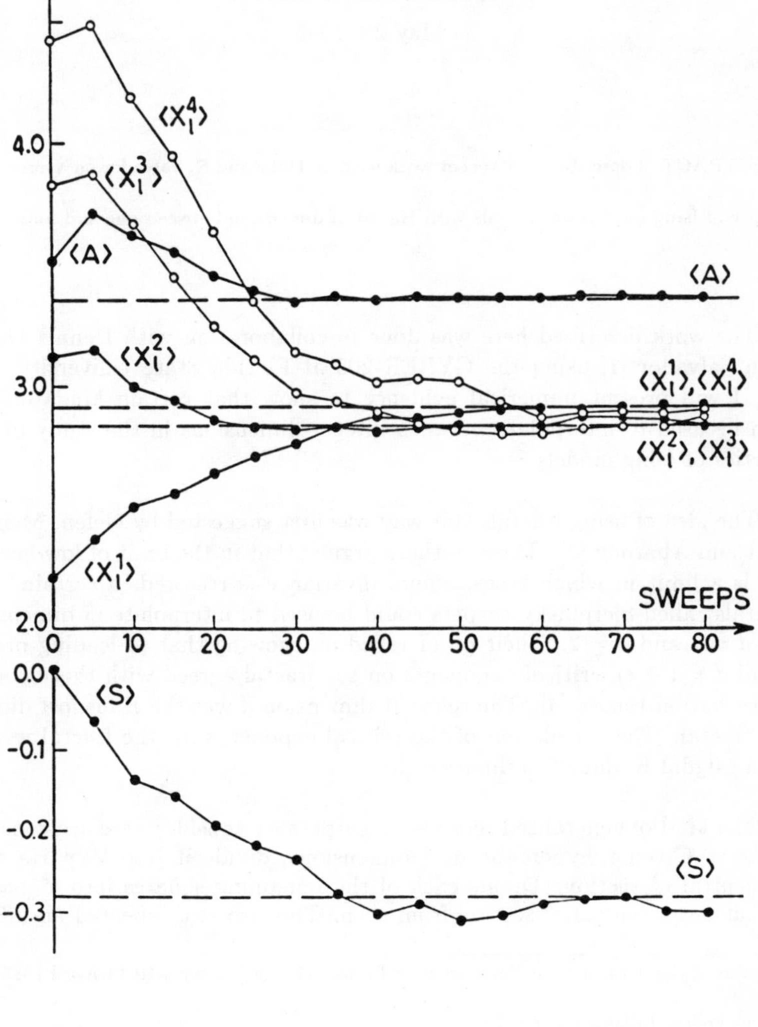

Critical properties of Ising models on Fractals of Dimension $1 < d < 4$[1]

G. BHANOT

Institute for Theoretical Physics,
University of California,
Santa Barbara, CA 93106

May 29, 1985

ABSTRACT. I describe some recent work with D. Duke and R. Salvador on a numerical study of Ising models on fractals with Hausdorf dimension between one and four.

The work described here was done in collaboration with Dennis Duke and Roman Salvador [1] using the CYBER-205 at Florida State University. In this paper, I will present numerical evidence to show that certain kinds of fractals [2] can be used to interpolate between integer dimensions in the study of critical properties of Ising models.

The idea of using fractals this way was first suggested by Gefen, Meir, Mandelbrot and Aharony [3]. These authors argued that in the limit of low lacunarity, which is a limit in which translational invariance is restored, a certain subclass of fractals called Sierpinsky carpets could be used to interpolate in dimension between $d = 1$ and $d = 2$. Their proof relied on showing that to leading order in ϵ (around $d = 1 + \epsilon$), critical exponents on the fractal agreed with those computed in perturbation theory [4]. The relevent dimension d was the Hausdorf dimension of the fractal. The calculation of the critical exponents for the fractal were done using a Migdal-Kadanoff decimation [5].

The kind of generalized Sierpinsky carpets we consider here are constructed as follows: Given a hypercube in δ dimensions, divide it into b^δ parts and cut out a central c^δ section. Divide each of the remaining squares into b^δ parts and again cut out a central c^δ section from each. This process, repeated ad-infinitum,

[1] Talk given at the workshop on " Advances in Lattice Gauge Theory", 10-14 April 1985, Florida State University, Tallahassee, Florida

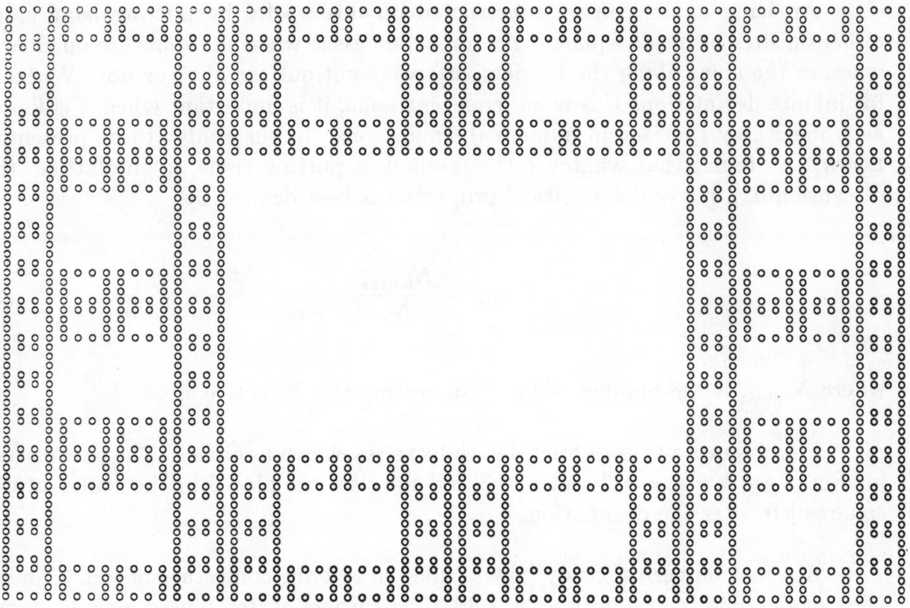

Figure 1. A Sierpinsky carpet with $b = 4$, $c = 2$ and $n = 3$. The circles show the location of spins which have been placed in the center of the squares left after decimation.

generates the fractal. As the number N_s of squares left after n decimations is $(b^\delta - c^\delta)^n$, and the linear size is b^n, the Hausdorf dimension d_h is given by,

$$(b^\delta - c^\delta)^n = (b^n)^{d_h}. \tag{1}$$

This gives,

$$d_h = \frac{\ln\,(b^\delta - c^\delta)}{\ln\, b}. \tag{2}$$

An Ising model can be constructed on this fractal by putting Ising spins either on sites or inside squares. An important issue when the spins are on sites is whether the spins along the borders that were cut out are alive or not. Whereas for infinite decimations this is an irrelevent issue, it is important when n is finite, as is inevitably the case in numerical simulations. In our study (to be presented below), we found that whatever the method of putting spins on the lattice, the dimension d_{nn} relevent for critical properties is best defined by,

$$d_{nn} = \frac{N_{bonds}}{N_s} \tag{3}$$

where N_{bonds} is the number of links connecting the N_s active sites.

Fig. 1 shows a typical fractal with live sites shown as circles. The parameters for Fig. 1 are $b = 4$, $c = 2$ and $n = 3$ and the sites are placed in the center of the squares left after the decimation.

A recent numerical study [6] was done to test the conjecture of Ref. 3 for an Ising model on a fractal with $d_h \approx 1.86$. The authors of Ref. 6 used a slight variation of the Sierpinsky carpet described above and measured the critical exponents γ, β and ν and the correlation function $< s_0 s_r >$. The strategy was to use the scaling laws [7] to compute the dimension d_c relevent to critical phenomena. The specific relationship used was:

$$d_c = \frac{\gamma + 2\beta}{\nu} \tag{4a}$$

d_c was also computed from the behavior of the two point function for large separations near the critical point:

$$< s_0 s_r >_{critical} \sim \frac{1}{r^{d_c - 2 + \eta}} = \frac{1}{r^{d_c - \gamma/\nu}} \tag{4b}$$

In addition, they also computed the so called "spectral dimension" d_{sp} from the behavior of the two point function for large r and far from the critical region using,

$$< s_0 s_r >_{non-critical} \sim \frac{e^{-mr}}{r^{(d_{sp}-1)/2}} \tag{4c}$$

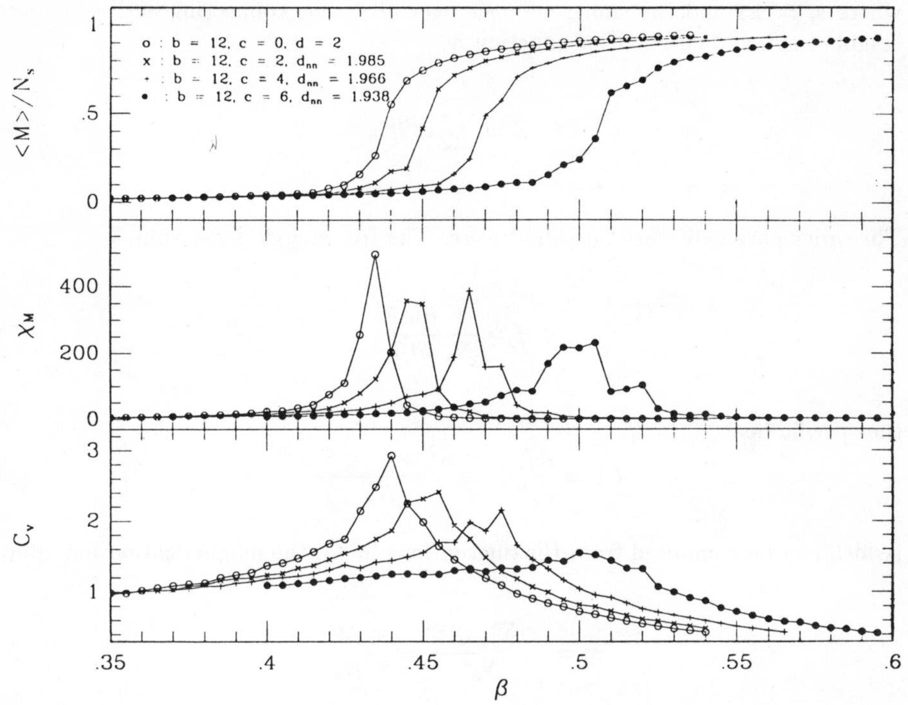

Figure 2. $< M >/N_s$, χ_M and C_v for four different fractals as functions of β. Each data point represents 50,000 measurements. The lines joining the points are for reference only.

where m is the mass (inverse correlation length). The values of d_{sp} and d_c were found to be consistent amongst themselves but disagreed with the Hausdorf dimension. This discrepancy was attributed to finite size effects. As the computer program used in Ref. 6 was highly optimized (using the microcanonical ensemble), it was not flexible. This meant that it was not possible to vary the dimension of the fractal.

In this paper, our aim is to study qualitative and quantitative features of Ising models on fractals of dimension between one and four. For all our fractals we imposed periodic boundary conditions. We worked in the usual canonical formulation and used the standard Metropolis [8] algorithm to generate configurations. The action was the simple nearest neighbor Ising action,

$$S = \sum_{i\mu} s_i s_{i+\mu} \tag{5}$$

where $s_i = \pm 1$ and the sum runs once over all bonds connecting active, nearest neighbor spins. The partition function is,

$$Z = \sum_s e^{\beta S(s)} \tag{6}$$

The order parameters we measured were: The free energy f per spin,

$$f = \frac{1}{N_s}\frac{\partial \ln Z}{\partial \beta}, \tag{7}$$

the specific heat

$$C_v = \frac{\partial f}{\partial \beta} = \frac{< S^2 > - < S >^2}{N_s} \tag{8}$$

(which can be computed from the fluctuations in f), the magnetization per spin,

$$m = \frac{< \sum_i s_i >}{N_s} = \frac{< M >}{N_s}, \tag{9}$$

and the magnetic susceptibility

$$\chi_M = \frac{< M^2 > - < M >^2}{N_s}. \tag{10}$$

The first dynamical question we studied was: How does the critical value of β vary with dimension? Fig. 2 is a typical study to determine this. In this figure, we plot m, χ_M and C_v as functions of β for four different fractals. The number of decimations and embedding dimension were two, the spins were in the center of squares, b was fixed to the value 12 and c took the values $0, 2, 4, 6$ as indicated in the figure. The lattice sizes were quite reasonable. For example, the $b = 12$, $c = 0$ system is a $d = 2$ Ising model on a 144^2 lattice while the $b = 12$, $c = 6$ system has 11660 active spins. The data shown represents $50,000$ measurements per point. The location of the critical point is clearly marked by the peaks in χ_M and C_v and the sudden drop in m. Note that the peaks in χ_M and C_v are almost coincident. Also note that β_c increases as the dimension decreases. This is also what one expects from the fact that the two dimensional Ising model has $\beta_c = 0.44068...$ and the $d = 1$ model has a critical point at $\beta_c = \infty$, implying that a system with dimension between two and one would have a critical point in the interval $(0.44068.., \infty)$.

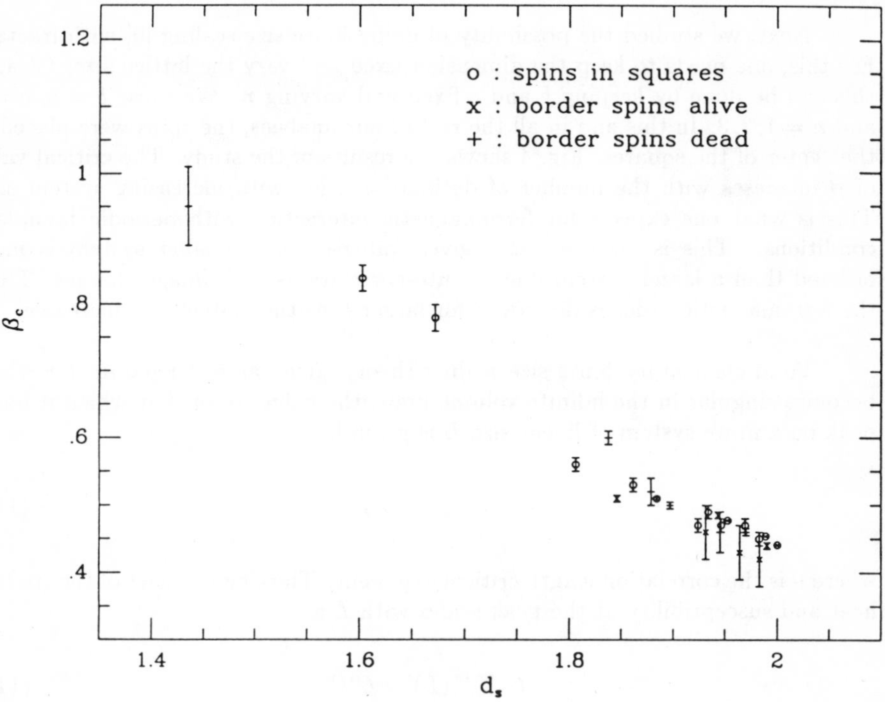

Figure 3. The critical value of beta as a function of d_h and d_{nn}.

In Fig. 3 results from studies such as the one of Fig. 2 are summarized. The data was obtained on 2-decimated fractals with b values ranging from 6 to 12. For $1 < d < 2$, the figure has data both from runs with spins at the center of squares and also from runs with spins on sites with border spins dead or alive. Note that this data plotted against d_h shows considerable scatter, depending on the way the spins are placed on the lattice. In contrast, when plotted against d_{nn}, the data falls on a unique curve. This suggests that for finite fractals, the relevent dimension is not d_h but d_{nn}. Fig. 3 also shows that there is a well defined relationship between the dimension of the fractal and the critical coupling as would be required if fractals were to smoothly interpolate in dimension between integer dimensions. Moreover, the qualitative features of Fig. 3 are also in agreement with expectations (β_c increasing as the dimension decreases).

Next, we studied the possibility of doing finite size scaling [9] with fractals. For this, one needs to keep the dimension fixed and vary the lattice size. Clearly, this can be done by keeping b and c fixed and varying n. We chose $b = 6$, $c = 2$ and $n = 1, 2, 3$. In this and in all the rest of our analysis, the spins were placed in the center of the squares. Fig. 4 shows the results of the study. The critical value of β increases with the number of decimations, ie. with increasing system size. This is what one expects for ferromagnetic interaction with periodic boundary conditions. This is because, at a given value of β, a smaller system is more ordered than a larger system due to auto-correlations with image charges. Thus, the ferromagnetic order is destroyed for larger β as the system size increases.

From elementary finite size scaling theory, given an order parameter which becomes singular in the infinite volume limit, the value β_L of β at which it has a peak on a finite system of linear size L is given by:

$$\beta_L{}^c = \beta_\infty{}^c + AL^{-1/\nu} \tag{11a}$$

where ν is the correlation length critical exponent. The singular part of the specific heat and susceptibility at the peak scales with L as:

$$C_v{}^{peak}(L) \sim L^{\alpha/\nu} \tag{11b}$$

$$\chi_M{}^{peak}(L) \sim L^{\gamma/\nu} \tag{11c}$$

As our system has $d < 2$ we expect $\alpha < 0$. That is, the specific heat does not diverge at the critical point in the infinite system. This means that as $L \to \infty$, the specific heat peak should become stable. However, γ is positive and so one expects the susceptibility peak to grow indefinitely as $L \to \infty$. These expectations are realized in Fig. 4 (note that the scale in Fig. 4 for the susceptibility is logarithmic while for the specific heat, it is linear). Estimating γ/ν from the results for $n = 2, 3$, we find,

$$\frac{\gamma}{\nu}(d_{nn} = 1.926) = 1.23 \tag{12}$$

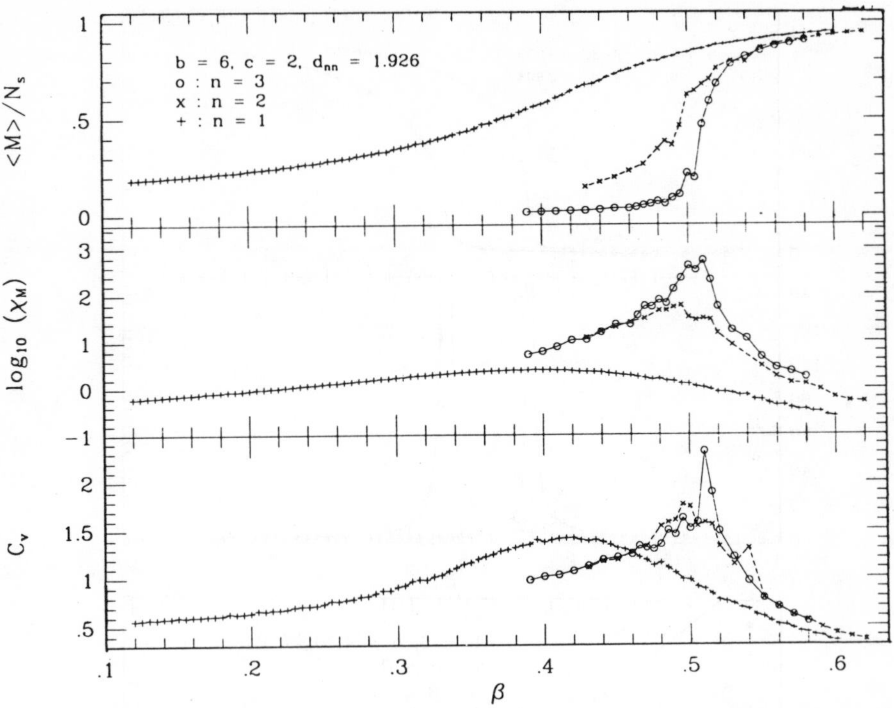

Figure 4. Results of the finite size scaling study. Each point represents an average over 20,000 measurements. Note the different scale for χ_M. The lines are again only decorative.

We do not give an estimate for $\beta_\infty{}^c$ or ν using Eq. 11a because we do not expect the data for $n = 1$ to be reliable for quantitative estimates. In summary, Fig. 4 shows that fractals give results in good qualitative agreement with expectations from finite size scaling theory.

Next, we did studies to measure the magnetic susceptibility exponent γ. In Fig. 5, a typical experiment to measure γ is summarized. The fractal used had $b = 6$, $n = 2$ and $c = 0, 4$ and was embedded in three dimensions. The data represent 50,000 measurements at each value of β. Once again, the qualitative features of the plot are in agreement with expectations. For the χ_M data, we did

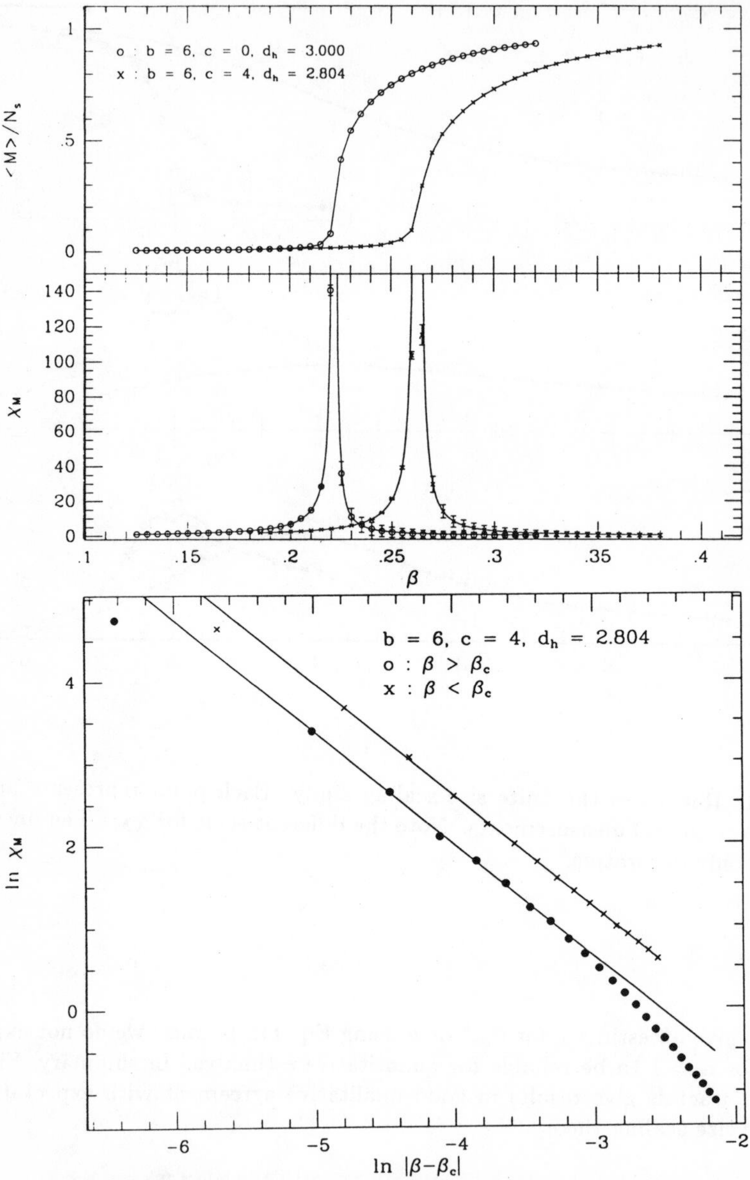

Figure 5. In the top figure, χ_M is shown as a function of β for $d = 3$ and $d_{nn} = 2.609$. Each data point represents $50,000$ measurements. The solid lines for the χ_M data show the fit as discussed in the text. The other two graphs show the fits in a log-plot.

least square fits to the expected form in the critical region:

$$\chi_M = A_{\pm}|\beta - \beta_c|^{-\gamma} \tag{13}$$

where $A_{\pm}$ refers to β values higher/lower than β_c.

These fits were made as follows: First, the critical point was approximately estimated from the peak in χ_M. A plot of lnχ versus ln$|\beta - \beta_c|$ was then made to determine the regions in which the form of Eq. 13 is valid. Finally, a least squres fit of lnχ to ln$A_{\pm} - \gamma$ln$|\beta - \beta_c|$ was made in these regions separately for the data above and below β_c. The final value of β_c was chosen as that for which the γ values from fits on either side of the critical point were the same.

The parameters obtained from the fit were:

$$d = 3.000 : \ \gamma = 1.26(3), \ \beta_c = 0.222(1) \tag{14a}$$

$$d_{nn} = 2.609 : \ \gamma = 1.37(4), \ \beta_c = 0.263(1) \tag{14b}$$

The errors in γ were estimated by computing the fluctuations in:

$$\gamma(est) = \frac{\ln(\chi_M^{(1)}/\chi_M^{(2)})}{\ln(|\beta^{(2)} - \beta_c|/|\beta^{(1)} - \beta_c|)} \tag{15}$$

where $\beta^{(1)}$ and $\beta^{(2)}$ are successive β values at which data was available in the fitted region.

The $d_h = 3$ value of γ obtained agrees well with the more accurate value $\gamma = 1.25$ [10]. The value of β_c in three dimensions also agrees with more accurate measurements [11]. We have done a series of studies for γ such as the one in Fig. 5 for fractals embedded in dimensions upto four. The summary of this presented in Fig. 6 where we plot γ as a function of d_{nn}. Note that the values of γ in non-integer dimensions interpolate smoothly between its values in integer dimensions.

Some of the questions we intend addressing in the near future are:

1) Do all critical exponents computed with the ϵ expansion agree with those measured on fractals for d near 4? This question can be answered using

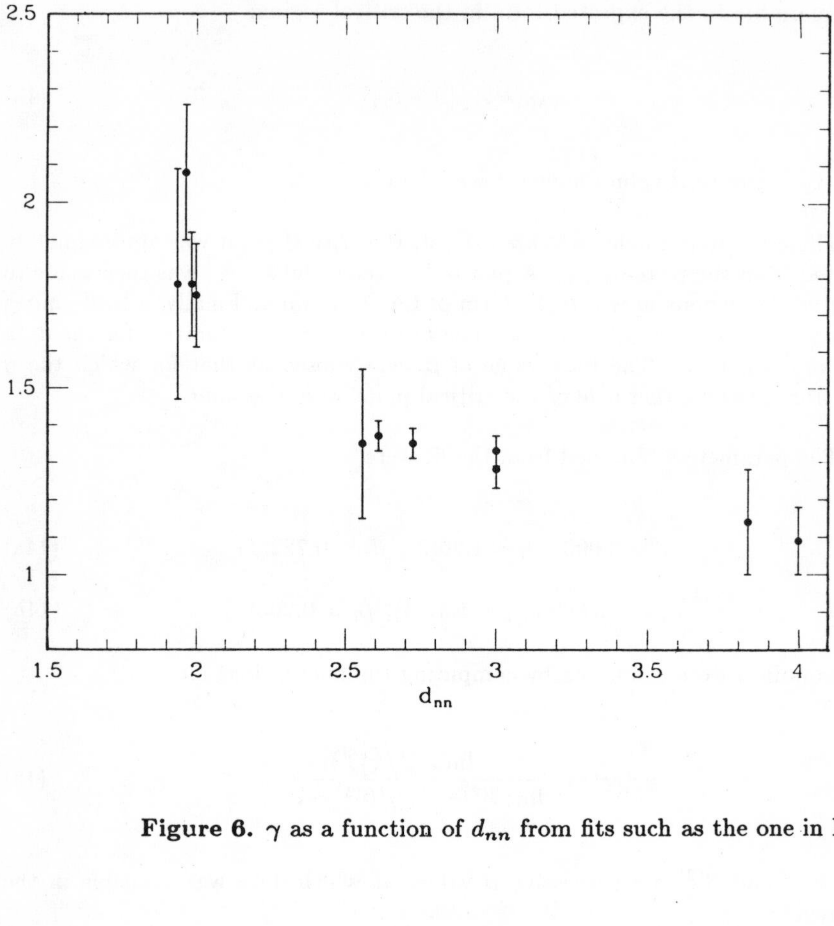

Figure 6. γ as a function of d_{nn} from fits such as the one in Fig. 5.

either the methods described above or by a Monte-Carlo renormalization group study.

2) Is ϕ^4 theory in $d = 4 - \epsilon$ interacting as $\epsilon \to 0$? Since in an appropriate limit, ϕ^4 theory becomes the Ising model, this question could first be studied on the Ising model by determining whether its critical exponents approach mean-field values smoothly as d_h approaches four. The advantage of using Ising models for numerical work is that very fast code can be written for it using multi-spin coding techniques.

3) Does all this also generalize to gauge theories? Note that it is straightforward

to define gauge theories on fractals.

In summary, we have presented a variety of studies of Ising models on fractals of the Sierpinsky carpet type. These studies indicate very strongly that these fractals can be used as systems that interpolate in dimension between integer dimensions in the sense of the ϵ expansion.

ACKNOWLEDGEMENTS

This work was supported by the National Science Foundation under grant number PHY82-17853 and supplemented by funds from the National Aeronautics and Space Administration. The data reported on in this paper was generated on the CYBER-205 at Florida State University during the time after its installation when it was undergoing testing.

REFERENCES

1) G. Bhanot, D. Duke and R. Salvador, 'Fractals and Interpolating Dimensions', Institute for Theoretical Physics, Santa Barbara preprint, NSF-ITP-85-34.

2) B. B. Mandelbrot, " Fractals : Form, Chance and Dimension " (Freeman, San Francisco, 1977) and " The Fractal Geometry of Nature " (Freeman, San Francisco, 1982).

3) Y. Gefen, Y. Meir, B. B. Mandelbrot and A. Aharony, Phys. Rev. Letts. **50** (1983) 145.

4) D. J. Wallace and R. K. P. Zia, Phys. Rev. Letts. **43** (1981) 808.

5) A. A. Migdal, Zh. Eksp. Teor. Fiz. **69** (1975) 1457.

L. P. Kadanoff, Ann. Phys. (N.Y.) **100** (1976) 359.

6) G. Bhanot, H. Neuberger and J. Shapiro, Phys. Rev. Letts. **53** (1984) 2277.

7) For a review and references to early literature, see S.-K. Ma, " Modern Theory of Critical Phenomena ", (Benjamin, 1976).

8) N. Metropolis, A. W. Rosenbluth, M. N. Rosenbluth, A. H. Teller, and E. Teller, J. Chem. Phys. **21** (1953) 1087.

9) For a recent review, see, E. Brézin, Journal de Physique **43** (1982) 15. Also see the paper by E. Brézin and J. Zinn-Justin, Saclay preprint, Saclay PhT/85-022.

10) E. Brézin, J. C. Le Guillou and J. Zinn-Justin in " Phase Transitions and Critical Phenomena ", eds. C. Domb and M. S. Green, volume VI (Academic Press, 1976).

11) G. S. Pawley, R. H. Swendsen, D. J. Wallace and K. G. Wilson, Phys. Rev. **B29**, (1984) 4030.

REVERSIBLE ISING DYNAMICS

Michael Creutz
Physics Department, Brookhaven National Laboratory
Upton, New York 11973
U.S.A.

ABSTRACT

I discuss a reversible deterministic dynamics for Ising spins. The algorithm is a variation of microcanonical Monte Carlo techniques and is easily implemented with simple bit manipulation. This provides fast programs to study non-equilibrium phenomena such as heat flow.

In this talk I will consider a variation on the microcanonical Monte Carlo algorithm that I discussed in last year's lattice gauge conference at Argonne. I introduced the present model in a recent preprint, to which you should refer for more details.

The microcanonical Monte Carlo method combines features of conventional Monte Carlo simulations of the Boltzmann distribution and molecular dynamics calculations. The algorithm involves the use of one or more auxiliary variables, called "demons," to transfer energy between the various degrees of freedom of a statistical system. In this way the combined energy of the system and the demons is held exactly constant. While this method can be useful for determining couplings in Monte Carlo renormalization group studies, its main virtue appears to be that for discrete systems, such as the Ising model, it can be programmed to run an order of magnitude faster than conventional Monte Carlo algorithms.

Here I wish to consider these demon variables more seriously as a part of the system. I will have one such variable for each site of an Ising lattice. This will then play the role of a momentum conjugate to the corresponding Ising spin. While still serving as temporary storage places for the lattice energy, the momenta can no longer transfer energy from one site to another. Instead, energy can only flow through the lattice bonds via the Ising interaction.

In this dynamics, as in molecular dynamics, temperature is a stochastic concept. The energies of the immobile demons fluctuate according to the Boltzmann distribution. The temperature is defined by inverting this distribution as observed on averages over a region of space or time, or both.

Before proceeding to a detailed description of the microscopic rules, let me discuss the obvious question of whether the dynamics is ergodic. Indeed, it is easily shown not to be. The algorithm commutes with translations of the lattice. Thus any state which is periodic will remain so. Of course, conventional molecular dynamics simulations have similar symmetries. The fraction of states with such symmetries should become insignificant as the volume of the system goes to infinity. It would, however, be interesting to know if the dynamics has any more subtle symmetries which can seriously further limit egodicity.

For simplicity I will consider the two dimensional model for the remainder of the talk. On each site of the lattice are four bits. The first of these is simply the Ising spin $S_i \in \{\pm 1\}$. associated with these spins is the energy

$$H_i = \sum_{\{ij\}} S_i S_j \tag{1}$$

where the sum is over all nearest neighbors. The next two bits on each site form a two bit integer $D_{2,i}, D_{1,i}$ $(D_{j,i} \in \{0,1\})$ representing the momentum conjugate to the spins. They are associated with the kinetic energy

$$H_K = 4 \sum_i (D_{1,i} + 2\, D_{2,i}). \tag{2}$$

The factor of 4 is inserted because flipping any spin in eq. (2) only changes the Ising energy by a multiple of 4, and we wish to keep this property for the kinetic term as well. The fourth bit on each site is its space-time parity. This is used to implement a checkerboard updating without violating the definition of cellular automata. At each time step we only consider changes of spins on that half of the sites that have a set parity bit. All parity bits are inverted for the next time step. Because of the rather trivial nature of this fourth bit, its value need not actually be stored.

The updating rule for this system is the microcanonical rule of ref. (1) on all spins with set parity bits. The resulting change in the Ising energy associated with the flip of a given spin is calculated. If this change can be absorbed by the kinetic variable on that site, then the spin is flipped and the corresponding change in the momentum made. Otherwise, the spin and its momentum retain their old values. Note that the dynamics is exactly reversed by an extra inversion of the parity bits. Hitting the red squares of our checkerboard twice returns the lattice to its original state.

Temperatures are determined from the distribution of kinetic energies. In particular, we expect

$$P(E_i) \propto \exp(-4\ \beta\ E_i). \tag{3}$$

where I have defined $E_i = D_{1,i} + 2\ D_{2,i}$ and $\beta = 1/T$. Thus the average value of E_i can be easily inverted to give the temperature.

Note that this dynamics is constructed so that the total energy

$$H = H_I + H_K \tag{4}$$

is exactly conserved. The only way the energy can flow around the lattice is through the bonds. Although the algorithm can simulate the equilibrium Ising model, it is not particularly good for this. In particular, we will see that a state with an initially non-uniform energy distribution can take a long time to equilibrate if the heat must flow long distances. On the other hand, the procedure allows numerical experiments which cannot be done with conventional Monte Carlo; in particular, one can study this heat flow.

In fig. (1) I show the results of a particular experiment of this type. Here the energy distribution of the initial lattice was not uniform; rather, the center half of the lattice was hotter than the remainder. The figure shows the temperature distribution as a function of distance through the lattice for varying times after the

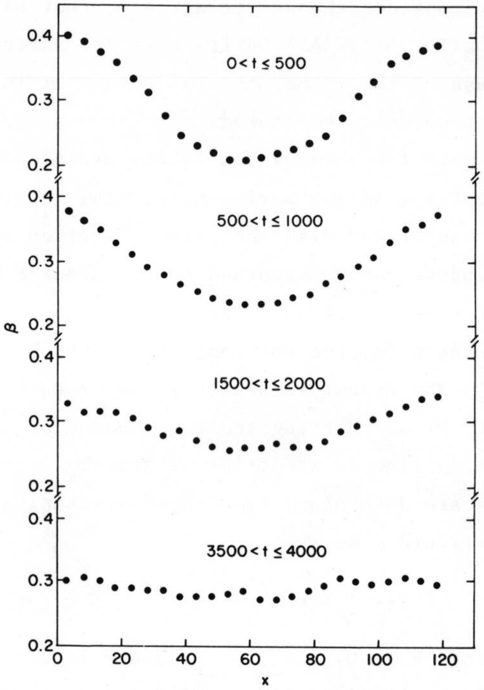

Fig. 1 The evolution of a thermal bump on a 120 × 120 lattice. The points are the temperature on a given row as obtained from an average over the given time range.

initial configuration. Note that the thermal peak gradually diffuses away and by 4000 iterations it is beginning to dissolve into thermal fluctuations. Thus we are studying a diffusion equation where the computer is only doing simple bit manipulations.

In fig. (2) I show the steady-state thermal profile of a periodic 120 by 120 lattice where heat is added to row 1 (=row 121) after each update and heat is simultaneously removed from row 61. In this experiment the heat was flowing at an average rate Q of 0.016 units per time step per site. Using the linear slope of 0.0055 for small beta in figure 4, we find the thermal conductivity at high temperatures behaves as

$$K \sim 3\beta^2 \tag{5}$$

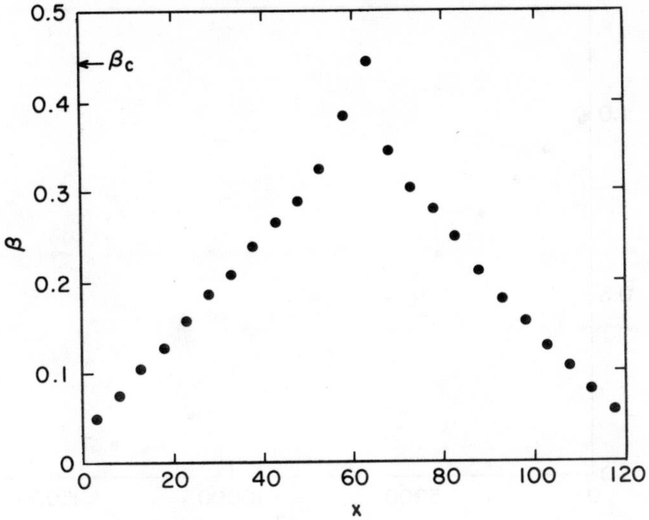

Fig. 2 The steady state thermal profile of a periodic 120 × 120 lattice heated at row 1 and cooled at row 61.

where K is defined in terms of the heat flow Q by

$$Q = - K \; \delta T / \delta x. \tag{6}$$

The slope in figure 2 increases rapidly as the critical temperature of the Ising model is approached. This indicates a sharp decrease of the thermal conductivity; indeed, in the ordered phase of this model, the thermal conductivity becomes extremely small, making convergence of this dynamics quite slow.

A good dynamics for statistical treatment should give a path through phase space which is quite sensitive to small disturbances. For our discrete Ising system, the correlation between the spins on two lattices gives a simple definition of a distance between two configurations. In fig. (3) I show the evolution of this correlation when the two lattices initially differ only in the value of a single spin. The correlation is plotted versus the square of the time to give a straight line at short times. This behavior is due to a region of decorrelated spins growing with time linearly in dimension about

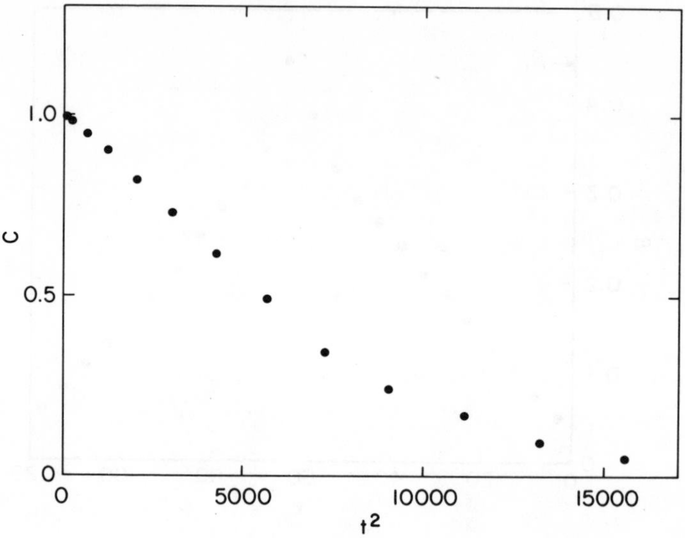

Fig. 3 As a function of time squared, the correlation between two lattices initially differing only in the value of a single spin. The quantity C represents the expectation of a spin on the first lattice times the corresponding spin on the second. The average inverse temperature of the lattices is $\beta = 0.409$.

the initial disturbed spin. An interesting experiment that I have not done is to see if this decorrelated region assumes a circular shape as it grows.

To conclude, I have presented a simple dynamics which is able to simulate a heat equation via an algorithm in which all bits used by the computer are of comparable importance. As the heat equation is a rather generic partial differential equation, the computational advantages of this bit manipulation may have considerably wider application[4].

This work supported under Contract No. DE-AC02-76CH00016 with the U.S. Department of Energy.

References

1. Creutz, M., Phys. Rev. Lett. 50, 1411 (1984); Proceedings of the Argonne National Laboratory Workshop on Gauge Theory on a Lattice (Argonne Nat. Lab., 1984); Bhanot, G., Creutz, M. and Neuberger, H., Nucl. Phys. B235[FS11], 417,(1984); Creutz, M., Mitra, P. and Moriarty, K.J.M., Comp. Phys. Comm. 33, 361 (1984); preprint BNL-35470 (1984).
2. Creutz, M., preprint # BNL 35875, (1985).
3. Creutz, M., Gocksch, A., Ogilvie, M. and Okawa, M., Phys. Rev. Lett. 53, 875 (1984).
4. Toffoli, T., Physica D10, 117 (1984).

SOME RECENT ADVANCES IN MONTE CARLO STUDIES IN STATISTICAL MECHANICS

D.P. Landau

Department of Physics and Astronomy
University of Georgia
Athens, Georgia 30602
U.S.A.

ABSTRACT

A brief review is given of some recent work involving Monte Carlo studies of problems in statistical mechanics. Several examples which may be of interest to researchers in lattice gauge theory are discussed.

1. INTRODUCTION

Monte Carlo methods have been used to study varied problems in statistical mechanics for many years now and it is clear that this approach can often be very useful for providing information and understanding which simply cannot be obtained in other ways. The methods most commonly used are documented and described elsewhere.[1] The very beautiful progress which has been made using Monte Carlo Renormalization Group methods is described elsewhere in these proceedings[2] so we shall not include this topic here. Our overview in the next section is not intended to be complete but should nonetheless give the flavor of recent activity.

2. OVERVIEW

Bulk phase transitions in simple models are relatively well understood and attention is turning towards the study of systems with surfaces/interfaces. One problem of interest is the determination of critical exponents at surfaces. These are not simply two dimensional exponents but are intrinsically dependent on the presence of the bulk. For sufficiently large values of surface exchange the surface may remain ordered at temperatures above the bulk T_c, and thus a multicritical point appears when surface and bulk fluctuations become simultaneously critical.[3] Since renormalization group studies as well as experimental results are beginning to appear the information supplied by the simulations should be quite useful. Monte Carlo calculations have also been used to study interface formation and surface tension in both two and three state models.[4] Similarly adsorbed layer/wetting transitions have been studied in a variety of models using Monte Carlo methods.[5]

One area which has grown substantially in interest has been that of systems far from equilibrium. Nucleation and spinoidal decomposition produced by quenching into a mixed-phase coexistence regime have been investigated over a wide range of times and concentrations in models for magnets, liquid-gas coexistence, binary alloys, and polymers.[6] Other non-equilibrium behavior such as diffusion limited aggregation, kinetic gelation, cluster-cluster aggregation have been studied in an attempt to distinguish between growth processes which give rise to compact or fractal matter.[7]

Monte Carlo methods have also been applied to a variety of models which are inhomogeneous. Spin glass models[8] have of course been studied for quite some time but more recent studies have concentrated on studying the degeneracies of metastable states and on the use of special purpose computers to increase the time scale over which the properties of the system could be followed. Several random field Ising models[9] have been examined with the general conclusion that $d_\ell=2$ is the lower critical dimensionality but the nature of the phase transition in three dimensions is unclear. Very recent work on granular superconductor models is also very promising.[10]

Kinetic critical behavior has been investigated in several models with the hope of shedding light on kinetic universality classes.[11] This work still has not in general achieved the necessary accuracy, although recent results for the Ising square lattice appear to be quite good. We still need accurate values for kinetic critical exponents which can be used to test the predicted relationship between linear and non-linear exponents and to examine dynamic universality and weak universality.

Monte Carlo methods have also been used to study models such as the ANNNI model which apparently have incommensurate phases.[12] The interpretation of the results is quite difficult and finite size effects are quite important since they not only limit the correlation length but also tend to "pin" the wave vector at values which are commensurate on the finite lattice.

Systems which have phase transitions to states with no long range order have been conjectured for many years. One such model which is now believed to possess such behavior is the two XY dimensional model which at low temperatures has a state with bound topological excitations (vortex-anti-vortex pairs) which unbind at T_c as the temperature is raised. The theoretical predictions of Kosterlitz and Thouless have been largely verified by Monte Carlo calculations[13] including the XY-like phase of the two dimensional anisotropic Heisenberg model[14] in the presence of a magnetic field. XY-like behavior has also been seen in systems with discrete local symmetry such as q-state clock models[15] for $q \geq 5$ and a competing interaction triangular Ising antiferromagnet.[16]

Substantial activity has existed in the area of Ising-lattice gas models[17] which are suitable for the study of adsorbed monolayers (2-dim. models) or binary alloys (3-dim. models). These studies have revealed a wide variety of phase diagrams and critical behavior which may arise due to competing interactions and external fields. One such example will be discussed in the next section.

3 GROUNDSTATE SYMMETRY AND UNIVERSALITY CLASSES

Early ideas associated with the concept of universality classes for critical behavior focused on the symmetry of the Hamiltonian largely without regard for the nature of the lattice upon which the spins reside. Thus a model with Ising spins, or Heisenberg spins with uniaxial anisotropy, would be expected to show Ising critical behavior. These ideas changed when it was realized that degeneracies in the ground state structures could lead to different critical properties. One striking example of such behavior is the triangular Ising model with antiferromagnetic nearest-neighbor coupling.[16] In zero applied field the triangular structure leads to "frustration" which prevents any transition from occurring. If ferromagnetic next-nearest-neighbor coupling is added an ordered (ferrimagnetic) phase appears at low temperature, but it is separated from the disordered phase by an XY-like line of critical points. The system thus shows non-universal Kosterlitz-Thouless behavior usually associated with two-dimensional continuous spin models. In the presence of an applied magnetic field the degeneracy is partially lifted, and the transition is now in the universality class of the 3-state Potts model. At sufficiently high fields a 3-state Potts tricritical point occurs beyond which the transition becomes first order. This picture can be changed dramatically if third-nearest-neighbor coupling is added.[18]

4. FINITE SIZE EFFECTS AT PHASE TRANSITIONS

The theory of finite size effects at 2nd order phase transitions is now well developed and tested.[19] In addition to describing "well behaved" transitions with an order parameter and power law singularities, finite size scaling has been extended to systems with Kosterlitz-Thouless behavior, i.e. no order parameter and exponential singularities. Thus for example for a system of linear dimension L and infinite lattice critical exponents α, β, γ, δ, ν, -- the singular part of the specific heat should be given by

$$C = L^{\alpha/\nu} f(x)$$

where $f(x)$ is the scaling function of the finite size scaling variable $x=tL^{1/\nu}$ where $t=|T-T_c/T_c|$. If $CL^{-\alpha/\nu}$ is plotted vs x, data in the critical regime should collapse onto a single curve $f(x)$ but only if the correct values of α, ν, T_c are chosen. In addition, for large x, $f(x)\rightarrow x^{-\alpha}$ so that for $L\rightarrow\infty$ the correct $t^{-\alpha}$ singular behavior is obtained. In Fig. 1 we show such a plot derived from a preliminary analysis of the charge 1 specific heat for the d=4 U(1) lattice guage model with negative charge 2 coupling.[20] This plot suggests that the transition is of 2nd order with exponents $\alpha \simeq 0.4$, $\nu \simeq 0.4$.

Of great recent interest is the finite size behavior at first order transitions.[21] In many cases the order of a transition can be determined only through examination of the finite size behavior. At a first order transition the susceptibility, for example, will vary as L^d, where d is the dimension of the lattice. The finite size behavior of the fourth order cumulants of energy or order parameter can also be used to distinguish between 1st and 2nd order transitions.[21,22]

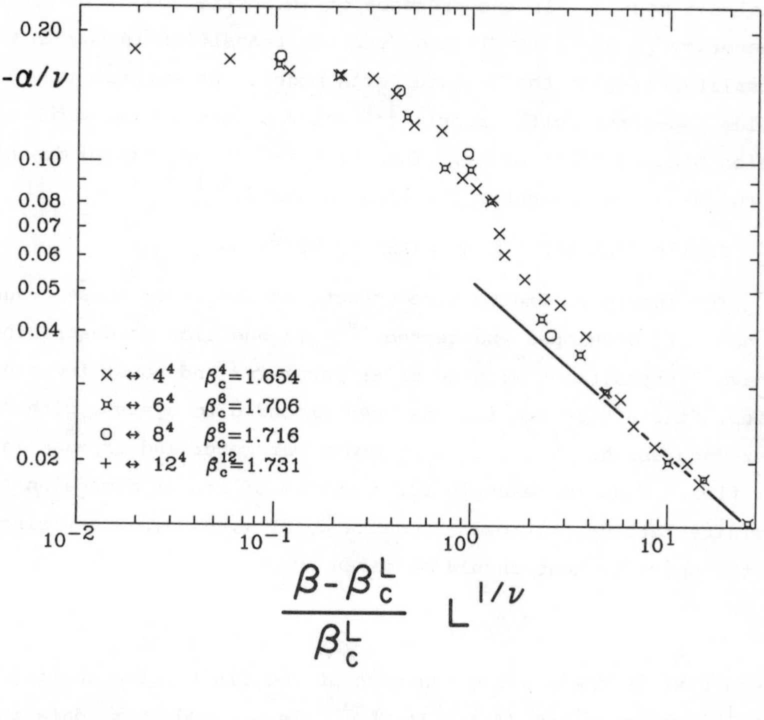

Fig. 1. Finite size scaling plot for charge 1 specific heat C_V. The plot was made with $\alpha=0.4$, $\nu=0.4$. The solid line has slope = α = 0.4.

5. REFERENCES

(1) Binder, K., in Monte Carlo Methods in Statistical Physics, ed. K. Binder (Springer Verlag, Heidelberg, 1979), and Applications of the Monte Carlo Method in Statistical Physics (Springer Verlag, Heidelberg, 1983).

(2) See the papers by R.H. Swendsen and R. Gupta in these proceedings.

(3) Binder, K. and Hohenberg, P.C., Phys. Rev. B9, 2194 (1974); Binder, K. and Landau, D.P., Phys. Rev. Lett. 52, 318 (1984).

(4) See for example, Binder, K., Phys. Rev. A25, 1699 (1982); Selke, W. and Pesch, W., Z. Phys. B47, 335 (1982); Selke, W. and Yeomans, J., J. Phys. A16, 2789 (1983); Mon, K.K. and Jasnow, D., Phys. Rev. A30, 670 (1984).

(5) Kim, I.M. and Landau, D.P., Surf. Sci. 110, 415 (1981); Ebner, C., Phys. Rev. A23, 1925 (1981); Binder, K. and Landau, D.P., J. Appl. Phys. 57, 3306 (1985).

(6) For a review see K. Binder and H. Muller-Krumbhaas in Monte Carlo Methods in Statistical Physics, ed. K. Binder (Springer Verlag, Heidelberg, 1979). See also Phani, M.K., Lebowitz, J.L., Kalos, M.H. and Penrose, O., Phys. Rev. Lett. 45, 366 (1980); Kaski, K., Yalabik, M.C., Gunton, J.D. and Sahni, P.C., Phys. Rev. B28, 5263 (1983); Grest, G.S., Srolovitz, D.J. and Anderson, M.P., Phys. Rev. Lett. 52, 1321 (1984).

(7) For an overview of work in this area see Kinetics of Aggregation and Gelation, eds. F. Family and D.P. Landau (North Holland, Ansterdam, 1984).

(8) For a review see K. Binder and D. Stauffer in Applications of the Monte Carlo Method in Statistical Physics, ed. K. Binder (Springer Verlag, Heidelberg, 1984); see also A.T. Ogielski.

(9) See for example Young, A.P. and Nauenberg, M., Phys. Rev. Lett. 54, 2426 (1985) and references therein.

(10) Jacobs, L., José, J.V. and Novotny, M.A., Phys. Rev. Lett. 53, 2177 (1984). See also Ebner, C. and Stroud, D., Phys. Rev. B25, 5711 (1982).

(11) See Williams, J.K., J. Phys. A18, 49 (1985); Novotny, M.A. and Landau, D.P., Phys. Rev. B (in press); and references therein.

(12) See e.g. W. Selke in Modulated Structure Materials, ed. T. Tsakules (NATO ASI Series: Martinus Nijhoff, 1984); Barber, M.N., and Selke, W., J. Phys. A15, L617 (1982); Morgenstern, I., Phys. Rev. B29, 1458 (1984); Rasmussen, E.B. and Jensen, S.J. Knak, Phys. Rev. B24, 2744 (1981).

(13) For a review, see D.P. Landau in Applications of Monte Carlo Methods in Statistical Physics, ed. K. Binder (Springer Verlag, Heidelberg, 1984).

(14) Landau, D.P. and Binder, K., Phys. Rev. B24, 1391 (1981).

(15) Challa, Murty S.S. and Landau, D.P., Phys. Rev. B (in press); Tobochnik, J., Phys. Rev. B26, 6201 (1982); Baltar, V.L.V., Carneiro, G.M., Pol, M.E. and Zagary, N., J. Phys. A18, 2017 (1985).

(16) Landau, D.P., Phys. Rev. B27, 5604 (1983).

(17) Landau, D.P. and Binder, K., Phys. Rev. B31, 5946 (1985).

(18) Subramaniam, R., Thesis, U. of Georgia, 1984 (unpublished).

(19) Fisher, M.E., in Critical Phenomena, ed. M.S. Green (Academic Press, New York, 1971); tests of finite size scaling using Monte Carlo data can be found in Landau, D.P., Phys. Rev. B13, 2997 (1976); B14, 255 (1976).

(20) Novotny, M.A., Gupta, R. and Landau, D.P. (unpublished).

(21) Binder, K. and Landau, D.P., Phys. Rev. B30, 1477 (1984); Privman, V. and Fisher, M.E., J. Stat. Phys. 33, 385 (1983).

(22) Challa, Murty S.S., Landau, D.P. and Binder, K. (to be published).

Noise in Monte Carlo[1]

A. D. KENNEDY
Institute for Theoretical Physics,
University of California,
Santa Barbara, CA 93106

May 23, 1985

ABSTRACT. I shall describe a new Monte Carlo method which allows us to use a "noisy" updating probability and yet obtain unbiased results. For example, it allows us to use stochastic methods to estimate the determinant ratio required for the inclusion of the effects of dynamical fermions without the penalty of either increased systematic or statistical errors in any measured physical quantities. The method is quite general, and amongst other examples I will show how it can be used for a stochastic "quenched" QCD calculation.

In this talk I want to introduce a new "noisy" Monte Carlo algorithm [**1**]. For the sake of definiteness, I shall describe it in terms of the problem of including the effects of dynamical quark fields into lattice QCD, but this should mislead us into thinking that this is the only possible application of the method. Indeed, as I shall mention at the end of this talk, I hope that "noisy" techniques will be useful for "improved action" calculations, as well as many other types of Monte Carlo problems.

Aim. The fundamental objective of a Monte Carlo calculation for lattice gauge theory is to evaluate a functional integral such as

$$\langle\Omega\rangle = \frac{1}{Z}\int d\mathcal{U}\ \det \mathcal{M}(\mathcal{U})\, e^{-\beta S_B(\mathcal{U})}\Omega(\mathcal{U}), \tag{1}$$

[1] Talk presented at the international conference *"Advances in Lattice Gauge Theory,"* Tallahassee, Florida, April 10–13 1985.

where $\mathcal{U}$ is a gluon lattice configuration, S_B is the bosonic action (the Wilson action), $\mathcal{M}$ is the fermion kernel, and Ω is some interesting operator (the Polyakov loop could be a typical example). As usual, Z denotes the partition function (or generating functional for disconnected Feynman diagrams if you prefer), and the configuration integral is with respect to Haar measure for SU(3).

Monte Carlo method. Outside the range of applicability of either strong- or weak-coupling perturbation theory, there are no reliable analytic methods to evaluate expressions such as Eq. 1. We therefore try to estimate the integrals numerically, which involves generating a sequence of gauge configurations $(\mathcal{U}_1, \mathcal{U}_2, \ldots)$ with probability distribution

$$P(\mathcal{U})\, d\mathcal{U} = \frac{1}{Z} \det \mathcal{M}(\mathcal{U}) e^{-\beta S_B(\mathcal{U})}\, d\mathcal{U}, \tag{2}$$

and then measuring the averages $\hat{\Omega} = \sum_{t=1}^{T} \Omega(\mathcal{U}_t)$. According to the law of large numbers and the central limit theorem, $\hat{\Omega} = \langle \Omega \rangle + O\left(1/\sqrt{T}\right)$ as $T \to \infty$, under some very general conditions.

Markov Processes. In order to generate such as sequence of configurations, we may use a Markov process to construct $\mathcal{U}_{t+1}$ from $\mathcal{U}_t$. If we can find an ergodic Markov chain which has the target distribution P as a fixed point, then it is guaranteed that the distribution of $\mathcal{U}_t$ approaches $P(\mathcal{U}_t)$ as $t \to \infty$, independent of the initial choice $\mathcal{U}_1$. Of course, the rate of convergence of the process, as well as the existence of metastable states and so forth, depend upon the exact nature of the Markov process and on the form of the distribution P — usually this reflects the physics of the underlying system. The Markov process which changes one configuration into the next may be conveniently broken down into a sequence of single-link steps; if each of these has P as a fixed point, and if any new configuration $\mathcal{U}$ is accessible after an entire "sweep" through the lattice (or maybe even a number of sweeps), then the composite Markov process will also preserve P and will be ergodic as required.

Detailed Balance. There are various methods of constructing such single-link updating procedures, but one of the simplest and most general is to find a method which satisfies the criterion of *detailed balance*

$$P(V \leftarrow U)P(U) = P(U \leftarrow V)P(V), \tag{3}$$

where U is the old value of the link being updated, V is its new value, and $P(V \leftarrow U)$ is the probability of changing U to V. The reason why this is a sufficient (but certainly not a necessary) condition for P to be a fixed point is that the probability of the link having value V after the updating step, given that initial value U occured with probability $P(U)$, is $\sum_U P(V \leftarrow U)P(U) = \sum_U P(U \leftarrow V)P(V) = P(V)$.

Metropolis algorithm. A clever trick [**2**] to generate V's which satisfy detailed balance is to choose a candidate V by selecting a "stepping" matrix $W = VU^{-1}$ from some suitable set[2] and then accepting V with probability[3]

$$P(V \leftarrow U) = \min\left(1, \frac{P(V)}{P(U)}\right). \tag{4}$$

This satisfies detailed balance (Eq. 3) because $\min(1, P(V)/P(U))\, P(U) = \min(1, P(U)/P(V))\, P(V)$, whether $P(U) > P(V)$ or $P(U) \leq P(V)$.

Fermions. If we are attempting to perform a Metropolis Monte Carlo calculation including the effects of dynamical quarks, almost all of our time (and money!) will be spent computing the ratio

$$R = \frac{P(V)}{P(U)} = \frac{\det \mathcal{M}(V)}{\det \mathcal{M}(U)}\, e^{-\beta \Delta S_B}. \tag{5}$$

At this point the question naturally arise as to why we should go to such a great effort to compute R *exactly* when all we are going to do with it is to compare

---

[2] Remembering that all we require is that detailed balance be satisfied and that eventually any V can be reached, it suffices that W and $W^\dagger$ be equiprobable, and that the set of possible W's "span" the SU(3) group manifold. A common choice for the set of stepping matrices is $\{e^{\pm i\theta\lambda_a} | a = 1, \ldots, 8\}$, with λ_a being a generator of SU(3) and the "step size" θ being selected at our discretion.

[3] There are many other possible choices for the acceptance probability which satisfy detailed balance too, for instance

$$P(V \leftarrow U) = \frac{P(V)/P(U)}{1 + P(V)/P(U)}.$$

it with a random number when it is used as an acceptance probability in the Metropolis algorithm? The fact that the determinant ratio is so costly to compute exactly is a symptom that we are computing more than we really need to. More generally we may describe this undesirable state of affairs as follows

> *Symptom: a negligible fraction of the computation time is spent in stochastic calculation (e.g., in generating pseudo-random numbers).*

What is the cure for this disease? What I would like to suggest is that we should only calculate R stochastically, that is we should generate an *unbiased estimator* $\hat{R}$ which we can find relatively cheaply, and then use an updating procedure which effectively averages $\hat{R}$ over the whole Markov chain. By an unbiased estimator I mean that $\hat{R}$ is to be chosen randomly (and independently) in such a way that $\langle \hat{R} \rangle = R$.

Linearity. In what sense can $\hat{R}$ be averaged over the Markov chain? All we know *a priori* about $\hat{R}$ is its mean value, so any updating algorithm which is to be correct cannot rely upon any other property of $\hat{R}$'s distribution: In other words, we are forced to use a *linear* updating algorithm if we desire unbiased results.

The requirement of linearity precludes the use of the Metropolis algorithm of Eq. 4, as we have no unbiased estimate of whether $R \geq 1$ given only $\hat{R}$. Indeed, the most general linear acceptance probabilty we can write down[4] is

$$P(V \leftarrow U) = \begin{cases} \lambda^+ + \lambda^- R & \text{if } U > V \\ \lambda^- + \lambda^+ R & \text{if } U \leq V, \end{cases} \qquad (6)$$

where $\lambda^\pm$ are constants, and U and V are to be compared using some arbitrary pairwise ordering. The reader will easily verify that detailed balance is satisfied by substituting Eq. 5 into the detailed balance equation for the cases $U > V$ and $U \leq V$ separately.

The ordering of the link values is quite arbitrary, except that it cannot depend upon R. For instance $U > V$ could mean $e^{-\beta S_B(U)} > e^{-\beta S_B(V)}$, or

---

[4] Assuming all we know is $\hat{R}$. If we compute an unbiased estimator of some function of R then other alternatives appear.

$\mathrm{Tr}(U) > \mathrm{Tr}(V)$, or $U_{ab} > V_{ab}$, etc.[5] The Metropolis procedure uses the ordering $R < 1 \Rightarrow U > V$, which cannot be determined from a knowledge of $\hat{R}$ alone.

Implementation. Although the acceptance probability of Eq. 6 depends on R, linearity allows us to implement it using only our unbiased estimator, namely for the case $U > V$ we accept V if $\lambda^+ + \lambda^- \hat{R} < r$, where r is a random number uniformly chosen from the unit interval, and likewise for the case $U \leq V$. There is, of course, a catch: This procedure is valid provided

$$0 \leq \lambda^+ + \lambda^- \hat{R} \leq 1, \tag{7}$$

as otherwise this quantity cannot be interpreted as a probability. This means that the noise in $\hat{R}$ must be bounded, but it does not mean that the noise must necessarily be small. Furthermore, it follows immediately that the acceptance rate for the linear algorithm falls only linearly as the allowed range for $\hat{R}$ grows.[6]

Example: Five State Model. In order to test whether this simple theoretical analysis works in practice let us consider the following toy model. It has only five states, which we shall label with an index i running from 0 through 4. Each state has an "energy" $E_i = i/10$, and a corresponding probability (or Boltzmann weight) $P_i \propto e^{-E_i}$. For simplicity we shall chose the parameters $\lambda^- = \frac{1}{2}$, $\lambda^+ = 0$, and we shall order the states according to their indices; the acceptance probability is, therefore,

$$P(k \leftarrow j) = \begin{cases} P_k/2P_j & \text{if } j > k \\ 1/2 & \text{if } j \leq k. \end{cases} \tag{8}$$

In this model it is a trivial matter to compute R exactly, since we know the probabilities P_i to start with, so we introduce the noise into the unbiased estimator

[5] These orderings all well-defined under normal circumstances, except perhaps on a set of measure zero.

[6] If you are familiar with using the Metropolis algorithm you will know that the acceptance rate also falls as the step size θ is increased. However, the cost of the computation grows as the inverse square of the step size (as, in some sense, we are taking a random walk through the space of gauge configurations), so in the usual case a high acceptance rate also means a large correlation between successive sweeps. This is not the case for the linear algorithm, where we adjust the acceptance rate by tuning the parameters $\lambda^{\pm}$ while keeping the step size fixed.

$\hat{R}$ artificially,

$$\hat{R} = \frac{P_k}{P_j} \pm \sigma, \tag{9}$$

where the signs are chosen equiprobably. This distribution gives the maximum amount of noise which does not fluctuate outside the range $R \pm \sigma$.

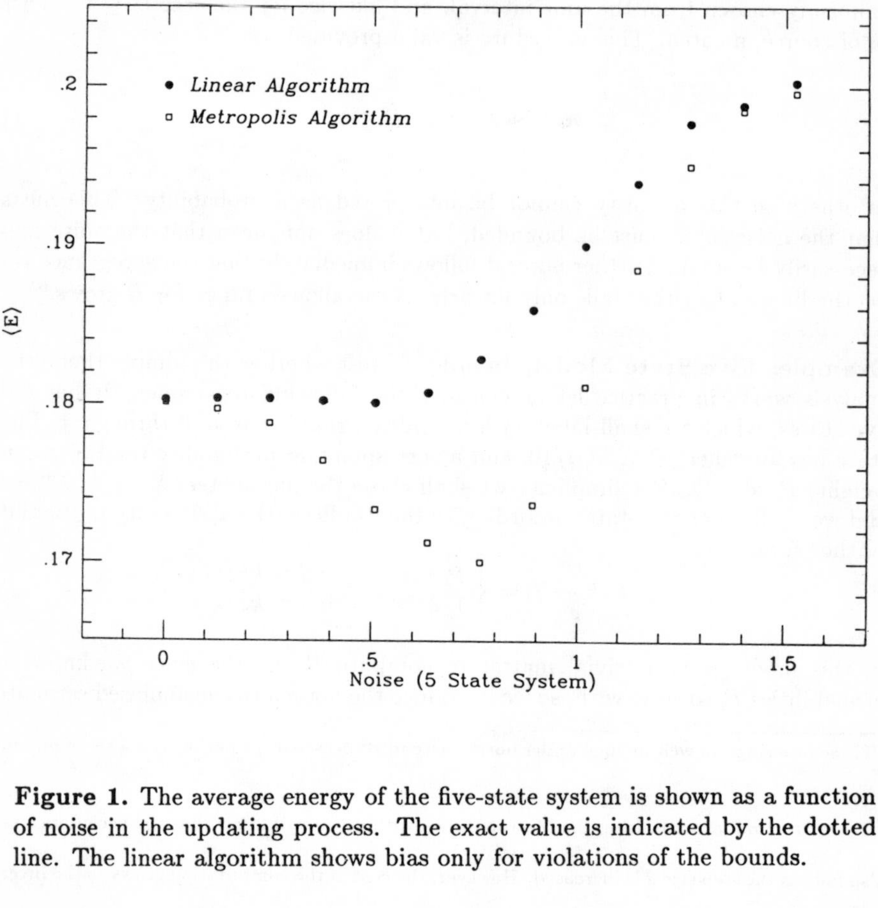

Figure 1. The average energy of the five-state system is shown as a function of noise in the updating process. The exact value is indicated by the dotted line. The linear algorithm shows bias only for violations of the bounds.

The results of Monte Carlo calculations using the linear method and the Metropolis method are shown in Fig. 1 and Fig. 2. Not only the average energy, but also *all* other physical quantities are given correctly by the new method provided only that the bounds of Eq. 7 are satisfied (which in this case means $\sigma \leq 0.508$). This follows from Fig. 2 which shows directly that the five states are visited with the correct relative frequencies for bounded noise. It is also to be noted that the Metropolis algorithm gives biased answers even for a small admixture of noise into the estimator of R. The amount of noise being put into the system can be judged by the fact that for $\sigma \approx 1.5$ all the states become equiprobable, indicating that the system is completely dominated by the noise and that the *a priori* probabilities no longer have any noticeable effect.

Stochastic Estimators. In order to utilize these ideas, we need to find cheap methods to compute $\hat{R}$, especially for fermions. Since [**3**]

$$\frac{\det \mathcal{M}(U+\delta U)}{\det \mathcal{M}(U)} = \det\left[1 + \mathcal{M}^{-1}(U)\delta\mathcal{M}(U)\right], \tag{10}$$

and $\delta\mathcal{M}$ is quasi-local, we need compute only stochastic estimates for a few components of $\mathcal{M}^{-1}$.[7] Various techniques for the noisy computation of $\mathcal{M}^{-1}$ have been discussed in the literature: I shall give only a very brief survey here.

von Neumann–Ulam random walks. [**4**] For Wilson or Kogut–Susskind (staggered) fermions the fermion kernel may be written as a unit matrix plus a hopping term $\mathcal{M} = 1 - \kappa\mathcal{W}$, so

$$\mathcal{M}^{-1}_{ij} = \delta_{ij} + \kappa\mathcal{W}_{ij} + \kappa^2\mathcal{W}^2_{ij} + \kappa^3\mathcal{W}^3_{ij} + \cdots. \qquad (\kappa < \kappa_c) \tag{11}$$

The locality of the fermionic action implies that $\mathcal{W}_{ij} = 0$ unless j is a neighbour of i, so we may estimate $\mathcal{M}^{-1}_{ij}$ as follows:

$$\begin{array}{ccccccccccc} \widehat{\mathcal{M}^{-1}_{ij}} & = & (1-\kappa)^{-1} & \xrightarrow{\kappa} & 1 & \xrightarrow{\kappa} & \mathcal{W}_{im} & \xrightarrow{\kappa} & \mathcal{W}_{mn} & \xrightarrow{\kappa} & \cdots, \\ & & \big\downarrow{\scriptstyle 1-\kappa} & & \big\downarrow{\scriptstyle 1-\kappa} & & \big\downarrow{\scriptstyle 1-\kappa} & & \big\downarrow{\scriptstyle 1-\kappa} & & \\ & & \delta_{ij} & & \mathcal{W}_{ij} & & \mathcal{W}_{mj} & & \mathcal{W}_{nj} & & \end{array} \tag{12}$$

where m, n, ... are randomly selected neighbours of i, m, ... respectively. The variance of the estimator may be reduced by taking a crowd of N such random

[7] Notice that the determinant is a linear function of the columns of $\mathcal{M}^{-1}$.

walks through the hopping parameter expansion (Eq. 11), in which case the standard deviation decreases asymptotically for large N as $1/\sqrt{N}$; however, for the reasons discussed previously, it is probably better to decrease the parameters $\lambda^{\pm}$ rather than to make N large. It may be beneficial, though, to walk more than once through the series. The restriction that $\kappa < \kappa_c$ is easily removed by modifying the matrix $\mathcal{M}$ suitably, for instance we could find the inverse $(\mathcal{M}^2)^{-1}$ and multiply it by $\mathcal{M}$. In practice, methods can be found which work well for heavy quarks, or for quarks at high temperature; for massless quarks there remains the problem that although we can arrange for the series we are summing stochastically to converge, it may still lead to an estimator whose variance is infinite.

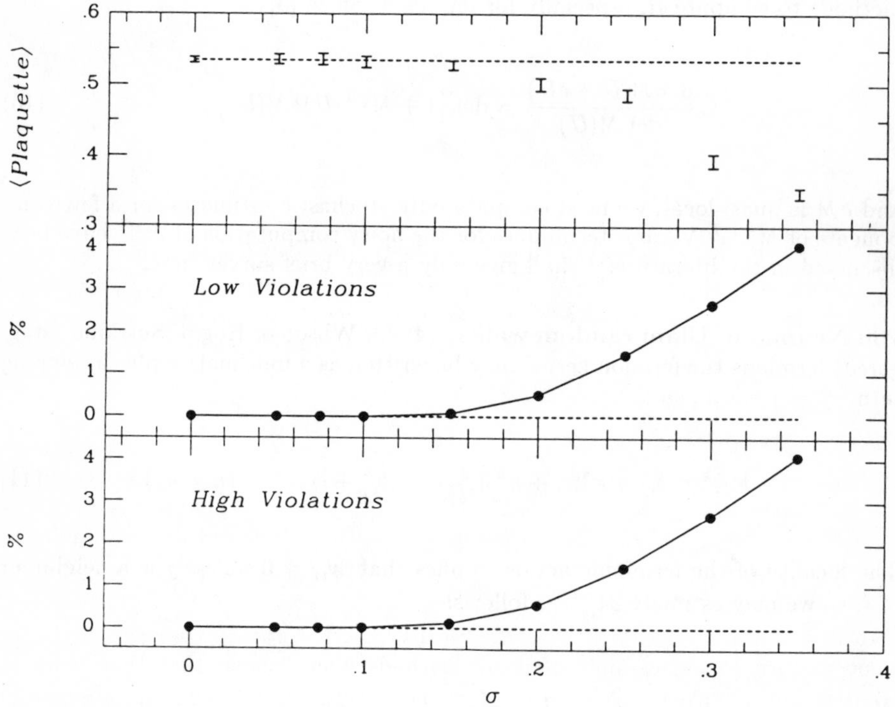

Figure 2. The relative frequencies of the five states are shown as the function of the noise. The dotted lines correspond to the exact results.

Pruning. [5] This method is somewhat similar in spirit to the straightforward random walk method described above, except that we solve for each column ψ_j of

$\mathcal{M}_{ij}^{-1}$ using the Jacobi[8] iteration scheme:

$$\psi_j^{(n+1)} = \delta_{ji} + \kappa \mathcal{W} \psi_j^{(n)} \qquad (\kappa < \kappa_c). \tag{13}$$

We perform this "wavefunction" updating step exactly, but we represent the components ψ_j of the wavefunction stochastically by applying the pruning operation $\psi_j \to \widehat{\psi_j}$ after each Jacobi iteration:

$$(\widehat{\psi_j})_i = \begin{cases} (\psi_j)_i & \text{if } \|(\psi_j)_i\| \geq \epsilon \\ \begin{cases} \frac{(\psi_j)_i}{\|(\psi_j)_i\|}\epsilon & \text{with probability } \frac{\|(\psi_j)_i\|}{\epsilon} \\ 0 & \text{with probability } 1 - \frac{\|(\psi_j)_i\|}{\epsilon} \end{cases} & \text{if } \|(\psi_j)_i\| < \epsilon \end{cases} \tag{14}$$

where ϵ is the "pruning parameter," which we can choose as we wish. The advantage of the pruning method over naïve random walks is that it deals more effectively with the sign cancellations which plague fermion calculations, but it has the disadvantage that there is a systematic error induced by truncating the Jacobi scheme, or equivalently the hopping parameter expansion, after a finite number of terms. Just as for the previous method, the variance also increases as the quark masses become small.

Pseudofermions. [6] The pseudofermion technique is not directly applicable for use in "noisy" Monte Carlo methods because in its usual formulation it produces estimators for $\mathcal{M}^{-1}$ which are biased as well as noisy. A possible technique to circumvent these difficulties is currently under investigation.

Example: 4–dimensional bosonic SU(3) lattice gauge theory. [7] I now want to turn to an example of noisy methods which shows that they work in realistic four dimensional systems. I want to consider QCD in the absence of quarks (the "quenched" approximation), but the bosonic action will be treated stochastically.

For the bosonic action we use the standard Wilson action, so for a single link update the change in the action ΔS depends only on the six plaquettes which include the link under consideration. In order to obtain an unbiased estimator for ΔS we shall choose at random one of these six plaquettes and multiply its value

[8] The Jacobi and Gauss–Seidel methods are the same for a nearest-neighbour hopping kernel.

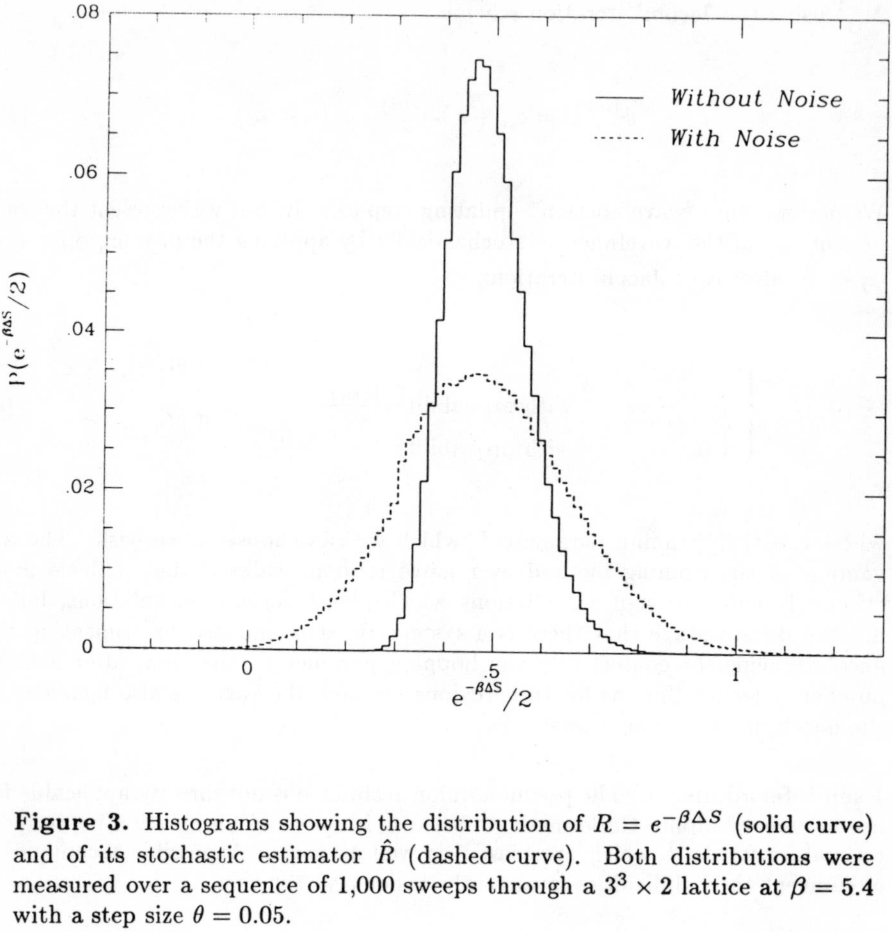

Figure 3. Histograms showing the distribution of $R = e^{-\beta\Delta S}$ (solid curve) and of its stochastic estimator $\hat{R}$ (dashed curve). Both distributions were measured over a sequence of 1,000 sweeps through a $3^3 \times 2$ lattice at $\beta = 5.4$ with a step size $\theta = 0.05$.

by six to get $\widehat{\Delta S}$. Notice that if this procedure is repeated we obtain a sequence of *independent* unbiased estimators, because $\langle \widehat{\Delta S}_1 \widehat{\Delta S}_2 \rangle = 0$.

Of course, $e^{-\beta\widehat{\Delta S}}$ is *not* an unbiased estimator for R, even though $\langle \widehat{\Delta S} \rangle = \Delta S$. We can produce an unbiased estimator for e^x given a set $\{\hat{x}_i\}$ of independent unbiased estimators for x.[9] First, write the series expansion for e^x in a suitable

[9]This method generalizes for any power series expansion.

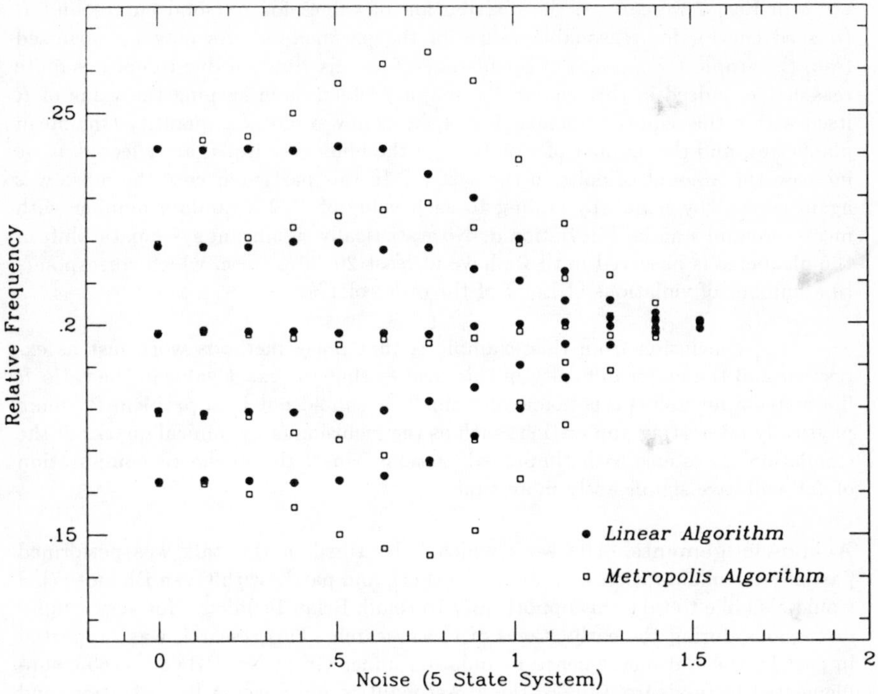

Figure 4. The average plaquette and the percentage of violations of Eq. 7 as a function of σ, as discussed in the text. The measurements were made over the last 9,000 of 10,000 sweeps through a $3^3 \times 2$ lattice at $\beta = 5.4$ starting from a configuration equilibrated with $\sigma = 0$. Because the step size $\theta = 0.05$ is small, successive sweeps are highly correlated.

factored form

$$\begin{aligned} e^x &= 1 + x + \frac{x^2}{2} + \frac{x^3}{3!} + \cdots \\ &= 1 + x + \tfrac{1}{2}\big(x^2 + \tfrac{1}{3}\big(x^3 + \cdots, \end{aligned} \tag{15}$$

hence

$$\langle \widehat{e^x} \rangle = \left\langle \begin{matrix} 1 + \hat{x}_1 & \xrightarrow{\frac{1}{2}} & + \hat{x}_1\hat{x}_2 & \xrightarrow{\frac{1}{3}} & + \hat{x}_1\hat{x}_2\hat{x}_3 & \xrightarrow{\frac{1}{4}} & \cdots \\ \big\downarrow \frac{1}{2} & & \big\downarrow \frac{2}{3} & & \big\downarrow \frac{3}{4} & & \end{matrix} \right\rangle = e^x. \tag{16}$$

On average this algorithm sums the first e terms of the series, but it is a simple matter to modify it to sum any mean number of terms we desire.

In Fig. 3 we can see the distribution of values for R (solid curve) and $\hat{R}$ (dashed curve), for reasonable values of the parameters. As may be surmized from the graph, the amount of broadening of the distribution due to noise is quite reasonable, indeed in this system the major problem is in keeping the value of R itself within the required bounds. Fig. 4 shows how a physical quantity (the mean plaquette), and the number of violations of the bounds of Eq. 7, are affected as we increase the amount of noise in the system. In this particular case the noise was again added "by hand" by adding to each value of $R/2$ a random number with mean zero and standard deviation σ. No statistically significant systematic shift in the plaquette is observed until we have at least 20–30% noise, which corresponds to a number of violations of Eq. 7 of the order of 1%.

The conclusion from this example is that noisy methods work just as expected, and the major difficulty in this case — that the exact value of the ratio R fluctuates a lot unless θ is made very small — should not be a problem for more physically interesting applications such as the inclusion of dynamical quarks or the simulation of systems with "improved" actions, where the stochastic computation of ΔS will save significantly more time.

Acknowledgements. The work which I described in this talk was performed partially in collaboration with Julius Kuti [**1**], and partly with Gyan Bhanot [**7**]. I would also like to take this opportunity to thank Brian Pendleton for very helpful discussions during the early stages of this research. This research was supported in part by the National Science Foundation under Grant No. PHY82–17853, supplemented by funds from the National Aeronautics and Space Administration, and by DOE under Grant No. DEAT0381ER40029.

REFERENCES

1. A. D. Kennedy and J. Kuti, NSF–ITP–85–21.

2. N. Metropolis, A. W. Rosenbluth, M. N. Rosenbluth, A. H. Teller, and E. Teller J. Chem. Phys. **21**, 1087 (1953) .

3. D. J. Scalapino and R. L. Sugar, Phys. Rev. Lett. **46**, 519 (1981) .

4. J. Kuti, Phys. Rev. Lett. **49**, 183 (1982) .

5. A. D. Kennedy, J. Kuti, B. J. Pendleton. In preparation.

6. F. Fucito, E. Marinari, G. Parisi, and C. Rebbi Nucl. Phys. B [FS2] **46**, 369 (1981) .

7. G. Bhanot and A. D. Kennedy NSF–ITP–85–24.

ANALYTIC CALCULATION OF MASS GAPS

A. Duncan and R. Roskies

Department of Physics & Astronomy
University of Pittsburgh
Pittsburgh, PA 15260

1. INTRODUCTION

The calculation of the spectrum of a nontrivial quantum field theory by explicit construction of the Schrödinger wavefunctions of the low-lying states may seem at first glance a hopeless task. The complexity of the problem is particularly acute in the case of asymptotically free field theories, where the long-distance physics is complicated, involving many degrees of freedom of the local fields. Nevertheless, much progress has been made using Hamiltonian techniques in problems of this sort in the areas of nuclear and many-body physics, so perhaps the situation is not quite so hopeless. Indeed, we would like to propose a particular approach[1)] to the construction of field-theory wavefunctions which appears to be sufficiently powerful to allow the accurate calculation of mass gaps in the deep weak-coupling region of several 2-dimensional field theories. In particular, we shall discuss results for the O(3) σ-model[2)] and the Gross-Neveu model[3)] in 1+1 dimensions. The technique used is to diagonalize the Hamiltonian of the field theory in the basis spanned by all linearly independent structures appearing in $H^n|\psi_0\rangle$, $0 \leq n \leq N$, where $|\psi_0\rangle$ is a suitably chosen trial wave-function (containing the basic qualitative features of the weak-coupling physics). We have obtained scaling results for N=2 in the O(3) σ-model and, with more than 3 flavors in the Gross-Neveu model, already at the N=1 level. The method is presently being extended to QCD in 3+1 dimensions, at the N=2 level.

2. THE O(3) σ-MODEL

Our Hamiltonian is

$$H = \frac{1}{a}\left[\beta \sum_n L_n^2 + \frac{1}{2\beta}\sum_n (1-\vec{\phi}_n\cdot\vec{\phi}_{n+1})\right] \tag{2.1}$$

where the sum is over the 1-dimensional spatial lattice sites, $\vec{\phi}_j$ is a unit vector at lattice site j, a is the lattice spacing, and

$$L_n^2 = \vec{\phi}_n \cdot \frac{\partial}{\partial\vec{\phi}_n}\left(\vec{\phi}_n\cdot\frac{\partial}{\partial\vec{\phi}_n} + 1\right) - \vec{\phi}_n\cdot\vec{\phi}_n \; \frac{\partial}{\partial\vec{\phi}_n}\cdot\frac{\partial}{\partial\vec{\phi}_n} \quad . \tag{2.2}$$

Small β is the weak coupling regime. The lattice is introduced to regulate the divergences of the continuum theory. We choose free boundary conditions $\vec{\phi}_0=\vec{\phi}_{L+1}=0$ where L is the number of lattice sites.

The original trial function in the scalar sector is

$$|\psi_s\rangle = e^{\lambda\sum_n \vec{\phi}_n\cdot\vec{\phi}_{n+1}}|0\rangle \tag{2.3}$$

where $|0\rangle$ is the strong-coupling vacuum defined by

$$L_j^2|0\rangle = 0 \quad \text{for all } j. \tag{2.4}$$

We find that

$$H|\psi_s\rangle \tag{2.5}$$

contains a linear combination of 4 additional topologically distinct states, providing a 5 dimensional basis for minimizing the ground state energy.

Similarly, in the triplet sector, our starting point is

$$|\psi_{t,\alpha}\rangle = (\sum_j \phi_j{}^{\alpha})\, e^{\lambda \sum_n \vec{\phi}_n \cdot \vec{\phi}_{n+1}} |0\rangle \qquad (2.6)$$

where α is an isospin index. $H|\psi_{t,\alpha}\rangle$ then contains a linear combination of 6 additional states.

At the N=2 level, the basis states contained in $H^2|\psi_s\rangle$ and $H^2|\psi_t\rangle$ are 26- and 60-dimensional respectively. The calculation consists of diagonalizing the Hamiltonian H in the 26 and 60 dimensional subspaces respectively, and interpreting the lowest eigenvalues as (over)estimates for the energies of the ground state and zero-momentum first excited states respectively. The mass gap is the difference between these two energies.

This is a completely analytic calculation (see Ref. 1), but the matrix elements of the Hamiltonian H are such complicated functions of λ that they are evaluated by a FORTRAN program, based on formulae produced by a REDUCE program. The numerical evaluation is by far the most time consuming. For example, for each value of λ, the 26x26 matrix evaluated on a 16 site lattice required 45 minutes on a RIDGE-32, while the 60x60 matrix required 20 hours. Each of these were evaluated at several values of λ, and the final estimate for the energies was obtained by searching for the minimum in λ by quadratic interpolation. Some details of the REDUCE calculations will be published elsewhere[4].

In tables 1,2 and 3, we present results for lattices of size 8, 12 and 16 indicating for each value of β in the weak coupling regime the energy estimate based on the basis generated by $H|\psi\rangle$ and by $H^2|\psi\rangle$.

Table 1

LOWEST LYING ENERGIES IN THE SINGLET AND TRIPLET SECTORS ON A LATTICE OF SIZE 8 AS A FUNCTION OF COUPLING. THESE ARE MINIMIZED IN SPACES OF DIMENSIONS 5 AND 26 FOR THE SINGLET AND DIMENSION 7 AND 60 FOR THE TRIPLET. THE MASS GAP IS THE DIFFERENCE BETWEEN TRIPLET AND SINGLET ENERGIES.

L=8

	Singlet		Triplet		Mass Gap	
β	5-dimsional basis	26-dimsional basis	7-dimsional basis	60=dimsional basis	7-5	60-26
.30	7.449	7.403	7.541	7.515	.091	.112
.31	7.374	7.332	7.480	7.456	.106	.123
.32	7.298	7.260	7.419	7.396	.121	.135
.33	7.221	7.187	7.358	7.335	.136	.148
.34	7.144	7.113	7.297	7.275	.153	.162
.35	7.065	7.038	7.236	7.215	.172	.177

Table 2

LOWEST LYING ENERGIES IN THE SINGLET AND TRIPLET SECTORS ON A LATTICE OF SIZE 12 AS A FUNCTION OF COUPLING. THESE ARE MINIMIZED IN SPACES OF DIMENSION 5 AND 26 FOR THE SINGLET AND DIMENSION 7 AND 60 FOR THE TRIPLET. THE MASS GAP IS THE DIFFERENCE BETWEEN TRIPLET AND SINGLET ENERGIES.

L=12

	Singlet		Triplet		Mass Gap	
β	5-dimensional basis	26-dimensional basis	7-dimensional basis	60-dimensional basis	7-5	60-26
.30	11.706	11.618	11.742	11.680	.036	.061
.31	11.587	11.507	11.639	11.581	.053	.074
.32	11.467	11.395	11.536	11.482	.069	.087
.33	11.345	11.281	11.433	11.382	.088	.101
.34	11.223	11.165	11.330	11.282	.107	.116
.35	11.099	11.048	11.226	11.181	.126	.133

Table 3

LOWEST LYING ENERGIES IN THE SINGLET AND TRIPLET SECTORS ON A LATTICE
OF SIZE 16 AS A FUNCTION OF COUPLING. THESE ARE MINIMIZED IN SPACES
OF DIMENSION 5 AND 26 FOR THE SINGLET AND DIMENSION 7 AND 60 FOR THE
TRIPLET. THE MASS GAP IS THE DIFFERENCE BETWEEN TRIPLET AND
SINGLET ENERGIES.

L=16

β	Singlet		Triplet			Mass Gap
	5-dim-sional basis	26-dim-sional basis	7-dim-sional basis	60-dim-sional basis	7-5	60-26
.30	15.9709	15.8414	15.9752	15.8710	.0043	.0296
.31	15.8075	15.6895	15.8307	15.7334	.0232	.0439
.32	15.6424	15.5355	15.6839	15.5942	.0415	.0587
.33	15.4756	15.3795	15.5373	15.4542	.0617	.0747
.34	15.3072	15.2214	15.3892	15.3129	.0820	.0915
.35	15.1378	15.0616	15.2407	15.1707	.1029	.1091

One can see immediately that for the larger lattices, for weak coupling, the small mass gap is the difference between two large numbers, which must therefore be evaluated quite precisely. (It is not even guaranteed that the mass gap turn out positive).

It should be emphasized that we are in the region of mass gaps very small compared to an inverse lattice spacing, so that we can hope for a good approximation of continuum results. However, we expect infrared effects, especially for the small lattices, to be present.

Several general features can be observed from the results. The increased basis reduces the singlet energy estimate more than the triplet estimate, thereby increasing the mass gap. At larger values of β, this effect is smaller so that the smaller basis already gives a good estimate of the mass gap. At small β, the increased basis is essential. We believe that at the very smallest value $\beta=0.30$, even the 26 and 60 dimensional bases do not quite suffice to accurately describe the mass gap.

We can compare our results to the continuum scaling behavior. The 1+1 dimensional σ model is an asymptotically free theory, and to three loops, the relation between the lattice spacing and coupling β is [5])

$$a = \frac{1}{\Lambda} e^{-\frac{1}{t} - \ln t - t/4} , \qquad t = \beta/\pi \quad . \tag{2.7}$$

Defining $m_\infty(\beta)$ to be the expected mass gap in an infinite lattice

$$a\, m_\infty(\beta) = (101) e^{-\frac{1}{t} - \ln t - t/4} \tag{2.8}$$

we plot in Figure 1, $m(\beta)/m_\infty(\beta)$ versus β where $m(\beta)$ is the observed mass gap. If $m(\beta)$ scaled, the graph would be flat. Note that although β ranges only between .30 and .35, $m_\infty(\beta)$ falls by a factor of more than 3 in this regime. It is clear that the $O(H^2)$ results for L=8,12 do not scale, and neither do the O(H) results for L=16. But the $O(H^2)$ results for L=16 show excellent agreement with scaling. This confirms the presence of infrared effects for small lattices, and the need for large bases in the weak coupling region for large lattices.

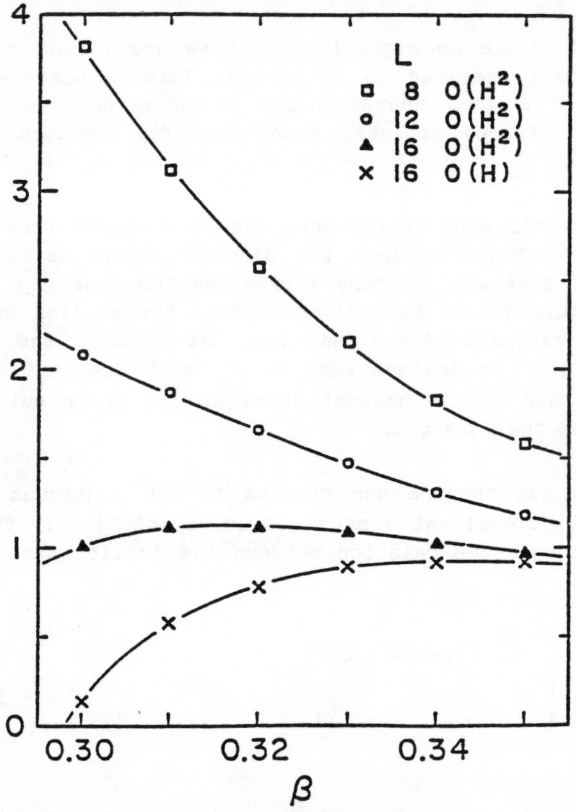

Fig. I: Plot of the ratio (calculated mass gap)/(three-loop continuum formula for the mass gap) vs. coupling on lattices of size 8, 12 and 16, in bases of 26 and 60 dimensions for the singlet and triplet. For a 16-lattice, the results for the smaller bases (5 and 7 dimensions) are also given. Asymptotic scaling implies that the ratio should be flat.

In Figure 2, we present the results somewhat differently. We plot $\ln m(\beta)$ versus $1/\beta$. If one-loop scaling were correct, this should be a straight line of slope $-\pi$. Such a line is shown on the plot, and gives a good approximation to the best straight line through the observed data points.

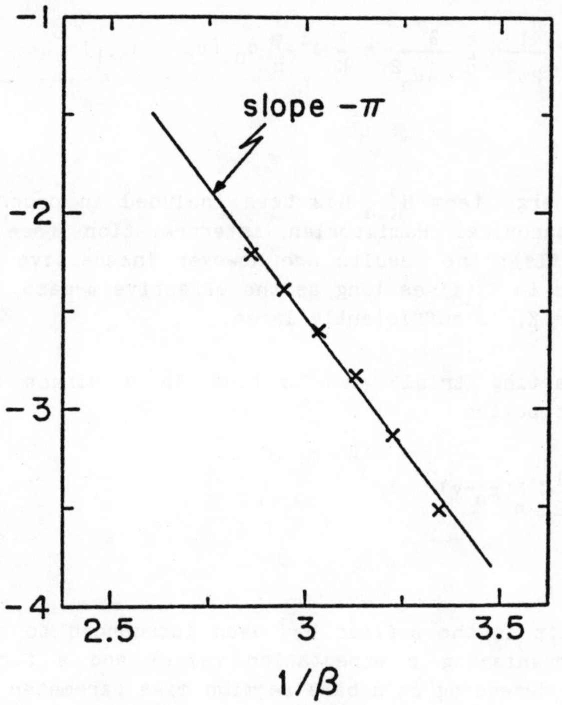

Fig. II: Plot of mass gap vs. coupling for a 16-lattice using bases of size 26 and 60. The solid line is the one-loop scaling result.

3. THE GROSS-NEVEU MODEL

A convenient representation of the Gross-Neveu Hamiltonian on a lattice is given as follows

$$aH = \frac{1}{2}\, \Sigma\, \sigma_n{}^2 + \sum_n \overline{\psi}_n(\psi_{n+1} - \psi_{n-1}) + ga \sum_n \overline{\psi}_n\, \psi_n(\sigma_n + \sigma_{n-1})$$

$$+\ aH^{\sigma}_{kin}(K) \qquad , \tag{3.1}$$

where

$$aH^{\sigma}_{kin} = -\frac{1}{2a^2}K\sum_n \frac{\partial^2}{\partial\sigma_n{}^2} + \frac{1}{K}a^2\sum_n \sigma_n(\sigma_n - \sigma_{n+1})$$

A kinetic energy term H^{σ}_{kin} has been included in order to give the σ-field a canonical Hamiltonian interpretation (see Ref. 3 for further details): the results are however insensitive to the choice of parameters in $H^{\sigma}_{kin}(K)$ as long as the effective σ-mass, determined by the parameter K, is sufficiently large.

The starting trial wave-function is a direct product of a σ-field wavefunction

$$|S\rangle \propto e^{-\frac{1}{2}C\sum_n(\sigma_n-v)^2} \qquad (3.2)$$

(where a shift in the σ-field has been introduced to accommodate an eventual nonvanishing σ expectation value) and a fermionic ground state $|0;\mu\rangle$, depending on a bare fermion mass parameter μ. The state $|0;\mu\rangle$ is obtained by diagonalizing the free massive fermion Hamiltonian - we thereby (partially) take into account the effects of chiral symmetry breaking. The parameters C, v, and μ will eventually be varied to bring the ground-state energy as low as possible (at C_{min}, v_{min}, μ_{min}).

The fermions in Eq. (3.1) are implemented following the Susskind prescription[6]. There are N_F fermion flavors at each site and one has

$$\bar{\psi}_n = (-)^{n+1}\,\psi_n{}^{\dagger}$$

$$\psi_n = \frac{1}{\sqrt{2a}}\,\phi_n \qquad n \text{ odd}$$

$$\frac{1}{\sqrt{2a}}\,\phi_n{}^{\dagger} \qquad n \text{ even} \qquad (3.3)$$

with

$$\{\phi_n, \phi_{n'}{}^{\dagger}\} = \delta_{nn'} \tag{3.4}$$

If $|0\rangle$ is the vacuum relative to the ϕ_n's,

$$\phi_n |0\rangle = 0 \tag{3.5}$$

then the fermionic starting state $|0;\mu\rangle$ is given by a Bogoliubov rotation

$$|0;\mu\rangle = U(\mu)|0\rangle \tag{3.6}$$

where

$$U^{\dagger}(\mu)\ \phi_n\ U(\mu) = \sum_{n'} (A_{nn'}\ \phi_{n'} + B_{nn'}\ \phi_{n'}{}^{\dagger}) \tag{3.7}$$

$A_{nn'}$, $B_{nn'}$ are explicitly known[3)] kinematical matrices. Thus, our starting state is $|\psi_0\rangle = |S\rangle \times |0;\mu\rangle$. In the 1 particle sector, one starts analogously from a trial wavefunction $|S'\rangle \times |p;\mu'\rangle$ (where p is a lattice momentum) involving new variational parameters C', v', μ'.

It is straightforward to write a REDUCE program which implements the operator algebra on $|\psi_0\rangle$. The calculations would be exceedingly tedious by hand, but are completely automated in REDUCE (see Ref. 4). At the N=1 level, in the vacuum sector, 6 additional states occur, so the ground-state energy E_0 is obtained by diagonalizing a 7×7 matrix.

The first matter for concern in the interpretation of the results is the dependence on the scalar field mass parameter K. For K large compared to 1, we are dealing with an inert scalar field whose mass is large in units of the inverse lattice spacing. Equivalently, the integration of the σ degrees of freedom leads to a fermionic action which is essentially local on the scale of the spatial lattice spacing. Thus, we should expect our results to be relatively insensitive to K in this regime. Quantities like the ground-state energy E_0 are of course additively renormalized by a change in K, and changing K will certainly change the values of the bare parameters C_{min}, V_{min}, μ_{min} (with C_{min} most sensitive) but physical quantities such as $\langle g\sigma \rangle$ should be relatively insensitive to K if our nonlocal Ansatz (3.2) is to be a good approximation to the local Gross-Neveu model.

Table 4

DEPENDENCE OF $\langle g\sigma \rangle$ ON SCALAR FIELD MASS PARAMETER K.

K	V_{min}	C_{min}	$\langle g\sigma \rangle$
3.0	1.38	.720	0.840
4.0	1.38	.590	0.840
5.0	1.36	.505	0.836
6.0	1.34	.455	0.832

The results of varying K are shown in Table 4. It is evident that the results for $\langle g\sigma \rangle$ are indeed insensitive to K as claimed. Indeed, variations of a factor of 2 in K (from K=3.0 to K=6.0), while leading to changes of a similar order in C_{min}, affect the ground-state expectation $\langle g\sigma \rangle$ by less than 1%. Accordingly, we shall henceforth adopt the value K=4.0 in presenting the results.

The results for the collective field expectation value $\langle g\sigma \rangle$ are displayed in Figs. 3,4.

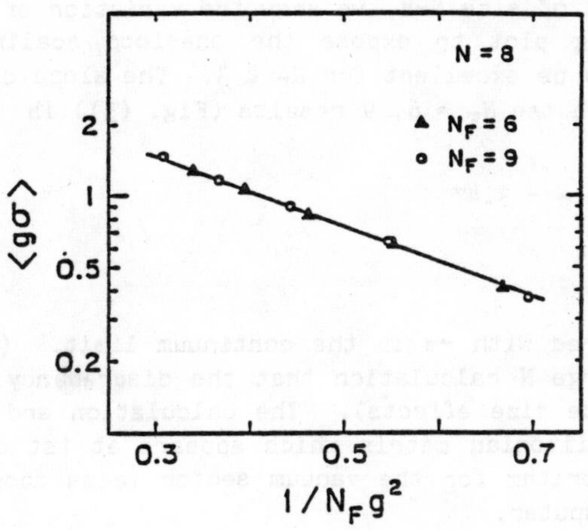

Fig. III: σ-expectation values on an N=8 lattice, N_F=6,9.

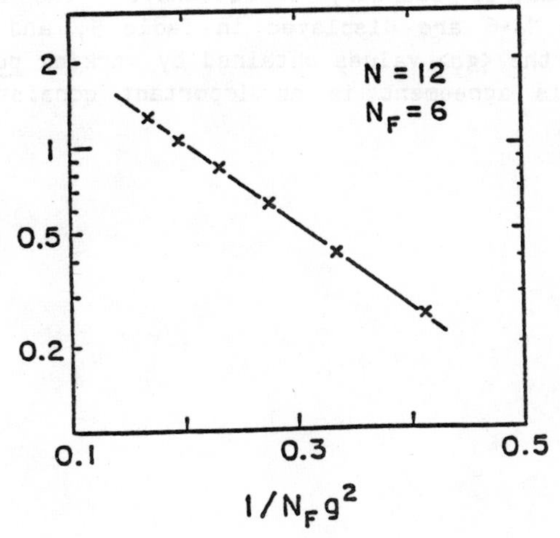

Fig. IV: σ-expectation values on an N=12 lattice, N_F=6.

On a lattice of size N=8, we show the variation of $\langle g\sigma\rangle$ with coupling on a semilog plot to expose the one-loop scaling behavior, which turns out to be excellent for $N_F \geq 3$. The slope of the straight-line drawn through the $N_F = 6, 9$ results (Fig. (3)) is

$$\frac{\Delta \ln\langle g\sigma\rangle}{\Delta\left(\frac{1}{N_F g^2}\right)} = -3.47 \qquad (3.8)$$

to be compared with $-\pi$ in the continuum limit. (It can be shown by analytic large N calculation that the discrepancy is almost entirely due to finite size effects). The calculation and diagonalization of the 7x7 Hamiltonian matrix which appears at 1st order $(H|s\rangle)$ in the Lanczos algorithm for the vacuum sector takes about 30 seconds on a RIDGE-32 computer.

The mass-gap has also been calculated directly by diagonalizing the Hamiltonian in the 12-dimensional basis obtained by isolating the independent structures in $H|p\rangle$. The energy E_1 is then minimized in this sector by varying C', μ', ν' and used to calculate the mass gap E_1-E_0 (for a particle of lattice momentum p=1/2, the minimum possible with antisymmetric boundary conditions). The results for a N=8 lattice with N_F=6 are displayed in Table 5, and are seen to agree closely with the $\langle g\sigma\rangle$ values obtained by working purely in the vacuum sector. This agreement is an important consistency check on our calculations.

Table 5

MASS GAP VS. COUPLING FOR N=8 LATTICE, N_F=6.

	g	E_0	E_1	E_1-E_0	$\langle g\sigma\rangle$
$N_F = 6$	0.5	-6.766	-6.218	0.548	0.45
	0.55	-7.565	-6.863	0.702	0.65
	0.60	-8.729	-7.854	0.875	0.86
	0.65	-10.228	-9.168	1.060	1.08
	0.70	-12.034	-10.777	1.257	1.31

REFERENCES

1. Duncan, A. and Roskies, R., Phys. Rev. D31, 364 (1985).

2. Duncan, A. and Roskies, R., Pitt Preprint PITT-02-85, May (1985), to be published in Phys. Rev..

3. Duncan, A., Nucl. Phys. B258, 125 (1985).

4. Duncan, A. and Roskies, R., 'Representations of Unusual Mathematical Structures in Scientific Applications of Symbolic Computation', Pitt preprint PITT-04-85, July 1985.

5. Brezin, E., J. Physique 43, 15 (1982).

6. Susskind, L., Phys. Rev. D16, 3031 (1977).

APPLICATION OF THE PROJECTOR MONTE CARLO METHOD TO THE TRANSFER MATRIX OF THE PLANAR SPIN MODEL

Daniel R. Stump

Department of Physics and Astronomy
Michigan State University
East Lansing, Michigan 48824

ABSTRACT

An application of the projector Monte Carlo method to the transfer-matrix eigenvalue problem is described, for the classical planar spin model. Importance sampling is used with a trial wave function derived from a mean-field approximation.

1. INTRODUCTION

A lattice quantum field theory or statistical system is defined by a partition function. In a lattice quantum field theory one dimension of the lattice is time, and the partition function is a discrete approximation of the path integral. For any lattice field theory a transfer matrix can be defined from the partition function, i. e. a matrix that defines the time-evolution of the probability amplitude in the space of fields. In the Kogut-Susskind limit, in which the time-like lattice spacing ε approaches zero, the transfer matrix approaches the matrix elements of the operator $\exp(-\varepsilon H)$ where H is the Kogut-Susskind Hamiltonian of the theory [1/].

Recently the projector Monte Carlo method has been applied to the Hamiltonian formulation of lattice gauge theories [2,3,4/]. In this method a diffusion process simulates iteration of an operator involving the Hamiltonian, for example the operator $1/(H+E)$ or the operator $\exp(-\varepsilon H)$ (in the limit $\varepsilon \to 0$). The iteration converges to the ground-state of the Hamiltonian. An interesting aspect of the method is the use of importance sampling with a trial function that is assumed to approximate the ground-state wave function.

The purpose of this paper is to describe a projector Monte Carlo calculation in which the transfer-matrix itself is iterated, rather than an operator involving the Hamiltonian. This transfer-matrix problem is equivalent to the partition function, whereas the Hamiltonian formulation differs from the partition function except in the limit of zero time-like lattice spacing. The calculation described here is for a spin system, but it may also be possible to apply this method to lattice gauge theories.

2. THE PLANAR SPIN MODEL

The classical planar spin model, or XY model, is a 2-dimensional square array of classical 2-dimensional spins with nearest neighbor interactions. If $\theta(x,y)$ is the angular position of the spin at (x,y) then the partition function is

$$Z = \int \prod_{x,y} \frac{d\theta(x,y)}{2\pi} \exp(-\beta S) , \tag{1}$$

where β is the inverse temperature and the interactions are

$$S = \sum_{x,y} \Big(1 - \cos[\theta(x+a,y) - \theta(x,y)] + 1 - \cos[\theta(x,y+a) - \theta(x,y)] \Big) . \tag{2}$$

This model is interesting and nontrivial. At small-β the spins are disordered; at large-β the partition function is dominated by harmonic spin waves. These two phases are separated by the subtle Kosterlitz-Thouless phase transition [5], which occurs around $\beta \simeq 1$. The partition function has been studied by the conventional Metropolis Monte Carlo method [6].

The transfer matrix (hereafter called T-matrix) of the XY model, which describes the evolution of the spins in the y direction, is a matrix connecting states of a one-dimensional chain of spins $[\theta(x): x=1,2,\ldots,N]$; it is defined by

$$T[\theta(x),\theta'(x)] = G[\theta,\theta'] \exp\big(-\beta V[\theta']\big) , \tag{3}$$

where

$$V[\theta(x)] = \sum_x \big(1 - \cos[\theta(x+a) - \theta(x)] \big) , \tag{4}$$

and

$$G[\theta(x),\theta'(x)] = \prod_x \exp\big(-\beta (1 - \cos[\theta(x) - \theta'(x)])\big) . \tag{5}$$

In the language of quantum mechanics, $V[\theta]$ is a diagonal operator, the potential energy of interaction along the chain; $G[\theta,\theta']$ is a non-diagonal operator that defines evolution of the spins in the y direction.

The relation between the partition function and the T-matrix is

$$Z = \mathrm{Tr}\, T^M = \int \prod_{x,j} \frac{d\theta_j(x)}{2\pi} T[\theta_1,\theta_2]\, T[\theta_2,\theta_3] \ldots T[\theta_M,\theta_1] \tag{6}$$

where M is the number of lattice sites in the y direction. In the limit $M \to \infty$, the operator T^M projects out the eigenstate of T with largest eigenvalue, i. e. the ground state. Thus the study of the partition function is equivalent to the eigenvalue problem

$$\psi[\theta(x)] = e^{\beta E} \int \prod_x \frac{d\theta'(x)}{2\pi} G[\theta,\theta'] \, e^{-\beta V[\theta']} \psi[\theta'] , \tag{7}$$

where the eigenvalue of T is $\exp(-\beta E)$. The ground-state energy E, which would be the eigenvalue of H in the Kogut-Susskind limit, is simply the free energy of statistical mechanics,

$$F = -\frac{1}{\beta} \ln Z = E M , \tag{8}$$

a result which follows from the connection between the partition function and the transfer matrix. The statistical average of a function of the spin variables is equal to an expectation value in the state $\psi[\theta(x)]$. For example the mean nearest-neighbor interaction is given by

$$\langle V\rangle = \frac{\int \prod_x d\theta(x) \; \psi^2[\theta] \; V[\theta] \; \exp\bigl(-\beta \, V[\theta]\bigr)}{\int \prod_x d\theta(x) \; \psi^2[\theta] \; \exp\bigl(-\beta \, V[\theta]\bigr)} \; . \tag{9}$$

The mean value of any operator that depends only on the spins at a fixed value of the y coordinate is given by an analogous expression.

A mean-field approximation of the eigenvalue problem (7) can be obtained by replacing the evolution operator $G[\theta,\theta']$, which propagates the spins independently, by an operator $G_m[\theta,\theta']$ that propagates independently the <u>difference</u> between neighboring spins. The resulting approximation of Eq. (7) factorizes into independent equations for the differences between neighboring spins. The mean-field approximation of the wave function is then

$$\psi_a[\theta(x)] = \prod_x u\bigl(\,\theta(x+a)-\theta(x)\,\bigr), \tag{10}$$

where the function $u(\eta)$ is the eigensolution of the one-dimensional equation

$$u(\eta) = e^{\beta\phi} \int \frac{d\eta'}{2\pi} \, g(\eta,\eta') \, e^{-\beta v(\eta')} \, u(\eta') , \tag{11}$$

where

$$v(\eta) = 1 - \cos\eta \; , \tag{12a}$$

$$g(\eta,\eta') = \exp\bigl(-\rho\,[1-\cos(\eta-\eta')]\bigr). \tag{12b}$$

The parameter ρ is chosen to make the mean-squared fluctuation of the spin $\theta(x)$ the same for $G[\theta,\theta']$ and $G_m[\theta,\theta']$; it can be shown that this condition implies

$$\frac{I_1(\rho)}{I_0(\rho)} = \frac{I_1^2(\beta)}{I_0^2(\beta)} \ . \qquad (12c)$$

The mean-field approximation of the ground-state energy E is

$$E = N \phi , \qquad (13)$$

where N is the number of spins in the x direction and ϕ is the eigenvalue in Eq. (11).

3. THE PROJECTOR MONTE CARLO CALCULATION

The projector Monte Carlo method (hereafter called pMC method) has been described in a number of papers, originally for application to quantum many-body problems [7] and recently to the Hamiltonian formulation of lattice gauge theories [2,4]. The method solves the eigenvalue equation (7) by iteration, by simulation of a diffusion process. The diffusion process acts on a weighted ensemble of points in the configuration space, $\{ \theta_\sigma(x): \sigma=1,2,3,\ldots,N_\sigma\}$. In the iteration, each point $\theta_\sigma(x)$ is reweighted by the factor $\exp(-\beta V[\theta_\sigma])$, and moved to a new point $\theta(x)$ by a stochastic process with probability distribution proportional to $G[\theta,\theta_\sigma]$. Thus the relation between the expected probability distribution of the new ensemble and the probability distribution of the previous ensemble is the same as iteration of the eigenvalue equation. The probability distribution of the ensemble of diffusing points converges to the ground-state eigenfunction.

The accuracy of the pMC calculation is improved by an importance sampling technique called biased diffusion, or guided random walk. A trial wave function, denoted here by $\psi_a[\theta(x)]$, which is supposed to approximate the ground-state eigenfunction, is used to generate the biased diffusion. The transfer-matrix eigenvalue problem is rewritten as an equation for the function

$$F[\theta(x)] = \psi_a[\theta] \ \psi[\theta] \ . \qquad (14)$$

The function $F[\theta(x)]$ obeys an eigenvalue equation of the form of Eq. (7), but with the Green's function $G[\theta,\theta']$ replaced by the biased Green's function

$$G_a[\theta,\theta'] = \psi_a[\theta] \ G[\theta,\theta'] \ / \ \psi_a[\theta']. \qquad (15)$$

The corresponding diffusion step $\theta_\sigma \rightarrow \theta$ is biased in favor of moves with a large value of $\psi_a[\theta]/\psi_a[\theta_\sigma]$. Technically, the diffusion step is a combination of a drift step $\theta_\sigma \rightarrow \theta'$, determined by the value of $\psi_a[\theta_\sigma]$, plus a random step $\theta' \rightarrow \theta$ with probability distribution proportional to $G[\theta,\theta']$. If $\psi_a[\theta]$ does approximate the ground-state wave function, then this biasing reduces the variance of estimates of expectation values by keeping the ensemble concentrated in the important region of configuration space. With importance sampling, the probability distribution of the pMC ensemble converges to $\psi_a[\theta]\psi[\theta]$, where $\psi[\theta]$ is the ground-state wave function.

In the calculation described here the trial function $\psi_a[\theta]$ is the mean-field approximation, Eq. (10).

Figure 1 shows the free energy per spin of the XY model, as a function of inverse temperature β. The curves are small- and large-β limits of E/N. The dots (•) are the mean-field approximation, Eq. (13). The crosses (+) are the results of the pMC calculation. The pMC values are computed from the growth estimate [2,4,7] of E; this estimate, which is a standard part of the pMC method, is derived from the fact that the change of the total weight of the ensemble of diffusing points over each iteration is related to the eigenvalue E. The pMC values differ little from the mean-field approximation, because the mean-field wave function $\psi_a[\theta]$ is constructed to describe accurately quantities like the free energy that do not depend on long-range correlations of the spins, but only on the mean nearest-neighbor correlation.

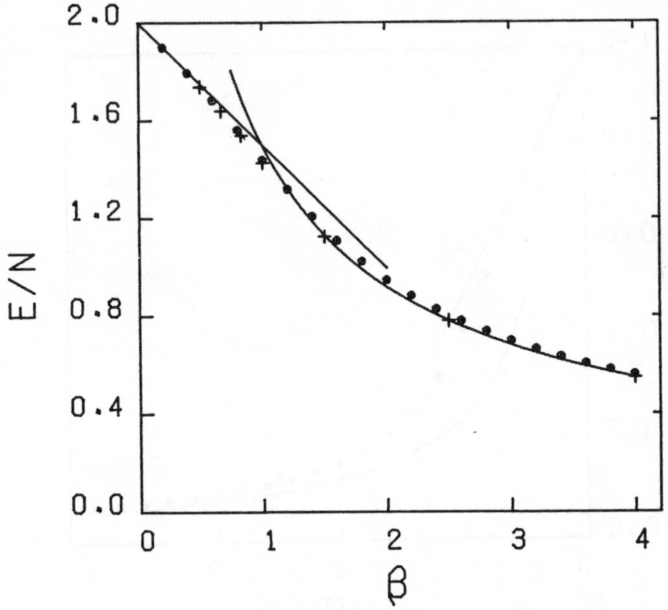

FIGURE 1. Free energy per spin vs inverse temperature. The curves are the small- and large-β limits. The dots (•) are the mean-field approximation; the crosses (+) are the projector Monte Carlo results.

Figure 2 shows the mean nearest-neighbor interaction per spin as a function of β. The curves and points have the same meaning as in Figure 1. As in Figure 1 the mean-field values and the pMC values both interpolate smoothly between the two asymptotic limits, and are nearly equal. The pMC values are computed from the mixed-expectation value[2,4,7]. This estimate is based on the approximation

$$\frac{\langle\psi\,|\,V\,|\,\psi\rangle}{\langle\psi\,|\,\psi\rangle} \simeq 2\,\frac{\langle\psi_a|\,V\,|\,\psi\rangle}{\langle\psi_a|\,\psi\rangle} - \frac{\langle\psi_a|\,V\,|\,\psi_a\rangle}{\langle\psi_a|\,\psi_a\rangle}\ . \qquad (16)$$

The first term on the right-hand side of Eq. (16) is called the mixed expectation value; the second term is simply the expectation value in the trial state. If the difference between ψ_a and ψ is of order ε, then the approximation in Eq. (16) is good to order ε^2. This estimate of the expectation value is valid if the trial function ψ_a is a good approximation of the eigenfunction, but otherwise there is some systematic order-ε^2 error. The crosses (+) in Figure 2 show just the mixed expectation values of V, which are computed from the average in the pMC ensemble.

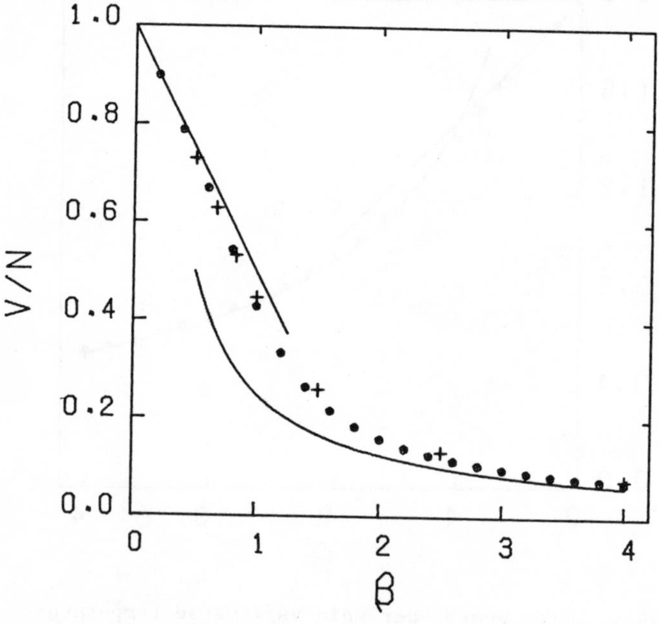

FIGURE 2. Mean nearest-neighbor interaction <u>vs</u> inverse temperature. The curves and points have the same meaning as in Figure 1.

By construction, the mean-field wave function $\psi_a[\theta]$ describes accurately the mean correlation between nearest-neighbor spins. For quantities that are insensitive to long-range correlations the pMC estimate will be approximately equal to the mean-field value. On the other hand, this mean-field approximation cannot describe accurately a quantity that depends on long-range correlations, because ψ_a lacks correlation beyond nearest neighbors. The pMC estimate of such a quantity should be different than the mean-field value, at least in the large-β limit where long-wavelength spin waves dominate the partition function.

Figure 3 shows the quantity γ, defined by

$$\gamma = \frac{\beta^2}{N}\left(\langle V^2\rangle - \langle V\rangle^2 \right)$$

$$= \frac{\beta^2}{N} \sum_{x,x'} \Big(\langle\cos(\theta(x'+a)-\theta(x'))\ \cos(\theta(x+a)-\theta(x))\rangle - \langle\cos(\theta(x'+a)-\theta(x'))\rangle \langle\cos(\theta(x+a)-\theta(x))\rangle \Big). \quad (17)$$

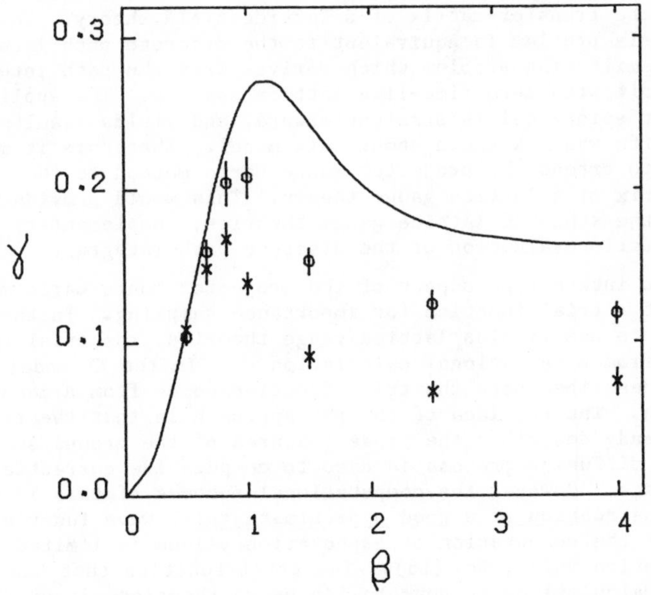

FIGURE 3. Mean-squared fluctuation of the interaction energy of spins with the same y coordinate vs inverse temperature. The curve is the mean-field approximation. The circles (o) and crosses (×) are the projector Monte Carlo results for the mixed expectation value for N=10 and N=20 spins.

The quantity γ is similar to the specific heat of the XY model. The specific heat is the mean-square fluctuation of the interaction energy of all neighboring spins, measured in units of temperature, by the fluctuation-dissipation theorem; but γ is the fluctuation of the interaction energy only of spins that have the same y coordinate. The quantity γ is sensitive to long-range correlations between spins, by the second equality in Eq. (17). In Figure 3, the curve is the mean-field estimate of γ; this is independent of the number of spins N. The points show the mixed expectation values of γ computed from the pMC ensemble average, with dots (•) for N=10 and crosses (×) for N=20. The pMC values agree with the mean-field approximation for small β, but disagree and show a significant N dependence for large β. These results are consistent with the fact that the Kosterlitz-Thouless transition occurs for $\beta \simeq 1$.

4. CONCLUSIONS

The projector Monte Carlo method has been applied to the Hamiltonian formulation of lattice gauge theories [2,4]. The present work was motivated in part to see whether this method can be applied directly to the transfer-matrix of a lattice field theory. The transfer-matrix problem is equivalent to the discrete path integral, unlike the Hamiltonian problem which derives from the path integral only in a limit with zero time-like lattice spacing. The application to the planar spin model is straightforward, and yields results consistent with what is known about this model. Therefore it may also be possible to extend the projector Monte Carlo method to the transfer-matrix of a lattice gauge theory. This would provide a new approach to the study of lattice gauge theories, complementary to the usual Monte Carlo evaluation of the discrete path-integral.

The most interesting aspect of the projector Monte Carlo method is the use of a trial function for importance sampling. In the applications to Hamiltonian lattice gauge theories, the trial function is obtained from a variational calculation [8]. In the XY model calculation described here the trial function comes from a mean-field approximation. The key idea of the pMC approach is that the trial function already describes the gross features of the ground state; then the pMC diffusion process is used to compute the correction to the trial state. However the computational success of this idea may depend on construction of a good approximate trial wave function, especially if the computation of expectation values is limited to the mixed expectation value, Eq. (16). The trial function that has been used in pMC calculations on non-Abelian gauge theories, is of the mean-plaquette form, i. e. a product of independent functions of the plaquette variables [8]. Accurate computation of quantities that depend on field variables with large separations may require the construction of a trial function with appropriate long-range correlations. The problem of constructing such a wave function may prove to be impossibly difficult for a non-Abelian lattice gauge theory.

REFERENCES

1. J. Kogut and L. Susskind, Phys. Rev. D11, 395 (1975); E. Fradkin and L. Susskind, Phys. Rev. D17, 2637 (1978).

2. D. W. Heys and D. R. Stump, Phys. Rev. D28, 2067 (1983); Phys. Rev. D30, 1315 (1984).

3. D. W. Heys and D. R. Stump, Phys. Rev. D29, 1784 (1984).

4. S. A. Chin, J. W. Negele, and S. E. Koonin, Ann. Phys. 157, 140 (1984); S. A. Chin, O. S. van Roosmalen, E. A. Umland, and S. E. Koonin, Caltech preprint; S. E. Koonin, invited talk at this conference.

5. J. M. Kosterlitz and D. J. Thouless, J. Phys. C6, 1181 (1973).

6. J. Tobochnik and G. V. Chester, Phys. Rev. B20, 3761 (1979).

7. M. H. Kalos, Phys. Rev. 128, 1791 (1962); Phys. Rev. A2, 250 (1970); M. H. Kalos, D. Levesque, and L. Verlet, Phys. Rev. A9, 2178 (1974); D. M. Ceperley and M. H. Kalos, in Topics in Current Physics, vol. 7, edited by K. Binder (Springer, Berlin, 1979).

8. D. W. Heys and D. R. Stump, Phys. Rev. D29, 1791 (1984); D. Horn and M. Karliner, Nucl. Phys. B235 [FS11], 135 (1984).

Using MACSYMA to Write Long FORTRAN Codes for Simplicial-Interpolative Lattice Gauge Theory *

Kevin Cahill and Randolph Reeder†

Department of Physics and Astronomy
University of New Mexico
Albuquerque, New Mexico 87131

Abstract

We describe a use of the symbol manipulator MACSYMA to write a long lattice-gauge-theory FORTRAN program. We used MACSYMA's Diff, Do, Ratcoef, Ratsimp, Subst, and Sum functions to calculate the principal formulas of the program. We then used MACSYMA's FORTRAN, Part, Print, Save, and Writefile functions to express these formulas in compilable FORTRAN code. The resulting 1,600-line code simulates SU_2 gauge theory in three dimensions by the simplicial interpolative method.

---

* Supported in part by the U.S. Department of Energy under grant DE-FG04-84ER40166.

† Address after August 1, 1985: AT-2, MS-H818, Los Alamos National Laboratory

The most important use of symbol manipulators in physics is to do calculations that cannot be done by hand. Many interesting physical problems involve infinitely many variables interacting nonlinearly. One can numerically simulate the behavior of such systems by reducing the number of variables to a finite but very large number. However because of the number of variables and of the nonlinearity of their interactions, it is seldom practical to write the necessary computer code without the aid of a symbol manipulator.

In such complex physical applications, one first uses the symbol manipulator to calculate the basic formulas of the numerical simulation. These formulas, which are the primary output of the symbol manipulator, then serve as the basis for the numerical computer program that actually performs the numerical simulation. The formulas that constitute this primary symbolic output are typically too long to edit manually. The usefulness of the symbol manipulator therefore depends upon whether it can automatically translate its primary output into statements of exactly the form required by the compiler of the computer being used for the numerical processing. This secondary output is as important to the writing of long error-free codes as is the primary output of the symbol manipulator.

We used MACSYMA [1] rather than SMP [2] or another symbol manipulator because it had a FORTRAN function that translates expressions from its own internal representation into FORTRAN statements.[1] Unfortunately, this command turns long internal expressions into long FORTRAN statements even though the maximum length of a FORTRAN statement is 20 lines. Many compilers run out of tree space after 10

---

1. SMP now also has a similar FORTRAN function.

sufficiently complex lines.

The FORTRAN statements produced by MACSYMA, in the complex problems for which MACSYMA is required, can easily run to hundreds or thousands of lines. It is not practical manually to edit such illegal FORTRAN statements. Fortunately, MACSYMA has a Part command which one can use to separate long expressions into small ones before using its FORTRAN function. However such use of the Part command can be tricky. Hopefully a future version MACSYMA will be able automatically to translate its symbolic expressions into compilable FORTRAN code. Such a capability ideally would provide an option in the FORTRAN function that would allow the user to specify the maximum number of lines in a given FORTRAN statement.[2]

In this article we describe an application of MACSYMA to lattice gauge theory. We used MACSYMA's Subst, Sum, Ratcoef, Diff, and Ratsimp functions to calculate the major formulas of an unconventional lattice-gauge-theory program. We then used MACSYMA's Part and FORTRAN functions to separate and translate these very long symbolic expressions into compilable FORTRAN code. The resulting 1,600-line code simulates SU_2 gauge theory in three dimensions by the simplicial interpolative method [3].

In the simplicial interpolative method, spacetime is filled with a cubic lattice each cube of which is tiled with six (tetrahedral) simplexes. The simplexes in each cube are oriented in the same way so that the simplexes of translated cubes are congruent. We use the temporal gauge so as to have only two gauge fields $A_n(x)^{\alpha}$, n=1 and 2, for each of the three group directions α. Each spacetime point x lying in a simplex with vertices v_i can be uniquely

2. The latest version of SMP has such an option.

expressed in the form

$$x = \sum_{i=0}^{3} \rho_i v_i \qquad (1)$$

in which the nonnegative weights ρ_i sum to one. We use this same formula to linearly interpolate the field $A_n(x)^\alpha$ at the point x from its values $A(\alpha,n,v_i)$ at the vertices v_i:

$$A_n(x)^\alpha = \sum_{i=0}^{3} \rho_i A(\alpha,n,v_i) \quad . \qquad (2)$$

Since the ρ's sum to one, ρ_0 is one minus the sum of the components of the 3-vector $\rho=(\rho_1,\rho_2,\rho_3)$. One may obtain this 3-vector by inverting the 3-by-3 matrix M defined by $M_{ji}=(v_i)_j-(v_0)_j$ and forming the matrix product $\rho=M^{-1}(x-v_0)$. The resulting formula (2) for the gauge field depends on the simplex the point x is in. For the simplex whose vertices are $v_0=(i,j,k)b$, $v_1=(i+1,j,k)b$, $v_2=(i+1,j+1,k)b$, and $v_3=(i+1,j+1,k+1)b$, where b is the lattice spacing, the interpolated gauge field at the point $x=(x,y,t)$ is

$$\begin{aligned} A_n(x)^\alpha = {} & [(i+1)b-x]A(\alpha,n,i,j,k) \\ & +[(j-i)b+x-y]A(\alpha,n,i+1,j,k) \\ & +[(k-j)b+y-t]A(\alpha,n,i+1,j+1,k) \\ & +(t-kb)A(\alpha,n,i+1,j+1,k+1). \end{aligned} \qquad (3)$$

Similar formulas obtain for the other five generic simplexes.

The field strength of the interpolated field $A_n(x)^\alpha$ is $F_{mn}(x)^\alpha = \partial_n A_m(x)^\alpha - \partial_m A_n(x)^\alpha + g\epsilon^{\alpha\beta\gamma} A_m(x)^\beta A_n(x)^\gamma$. The field A is continuous, but its derivatives have step-function discontinuities on the boundaries between simplexes. The field strength F inherits these integrable singularities. A key feature of the present method is that the interpolated fields are defined throughout spacetime. It is therefore possible to use the action of the continuum theory unaltered apart from the granularity of the simplicial lattice. Thus we define the euclidean action of the interpolated field $A_n(x)^\alpha$ to be the integral over spacetime of the sum of the squares of the interpolated field strengths:

$$S(A) = \int d^3x \, \tfrac{1}{4}[F_{mn}(x)^\alpha]^2 \,. \tag{4}$$

Because the interpolation is linear in the field variables A, the jacobian $\det[\partial A_n(x)^\alpha / \partial A(\beta,m,i,j,k)]$ cancels in ratios of path integrals. By restricting spacetime to a finite volume, one thus approximates the vacuum expected value of a euclidean operator **Q**(**A**) by a multiple integral over the $A(\alpha,n,i,j,k)$'s:

$$\langle 0|\mathbf{Q}(\mathbf{A})|0\rangle \approx \frac{\int \prod dA(\alpha,n,i,j,k)\; e^{-S(A)}\, Q(A)}{\int \prod dA(\alpha,n,i,j,k)\; e^{-S(A)}} \tag{5}$$

where Q(A) is obtained from **Q**(**A**) by replacing the operator **A**(x) with the interpolated field A(x).

We used a 10^3 lattice and a 16^2-by-32 lattice. For the smaller lattice with 1,000 vertices, the action is a quartic polynomial in 6,000 variables, the $A(\alpha,m,i,j,k)$; for the larger lattice with 8,192 vertices, the action involves 49,152 A's. However because each field variable $A(\alpha,m,i,j,k)$ influences the action in only 24 simplexes, it is coupled to only 90 of the 6,000 or 49,152 variables.

The antisymmetry of $F_{mn}(x)^\alpha$ implies that no field variable occurs more than quadratically in the action. That is, the dependence of the action upon any single field variable A, if all the other A's are held fixed, is only quadratic. We used this fact to implement a fast heat-bath method that only requires knowledge of the first and second derivatives of the action with respect to each field variable. The method constructs the parabola that describes the dependence of the action on each A, goes to the minimum of the parabola, and then adds noise at the inverse temperature β. In terms of the dimensionless scaled field variable $a(\alpha,n,i,j,k) \equiv bgA(\alpha,n,i,j,k)$, the formula for da is

$$da = -S'/S" + 4x(\beta/S")^{\frac{1}{2}} \tag{6}$$

in which S' and S" are the first and second derivatives of the action with respect to the field variable a and x is a random number normally distributed on $(-\infty,\infty)$. We used Knuth's Algorithm P [4], Moler's URAND [5], and some tricks to write a fast subroutine [6] that generates a few thousand x's per call. The update da of each field variable is as big as required, and thermalization is very fast.

We used MACSYMA to calculate the derivatives S' and S", and to express them in legal FORTRAN. We did this in several steps. We first used MACSYMA to construct the action in each of the six generic simplexes into which we divided each cube of the lattice. We then used MACSYMA to translate these simplexes from one cube to another and to differentiate the action with respect to any field variable at an arbitrary vertex. Finally we used MACSYMA to write these derivatives in compilable FORTRAN.

We constructed the action in each of the six simplexes in five steps. The first step was to define the field in each simplex. We wrote six MACSYMA programs f*.mac in block form to do this in the six simplexes. The program f1.mac and several of our other MACSYMA programs are listed in the Appendix. The next step was to construct the color-electric field in each simplex. We wrote six programs fe*.mac in block form, each of which called the corresponding field program f*.mac. We wrote one block program sc.mac to make the structure constants for the group SU(2). We wrote six block programs fm*.mac that construct the color-magnetic field in each simplex. These programs call sc.mac and the f*.mac's. Finally we wrote six block programs s*.mac that compute the action in simplex *. These call fe*.mac and fm*.mac. In each simplex the action is the integral over the simplex of the sum of the squares of the color-electric and color-magnetic fields. In the temporal gauge the color-electric part is trivial. The color-magnetic fields are quadratic in the gauge fields, and also in the spacetime coordinates x, y, and t. S*.mac uses the Ratcoef function to compute the coefficient of each monomial $x^p y^q t^r$. Next, using the Sum function inside a Do loop, it multiplies products of the coefficients by integrals of products of the monomials over the simplexes, in the way appropriate to the integration of perfect squares.

We wrote a block program d.mac to calculate the derivatives of the action with respect to the field variable $a(\alpha,n,i,j,k)$. This program calls the s*.mac's and uses the Subst, Diff, and Ratsimp functions to calculate the first and second derivatives fd and sd, and to express them in terms of the 90 field variables b(k) with which $a(\alpha,n,i,j,k)$ actually interacts. It uses the Save function to create a LISP file d.l containing fd and sd. The program d.mac actually only calculates the derivatives with respect to a(1,1,i,j,k); the FORTRAN program uses the symmetry of the action under spatial and SU(2) rotations to compute the other five derivatives by permuting the b(k)'s. Finally the program d.mac calls programs that express the derivatives fd and sd in compilable FORTRAN. We shall now describe these formatting programs in some detail.

The formatting program terms.mac separates the derivatives into terms that the programs for1.mac and for2.mac can format. The program d.mac uses the labels fd and sd for the first and second derivatives of the action with respect to a(l,l,i,j,k). When the overall sign of fd (or sd) is positive; then its 0th part is /, its 1st part is its numerator, a polynomial in the a's, and its 2nd part is its denominator, an integer. However, when the overall sign of fd or sd is negative, then the 0th part is - and the first part is the whole expression, -fd or -sd. Therefore terms.mac first determines this sign. If it is negative, terms.mac reverses it and sets the flag sigfd or sigsd. The programs for*.mac are sensitive to these flags and reverse any sign changed by terms.mac. Terms.mac then uses the Part function to extract the denominators of fd and sd, which it labels fden and sden, and their numerators, fnum and snum. It next uses the function Nterms to determine the number of terms in fnum and snum, which it calls fnt and snt. The numerators fnum and snum are sums of simple polynomials in the basic field variables. Terms.mac makes arrays labeled fterm and sterm in which the ith element is the ith part of fnum or snum. It also makes the arrays nfterm and nsterm of which the ith element is the number of terms in fterm[i] or sterm[i]. In the same Do loops, it counts the number of parts in fnum and snum, which it calls cfp and csp, these being also the dimensions of the arrays fterm and sterm. Finally terms.mac uses the Save and Writefile functions to create LISP and ASCII files containing these arrays and various numerical characteristics of fd and sd.

The program forl.mac uses Writefile to create the ASCII file containing the FORTRAN statements that define the first derivative fd. Forl.mac first uses the FORTRAN function to initialize fd to 0. Next it checks the the number of terms nfterm[i] in successive terms fterm[i]. If nfterm[i] is less than 130, then it uses the FORTRAN function to express fterm[i] in fewer than 21 lines. When nfterm[i] is greater than or equal to 130, fterm[i] is a product of two terms of which the first, part(fterm[i],1), is a polynomial and the second, part(fterm[i],2), is a monomial. In this case, after fortraning the initializing statement "thing=0," forl.mac gathers enough terms, part(fterm[i],1,k), to form a suitably (ntj<122) long term, which it calls sfterm[i,j], and fortrans the statement "thing=thing+sfterm[i,j]." It continues to form such terms and to add them to thing until it exhausts part(fterm[i],1). It then fortrans the statement "fd=fd+thing*part(fterm[i],2)." Next it proceeds to fterm[i+1], etc., until it has analyzed all cfp terms of fnum. Finally, it examines the flag sigfd, and if necessary fortrans the statement "fd=-fd."

For2.mac uses Writefile to record the FORTRAN statements that it uses to express the second derivative sd. The terms sterm[k] are usually too small to form efficient FORTRAN statements by themselves. So for2.mac gathers them into larger terms that it calls cterm[i], continuing to add sterm[k]'s as long as nterms(cterm[i])<100. It then fortrans the equation "sd=sd+cterm[i]." It so continues until it has exhausted all csp terms in snum. Finally it examines the flag sigsd and acts appropriately.

The program forl.mac does produce a tail of 20 very short FORTRAN statements due to the fact that some fterm[i]'s are monomials. We used a text editor to gather these monomials into a single FORTRAN statement.

We used MACSYMA to write the derivatives fd and sd in terms of 90 variables b1 through b90, instead of the 90-element array b. The resulting FORTRAN code proved to be slightly slower on the Ridge 32C. A portion of the derivative fd is displayed in the Appendix. Some subexpressions contain common integral factors. We factored these subexpressions by using the Factor function on part(fterm[i],1,k) and on sterm[k] in our formatting programs; this change also slowed the code slightly.

We tested the final FORTRAN program by computing the action in artificial cases in which we knew its exact value; the program agreed with these values to within 1 part in 10^7. We calculated the weak- and strong-coupling limits of the mean value of the action per cube; the program agreed with these predictions to within 1%. We shall describe these and other results in a future publication.

Although symbol manipulators are useful in the treatment of very complex physical problems, there are some difficulties of which the novice user should be aware. MACSYMA and SMP jobs often require many hours of cpu time and many MB of real or virtual memory. One typically must repeatedly reformulate one's problem in order to fit it within the limits of one's computer. The form of the output of a good symbol manipulator is so regular that it may seduce one into believing that it is correct. But symbol manipulators have bugs and sometimes interpret programs in unexpected ways. One should always test the output. Finally, symbol manipulators are not yet available for all computers. Symbolics Corp. supports MACSYMA on its

LISP machines and on VAXes. Fortunately, it is the policy of Inference Corp. to port SMP to the widest possible range of computers. SMP now is available, for example, on the Ridge 32. This policy of widely porting SMP may make it the most important symbol manipulator and the best one for a beginner to learn.

Acknowledgments

The authors are particularly grateful to Stanly Steinberg for many useful conversations; they also wish to thank Michael Creutz, Anthony Kennedy, and Brent Richert for helpful suggestions. This work was supported by the Department of Energy under grant DE-FG04-84ER40166.

Appendix: Macsyma Codes and Output

Fl.mac

```
/*  Fl.mac creates the gauge field in the first simplex  */
/*  of the cube (i,j,k).  */
fl() := block(array(x,3),
b(d,n,x,i,j,k) := a(d,n,i,j,k)
        +(x[l]-i)*(a(d,n,i+l,j,k)-a(d,n,i,j,k))
        +(x[2]-j)*(a(d,n,i+l,j+l,k)-a(d,n,i+l,j,k))
        +(x[3]-k)*(a(d,n,i+l,j+l,k+l)-a(d,n,i+l,j+l,k)),
fl : b(d,n,x,i,j,k))$
```

Fel.mac

```
/*  Fel.mac makes the color-electric fields  */
/*  in the first simplex.  */
fel() := block(load(fl), fl(), array(fel,3,2),
for dq:l thru 3 do
for nq:l thru 2 do
    (fel[dq,nq]:diff(subst([d=dq,n=nq],fl),x[3])))$
```

Sc.mac

```
/*  Sc.mac creates the SU(2) structure constants.  */
sc() := block(sc[l,2,3]:l,
```

```
for k: 1 thru 2 do
(sc[1,2,k]:0),
for c: 1 thru 3  do
for b: 1 thru 3  do
for a: 1 thru 3  do
    (if a=b or b=c or c=a then sc[a,b,c]:0),
for c: 1 thru 3  do
for b: 1 thru c  do
for a: 1 thru b  do
    (sc[a,c,b]:sc[b,a,c]:sc[c,b,a]:-sc[a,b,c],
     sc[b,c,a]:sc[c,a,b]:sc[a,b,c]))$
```

Fml.mac

```
/*  Fml.mac creates the color-magnetic field  */
/*  in the first simplex.  */
fml() := block(load(fl), fl(), load(sc), sc(),
array(x,3), array(v,3), array(fml,3),
fl : subst([x[1]-i=v[1],x[2]-j=v[2],x[3]-k=v[3]],fl),
for dq: 1 thru 3 do
   (fml[dq] : diff(subst([d=dq,n=1],fl),v[2])
               - diff(subst([d=dq,n=2],fl),v[1])
              +sum(sum(sc[dq,dp,dpp]*subst([d=dp,n=1],fl)
                            *subst([d=dpp,n=2],fl),
                                 dp,1,3),dpp,1,3)))$
```

Sl.mac

```
/*  Sl.mac makes the action of the first simplex.  */
/*  The quantity calculated is S(1)*(4*b*g^2).  */
sl() := block(load(fel), fel(),
kill(labels), kill(fl),
som:0,
for d:1 thru 3 do
for n:1 thru 2 do
      (som : som + 2*fel[d,n]^2),
som : som/6,
kill(fel),
load(fml), fml(),
kill(labels),
for d:1 thru 3 do
      (for n1:0 thru 2 do
        (cf1:ratcoef(fml[d],v[1],n1),
         for n2:0 thru 2-n1 do
         (cf2:ratcoef(cf1,v[2],n2),
          for n3:0 thru 2-n1-n2 do
          (rcm[n1,n2,n3]:ratcoef(cf2,v[3],n3)),
           cf2:0), cf1:0),
       fml[d]:0,
       for n1:0 thru 2 do
       for n2:0 thru 2-n1 do
```

```
      for n3:0 thru 2-n1-n2 do
            (som : som + 2*rcm[n1,n2,n3]*(
                   2*sum(sum(sum(rcm[m1,m2,m3]/
                   ((n3+m3+1)*(n3+m3+n2+m2+2)*
                   (n3+m3+n2+m2+n1+m1+3)),
                   m3,0,2-m1-m2),m2,0,2-m1),m1,0,n1-1)
                   + 2*sum(sum(rcm[n1,m2,m3]/
                   ((n3+m3+1)*(n2+m2+n3+m3+2)*
                   (n3+m3+n2+m2+2*n1+3)),
                   m3,0,2-n1-m2),m2,0,n2-1)
                   + 2*sum(rcm[n1,n2,m3]/
                   ((n3+m3+1)*(n3+m3+2*n2+2)*
                   (n3+m3+2*n2+2*n1+3)),
                   m3,0,n3-1)
                   + rcm[n1,n2,n3]/
                   ((2*n3+1)*(2*(n2+n3)+2)*
                   (2*(n1+n3+n2)+3)))),
      kill(rcm)), kill(fm1), kill(labels))$
```

D.mac

```
/*  D.mac calculates the first and second derivatives  */
/*  of S*4*b*g^2 with respect to a(1,1,i,j,k).  */
d() := block( load(s1), s1(),
kill(allbut(som)),
dif:ratsimp(diff(som,a(1,1,i,j,k),1)),
dif2:ratsimp(diff(dif,a(1,1,i,j,k),1)),
kill(labels),
som:subst(i=i-1,som),
df:diff(som,a(1,1,i,j,k),1),
df2:diff(df,a(1,1,i,j,k),1),
 dif:ratsimp(dif+df),
 dif2:ratsimp(dif2+df2),
kill(labels), kill(df,df2),
som:subst(j=j-1,som),
df:diff(som,a(1,1,i,j,k),1),
df2:diff(df,a(1,1,i,j,k),1),
 dif:ratsimp(dif+df),
 dif2:ratsimp(dif2+df2),
kill(labels), kill(df,df2),
som:subst(k=k-1,som),
df:diff(som,a(1,1,i,j,k),1),
df2:diff(df,a(1,1,i,j,k),1),
 dif:ratsimp(dif+df),
 dif2:ratsimp(dif2+df2),
fd1:dif, sd1:dif2,
kill(s1,som,labels), kill(df,df2),
/*  Similar steps compute the contributions  */
/*  of the other simplexes.  */
kill(allbut(fd1,fd2,fd3,fd4,fd5,fd6,
sd1,sd2,sd3,sd4,sd5,sd6)),
```

```
fd:ratsimp(fd1+fd2+fd3+fd4+fd5+fd6),
kill(fd1,fd2,fd3,fd4,fd5,fd6),
sd:ratsimp(sd1+sd2+sd3+sd4+sd5+sd6),
kill(sd1,sd2,sd3,sd4,sd5,sd6),
load(terms), terms(),
load(for1), for1(),
load(for2), for2(),);
```

Terms.mac

```
/*  Terms.mac splits fd and sd into fterms and sterms.  */
terms():= block( writefile("terms"),
/*  Making fd and sd positive & setting the flags sig*d:  */
sigfd:0, sigsd:0,
if part(fd,0)="-"
   then (fd:-fd, sigfd:1, print("fd: MINUS!!")),
if part(sd,0)="-"
   then (sd:-sd, sigsd:1, print("sd: MINUS!!")),
/*  Extracting the denominators fden and sden:  */
fden:part(fd,2),  sden:part(sd,2),
/*  And the numerators fnum and snum:  */
fnum:part(fd,1),  snum:part(sd,1),
kill(fd,sd,labels),
print("fden=",fden),  print("sden=",sden),
/*  Determining the lengths of fnum and snum:  */
fnt:nterms(fnum),  snt:nterms(snum),
print("fnt=",fnt),  print("snt=",snt),

/*  Splitting fnum into fterm[i]'s:  */
cfnt:0,  cfp:0,
for i:1 while cfnt<fnt do
(fterm[i]:part(fnum,i),
 nfterm[i]:nterms(fterm[i]),
 cfnt:cfnt+nfterm[i],
 cfp:cfp+1),
print("# of parts in fnum=", cfp),

/*  Splitting snum into sterm[i]'s:  */
csnt:0,  csp:0,
for i:1 while csnt<snt do
(sterm[i]:part(snum,i),
 nsterm[i]:nterms(sterm[i]),
 csnt:csnt+nsterm[i],
 csp:csp+1),
print("# of parts in snum=", csp),

/*  Recording some properties of fterm & sterm:  */
save("terms.1",all),
for i:1 thru cfp do
(fop[i]:part(fterm[i],0),
 l[i]:length(fterm[i]),
```

```
 print("nfterm[",i,"]=",nfterm[i],
       "  fop=",fop[i],"  l=",l[i]),
 for j:1 thru l[i] do
 (size[i,j]:nterms(part(fterm[i],j))),
  print("size[",i,"1]=",size[i,1],
     "  size[",i,"2]=",size[i,2])),
closefile("terms"),
kill(fnum,snum));
```

For1.mac

```
/*  For1.mac writes fd in legal FORTRAN.  */
for1():= block( writefile("fd.der"),
fortran(fd=0),
/*  Examining successive fterm[i]'s:  */
for i:1 thru cfp do
(print("c    term[",i,"]:"),
/*  Formatting fterm[i] if it's small enough:  */
 if nfterm[i]<130 then fortran(fd=fd+fterm[i])
/*  Otherwise splitting fterm[i] into sfterm[i,j]'s:  */
 else (fortran(thing=0),
       np:1,  nt:0,
       for j:1 while nt<nfterm[i] do
       (sfterm[i,j]:0, ntj:0,
        for k:np while (nt<nfterm[i] and ntj<122) do
        (sfterm[i,j]:sfterm[i,j]+part(fterm[i],1,k),
         np:np+1,  ntj0:ntj,
         ntj:nterms(sfterm[i,j]),
         nt:nt+ntj-ntj0),
/*  Adding the sfterm[i,j]'s into thing:  */
        fortran(thing=thing+sfterm[i,j])),
/*  Adding the reconstructed fterm[i] to fd:  */
       fortran(fd=fd+thing*part(fterm[i],2)))),
/*  Reverting to fd's original sign:  */
if sigfd=1 then fortran(fd=-fd),
fortran(fd=fd/fden),
closefile("fd.der"));
```

For2.mac

```
/*  For2.mac writes sd in legal FORTRAN.  */
for2():= block( writefile("sd.der"),
nt:0,  np:1,
fortran(sd=0),
/*  Gathering sterm[k]'s to form cterm[i]'s:  */
for i:1 while (nt<snt and np<=csp) do
(print("c    term[",i,"]:"),
 cterm[i]:0,  nti:0,
 for k:np while (nt<snt and np<=csp and nti<100) do
```

```
 (cterm[i]:cterm[i]+sterm[k],
  np:np+1,  nti0:nti,
  nti:nterms(cterm[i]),
  nt:nt+nti-nti0),
/*  Adding cterm[i] to sd:  */
 fortran(sd=sd+cterm[i])),
if sigsd=1 then fortran(sd=-sd),
fortran(sd=sd/sden),
closefile("sd.der"));
```

A Portion of the Derivative fd Produced by Macsyma

```
      thing = 0
      thing = thing+(8*b(14)+8*b(10)+8*b(9)+72*b(8)+48*b(7)+8*b(3)+8*b(2
     1   )+8*b(1))*b(82)+(8*b(13)+8*b(11)+8*b(9)+72*b(8)+48*b(6)+8*b(4)+
     2   8*b(2)+8*b(1))*b(81)+(8*b(12)+8*b(11)+8*b(10)+72*b(8)+48*b(5)+8
     3   *b(4)+8*b(3)+8*b(1))*b(80)+(8*b(11)+48*b(8)+8*b(6)+8*b(5)+32*b(
     4   4)+8*b(1))*b(79)+(8*b(10)+48*b(8)+8*b(7)+8*b(5)+32*b(3)+8*b(1))
     5   *b(78)+(8*b(9)+48*b(8)+8*b(7)+8*b(6)+32*b(2)+8*b(1))*b(77)+(72*
     6   b(8)+8*b(7)+8*b(6)+8*b(5)+8*b(4)+8*b(3)+8*b(2)+48*b(1))*b(76)+2
     7   52*b(75)+252*b(74)+252*b(73)-252*b(72)-252*b(71)-252*b(70)+756*
     8   b(69)+252*b(67)+252*b(66)-756*b(65)-252*b(64)-252*b(63)+252*b(6
     9   2)-252*b(61)+(-24*b(45)-4*b(44)-4*b(43)-4*b(42)-4*b(41)-4*b(40)
     :   -4*b(39)-36*b(38))*b(60)+(-4*b(45)-16*b(44)-4*b(40)-4*b(39)-24*
     ;   b(38)-4*b(37))*b(59)+(-4*b(45)-16*b(43)-4*b(41)-4*b(39)-24*b(38
     <   )-4*b(36))*b(58)+(-4*b(45)-16*b(42)-4*b(41)-4*b(40)-24*b(38)-4*
     =   b(35))*b(57)+(-4*b(45)-4*b(43)-4*b(42)-24*b(41)-36*b(38)-4*b(36
     >   )-4*b(35)-4*b(34))*b(56)+(-4*b(45)-4*b(44)-4*b(42)-24*b(40)-36*
     ?   b(38)-4*b(37)-4*b(35)-4*b(33))*b(55)+(-4*b(45)-4*b(44)-4*b(43)-
     @   24*b(39)-36*b(38)-4*b(37)-4*b(36)-4*b(32))*b(54)+(-36*b(45)-24*
     1   b(44)-24*b(43)-24*b(42)-36*b(41)-36*b(40)-36*b(39)-576*b(38)-36
     2   *b(37)-36*b(36)-36*b(35)-24*b(34)-24*b(33)-24*b(32)-36*b(31))*b
     3   (53)
      thing = thing+(-4*b(44)-4*b(40)-4*b(39)-36*b(38)-24*b(37)-4*b(33)-
     1   4*b(32)-4*b(31))*b(52)+(-4*b(43)-4*b(41)-4*b(39)-36*b(38)-24*b(
     2   36)-4*b(34)-4*b(32)-4*b(31))*b(51)+(-4*b(42)-4*b(41)-4*b(40)-36
     3   *b(38)-24*b(35)-4*b(34)-4*b(33)-4*b(31))*b(50)+(-4*b(41)-24*b(3
     4   8)-4*b(36)-4*b(35)-16*b(34)-4*b(31))*b(49)+(-4*b(40)-24*b(38)-4
     5   *b(37)-4*b(35)-16*b(33)-4*b(31))*b(48)+(-4*b(39)-24*b(38)-4*b(3
     6   7)-4*b(36)-16*b(32)-4*b(31))*b(47)+(-36*b(38)-4*b(37)-4*b(36)-4
     7   *b(35)-4*b(34)-4*b(33)-4*b(32)-24*b(31))*b(46)-252*b(30)-252*b(
     8   29)+252*b(28)-252*b(27)+252*b(26)-756*b(25)+252*b(24)-252*b(22)
     9   +756*b(21)-252*b(20)+252*b(19)-252*b(18)+252*b(17)+252*b(16)
      fd = b(83)*thing+fd
```

References

1. MACSYMA was developed at MIT and is marketed by Symbolics, Inc. (617-577-7350). In the computations described in this article, we used beta-test release 308.

2. SMP was developed largely by Stephen Wolfram; it is enhanced, ported, and marketed by Inference Corp. (213-417-7997).

3. K. Cahill, S. Prasad, and R. Reeder, Physics Lett. 149B (1984) 377.

4. D. E. Knuth, The Art of Computer Programming, Vol. II (Addison-Wesley, second ed., 1981), p. 117.

5. G. Forsythe, M. Malcolm, and C. Moler, Computer Methods for Mathematical Computations (Prentice-Hall, 1977), p. 246.

6. A copy of this subroutine, snrand, adapted to the Ridge 32C is available from Ridge Computers, Santa Clara, CA.

THE GF11 SUPERCOMPUTER

J. Beetem, M. Denneau, and D. Weingarten

IBM T. J. Watson Research Center
Yorktown Heights, NY 10598

Abstract

GF11 is a parallel computer currently under construction at the IBM Yorktown Research Center. The machine incorporates 576 floating- point processors arranged in a modified SIMD architecture. Each has space for 2 Mbytes of memory and is capable of 20 Mflops, giving the total machine a peak of 1.125 Gbytes of memory and 11.52 Gflops. The floating-point processors are interconnected by a dynamically reconfigurable non-blocking switching network. At each machine cycle any of 1024 pre-selected permutations of data can be realized among the processors. The main intended application of GF11 is a class of calculations arising from quantum chromodynamics.

A detailed treatment appears in the Proceedings of the Twelfth International Symposium on Computer Architecture, Boston, June, 1985.